Principles of Terahertz Science and Technology

Yun-Shik Lee

Principles of Terahertz Science and Technology

Springer

Yun-Shik Lee
Physics Department
Oregon State University
Corvallis, Oregon 97331
USA

ISBN 978-0-387-09539-4 e-ISBN 978-0-387-09540-0

Library of Congress Control Number: 2008935382

Printed on acid-free paper.

9 8 7 6 5 4 3 2 1

springer.com

To my parents Su-Ho Lee and Soon-Im Shin

Preface

Over the last two decades, THz technology has ripened enough that a thorough summary and review of the relevant topics is in order. Many different disciplines such as ultrafast spectroscopy, semiconductor device fabrication, and bio-medical imaging involve the recent development of THz technology. It is an important task to lay down a common ground among the researchers, so that they can communicate smoothly with one another. Besides, the THz community is growing fast and the THz technology is in a transitional period. The THz research activities have mainly focused on generation and detection until lately, but the focal point has shifted to the practical applications such as high-speed communication, molecular spectroscopy, security imaging, and medical diagnosis, among many others.

This book covers a broad range of topics and fundamental issues. Individuals from distinct disciplines have helped developing new THz technologies, and in order to reach the next level, i.e., practical applications, the technology requires its researchers to understand and communicate with one another. This book serves this general purpose by providing the researchers with a common reference, thus bridging "the THz gap".

I have tried to elucidate the fundamentals of THz technology and science for their potential users. This book surveys major techniques of generating, detecting, and manipulating THz waves. It also discusses a number of essential processes where THz waves interact with physical, chemical, and biological systems. Scientists and engineers of various disciplines realize that the THz gap in the electromagnetic spectrum is now accessible thanks to the recent advances in THz source and detection technologies. Many are seeking ways by which they can incorporate the new technologies into their expertise and research agenda. Younger researchers, who wish or are to join THz research groups, would also find this new field challenging due to many barriers, the lack of comprehensive introduction and/or instruction among them. Potential users of THz technology should be prepared in the essential concepts and techniques of THz science and technology; I hope this book be an introductory guide for the new comers.

During the process of writing this book, many colleagues, friends, and students gave me worthy criticism and introduction. Although it is impossible to acknowledge all scientific contributions, I am deeply grateful of those whose works I use in this book. I am much obliged to Joe Tomaino, Andy Jameson, and Jeremy Danielson for their invaluable advice. I am indebted to the National Science Foundation and the Alexander von Humboldt-Foundation for their generous support. Finally, I thank my wife, JungHwa, for her support in every possible ways.

Corvallis
September 2008 *Yun-Shik Lee*

Contents

1

Introduction

Terahertz (THz) radiation is electromagnetic radiation whose frequency lies between the microwave and infrared regions of the spectrum. We cannot see THz radiation but we can feel its warmth as it shares its spectrum with far-infrared radiation. Naturally occurring THz radiation fills up the space of our everyday life, yet this part of the electromagnetic spectrum remains the least explored region mainly due to the technical difficulties involved in making efficient and compact THz sources and detectors. The lack of suitable technologies led to the THz band being called the "THz gap". This technological gap has been rapidly diminishing for the last two decades. Optical technologies have made tremendous advances from the high frequency side, while microwave technologies encroach up from the low frequency side. This chapter gives a brief perspective on the basic properties of THz radiation and its interaction with materials, which lays down the foundation to discuss the progress of THz science and technology in subsequent chapters.

1.1 Terahertz Band

"Terahertz radiation" is the most common term used to refer to this frequency band, analogous to microwaves, infrared radiation, and x-rays. It is rather awkward to use a frequency unit for naming a spectral band. Nevertheless, as "terahertz" has become a symbolic word designating the entire field, we will universally use this term throughout this book. An alternative and seemingly better terminology is "T-rays", where "T" stands for terahertz. It was initially coined for an imaging technique, and will be used in Chapter 7, which is dedicated to T-ray imaging technologies and applications.

Until quite recently, THz technologies had been independently developed by researchers from several different disciplines. In practice, different communities use different units to describe the spectrum of THz radiation. We will use THz (10^{12} Hz) as the universal unit in this book, but other units will also be used when they are appropriate. Frequently used units and their conversions at 1 THz are as follows:

- Frequency: $\nu = 1$ THz $= 1000$ GHz
- Angular frequency: $\omega = 2\pi\nu = 6.28$ THz
- Period: $\tau = 1/\nu = 1$ ps
- Wavelength: $\lambda = c/\nu = 0.3$ mm $= 300$ μm
- Wavenumber: $\bar{k} = k/2\pi = 1/\lambda = 33.3$ cm^{-1}
- Photon energy: $h\nu = \hbar\omega = 4.14$ meV
- Temperature: $T = h\nu/k_B = 48$ K

where c is the speed of light in vacuum, h is Plank's constant, and k_B is Boltzmann's constant. Physicists tend to use μm and meV as units of photon wavelength and energy, respectively; chemists use cm^{-1} as a unit of wavenumber; engineers use mm and GHz as units of wavelength and frequency, respectively. In physics, angular wavenumber ($k = 2\pi/\lambda$) is usually abbreviated as wavenumber. In this book, we will use the abbreviated notation when it is clearly defined.

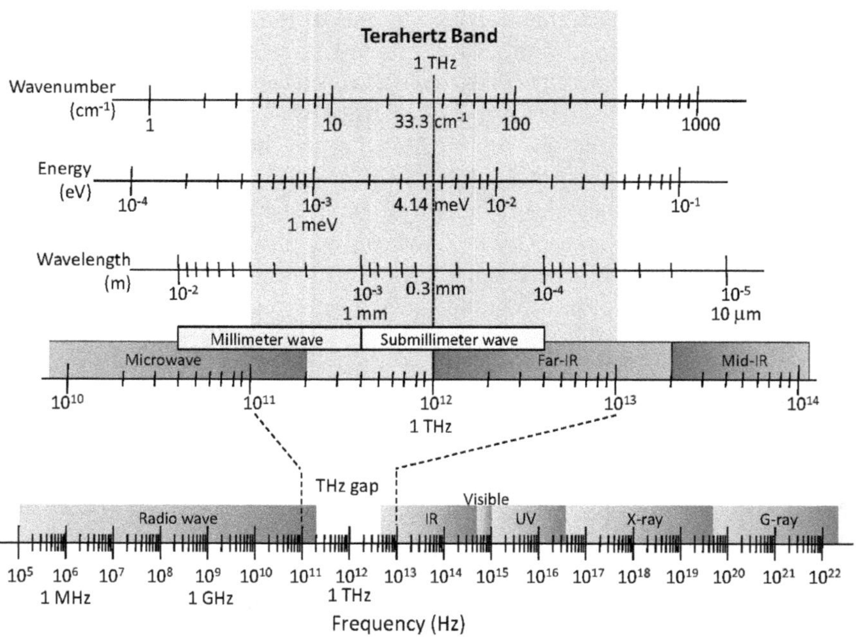

Fig. 1.1. Terahertz band in the electromagnetic spectrum

The THz band does not have a standard definition yet. Commonly used definitions are included in the spectral region between 0.1 and 30 THz. The range of 10-30 THz, however, exceeds the far-IR band and intrudes on the mid-IR band, where well established optical technologies exist. Unless we deal with ultrabroadband THz pulses, we will use 0.1-10 THz as a universal definition of the THz band. Figure 1.1 illustrates the THz band in the electromagnetic spectrum. The THz band merges into neighboring spectral bands such as the millimeter-wave band, which is the highest radio frequency band known as Extremely High Frequency (EHF), the submillimeter-wave band, and the far-IR band. The definitions of these bands are as follows:

- Millimeter wave (MMW): 1-10 mm, 30-300 GHz, 0.03-0.3 THz
- Submillimeter wave (SMMW): 0.1-1 mm, 0.3-0.3 THz
- Far infrared radiation (Far-IR): (25-40) to (200-350) μm, (0.86-1.5) to (7.5 to 12) THz
- Sub-THz radiation: 0.1-1 THz

These bands are also distinguished by their characteristic technologies. Millimeter wave emitters and sensors are solid-state devices based on microwave technologies. Traditionally, far-IR applications rely on optical and thermal devices.

1.2 Terahertz Generation and Detection

Technological advances in optics and electronics have resulted in many different types of THz sources and sensors. Chapters 3 and 4 are devoted to broadband and continuous-wave (CW) THz technologies, which are classified by similarities in radiation characteristics. In this section, we make brief descriptions of the schemes used for THz generation and detection, grouped by operational concepts.

1.2.1 Terahertz Sources

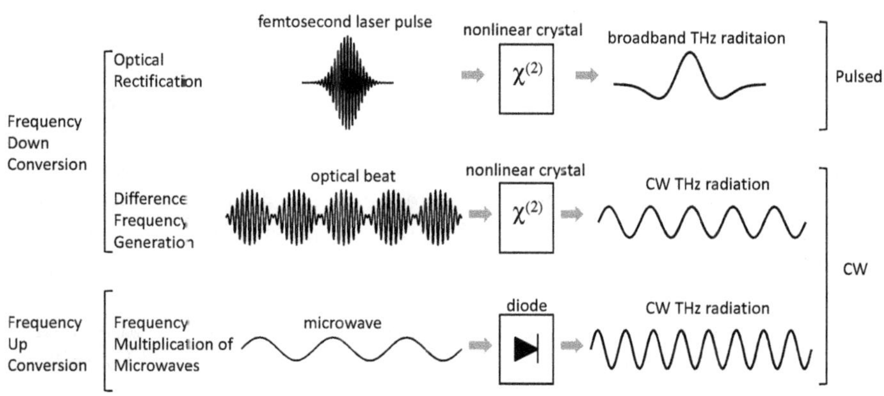

Fig. 1.2. Terahertz generation in nonlinear media

One way of generating THz radiation is to exploit a nonlinear medium in which incident electromagnetic waves undergo nonlinear frequency conversion (Fig. 1.2). Optical rectification and difference frequency generation (DFG) are second order nonlinear optical processes in which a THz photon at frequency ω_T is created by interaction of two optical photons at frequencies ω_1 and ω_2 with a nonlinear crystal, such that $\omega_T = \omega_1 - \omega_2$. Femtosecond laser pulses with a broad spectrum (bandwidth$\sim$10 THz) generate broadband THz pulses, whose shape resembles the optical pulse envelope, via optical rectification. Two CW optical beams produce CW THz radiation by DFG. Solid-state THz sources based on microwave technology convert incoming microwaves into their harmonic waves utilizing diodes with strongly nonlinear I-V characteristics.

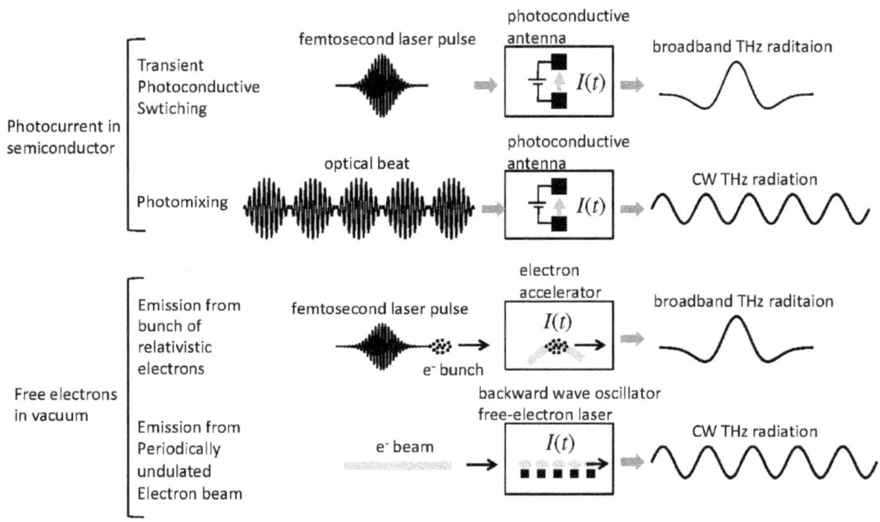

Fig. 1.3. THz radiation from accelerating electrons

Accelerating charges, and time-varying currents, radiate electromagnetic waves (Fig. 1.3). THz radiation can be generated from a biased photoconductive (PC) antenna excited by laser beams. A PC antenna consists of two metal electrodes deposited on a semiconductor substrate. An optical beam, illuminating the gap between the electrodes, generates photocarriers, and a static bias field accelerates the free carriers. This photocurrent varies in time corresponding to the incident laser beam intensity. Consequently, femtosecond laser pulses produce broadband THz pulses. Mixing two laser beams with different frequencies forms an optical beat, which generates CW THz radiation at the beat frequency. This technique is called photomixing.

Electron accelerators produce extremely bright THz radiation using relativistic electrons. A femtosecond laser pulse triggers an electron source to generate a ultrashort pulse of electrons. After being accelerated to a relativistic speed, the electrons are smashed into a metal target, or are forced into circular motion by a magnetic field. Coherent THz radiation is generated by this transient electron acceleration.

Backward wave oscillators (BWOs) are laboratory-size equipment, and free-electron lasers (FELs), small scale electron accelerators, are large facilities. In spite of the huge difference in size, there is a similarity in their THz generation mechanism. In both, an electron beam is undulated by a periodic structure: a BWO has a metal grating and a FEL consists of a magnet array. CW THz radiation is produced by the periodic acceleration of electrons.

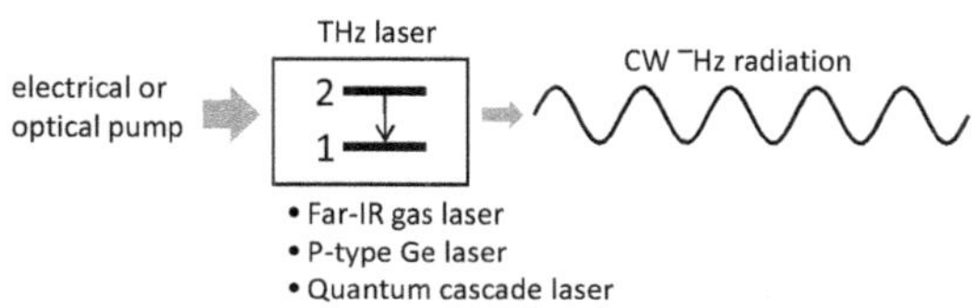

Fig. 1.4. THz emission from laser

Laser action requires a population-inverted two-level quantum system (Fig. 1.4). Far-IR gas lasers utilize molecular rotation energy levels, whose transition frequencies fall into the THz region. P-type germanium lasers are electrically pumped solid-state lasers. Their lasing action relies on the population inversion of two Landau levels formed by hot-carriers submerged in crossed electric and magnetic fields. Quantum cascade lasers (QCLs) are semiconductor heterostructure lasers consisting of periodically alternating layers of dissimilar semiconductors. Transitions between subbands of these semiconductor nanostructures involve THz photons. In a QCL, electrons undergo successive intersubband transitions to generate coherent THz radiation.

1.2.2 Terahertz Detectors

THz detection schemes are largely classified as either coherent or incoherent techniques. The fundamental difference is that coherent detection measures both the amplitude and phase of the field, whereas incoherent detection measures the intensity. Coherent detection techniques are closely associated with generation techniques in that they share underlying mechanisms and key components. In particular, optical techniques utilize the same light source for both generation and detection.

Figure 1.5 illustrates the commonly used coherent detection schemes. Free-space electro-optic (EO) sampling measures the actual electric field of broadband THz pulses in the time domain by utilizing the Pockels effect, which is closely related to optical rectification. A THz field induces birefringence in a nonlinear optical crystal which is proportional to the field amplitude. The entire waveform is determined by a weak optical probe measuring the field-induced birefringence as a function of the relative time delay between the THz and optical pulses.

Sensing with a PC antenna also measures broadband THz pulses in the time domain. In the absence of a bias field, a THz field induces a current in the photoconductive gap when an optical probe pulse injects photocarriers. The induced photocurrent is proportional to the THz field amplitude. The THz pulse shape is mapped out in the time domain by measuring the photocurrent while varying the time delay between the THz pulse and the optical probe.

A combined setup of broadband THz generation and detection measures changes in both the amplitude and phase of THz pulses induced by a sample, which provides enough information to simultaneously determine the absorption and dispersion of the sample. This technique is named THz time-domain spectroscopy, or, in short, THz-TDS.

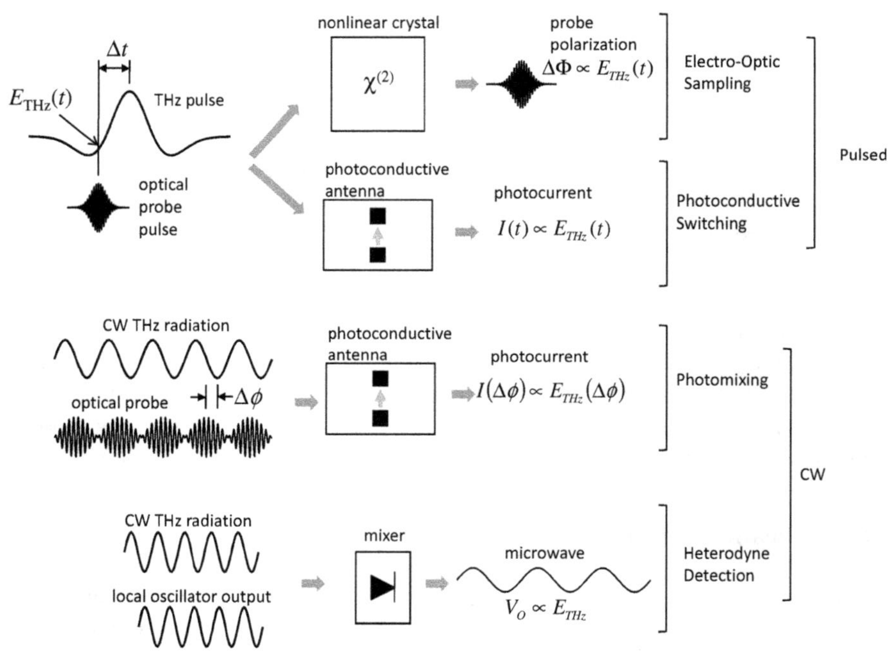

Fig. 1.5. Coherent detection of THz radiation

Photomixing measures CW THz radiation by exploiting photoconductive switching. In this case, the photocurrent shows sinusoidal dependence on the relative phase between the optical beat and the THz radiation.

Heterodyne detection utilizes a nonlinear device called a "mixer". Schottky diodes are commonly used as mixers. The key process in a mixer is frequency downconversion, which is carried out by mixing a THz signal ω_s with reference radiation at a fixed frequency ω_{LO}. The mixer produces an output signal at the difference frequency called the "intermediate frequency", $\omega_D = |\omega_S - \omega_{LO}|$. The amplitude of the output signal is proportional to the THz amplitude. Unlike the optical techniques, heterodyne detection is usually used to detect incoherent radiation.

Commonly used incoherent detectors are thermal sensors such as bolometers, Golay cells, and pyroelectric devices. A common element of all thermal detectors is a radiation absorber attached to a heat sink. Radiation energy is recorded by a thermometer measuring the temperature increase in the absorber. Each type of thermal detector is distinguished by its specific scheme used to measure the temperature increase. Bolometers are equipped with an electrical resistance thermometer made of a heavily doped semiconductor such as Si or Ge. In general, bolometers operate at cryogenic temperature. Pyroelectric detectors employ a pyroelectric material in which temperature change gives rise to spontaneous electric polarization. In a Golay cell, heat is transferred to a small volume of gas in a sealed chamber behind the absorber. An optical reflectivity measurement detects the membrane deformation induced by the pressure increase. These thermal detectors respond to radiation over a very broad spectral range. Because a radiation absorber must reach to thermal equilibrium for a temperature measurement, detection response is relatively slow compared with typical light detectors.

1.3 Terahertz Applications

The THz region is crowded by innumerable spectral features associated with fundamental physical processes such as rotational transitions of molecules, large-amplitude vibrational motions of organic compounds, lattice vibrations in solids, intraband transitions in semiconductors, and energy gaps in superconductors. THz applications exploit these unique characteristics of material responses to THz radiation.

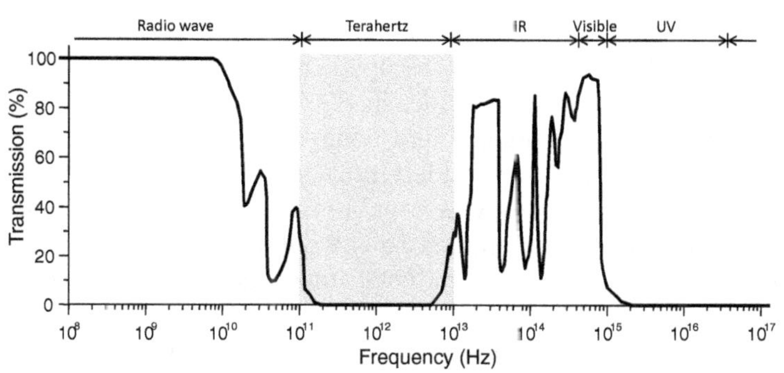

Fig. 1.6. Atmospheric transmission spectrum of electromagnetic waves

Compared with the neighboring regions of radio waves and infrared radiation, the THz band shows exceedingly high atmospheric opacity due to the rotational lines of constituent molecules (Fig. 1.6). In particular, absorption by water vapor is the predominant process of atmospheric THz attenuation. Figure 1.7 shows a high-resolution transmission spectrum of water vapor. In practice, water absorption is an important factor to be considered when designing an operation scheme for a THz application.

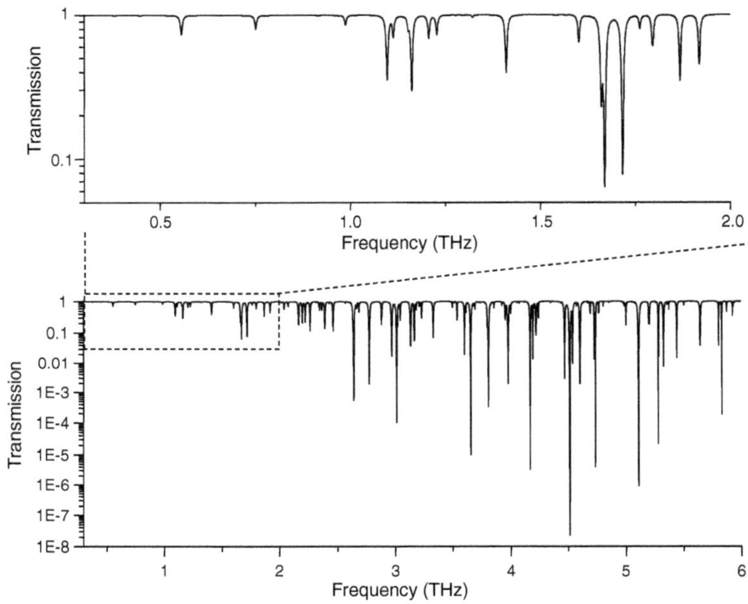

Fig. 1.7. Water vapor transmission spectrum from 0.3 to 6 THz

Distinctive line structures of each molecular species can be utilized to identify it in an unknown specimen. Furthermore, spectral line-shapes provide crucial information concerning microscopic mechanisms of molecular collisions. High-resolution THz spectroscopy is being used to monitor the Earths atmosphere and to observe molecules in the interstellar medium.

Spectral signatures of organic and biological molecules in the THz region are associated with large amplitude vibrational motions and inter-molecular interactions. THz spectroscopy is capable of analyzing these molecular dynamics, it can therefore be applied to the detection of explosives and illicit drugs, testing pharmaceutical products, investigating protein conformation, etc.

Based on optical properties at THz frequencies, condensed matter is largely grouped into three types: water, metal, and dielectric. Water, a strongly polar liquid, is highly absorptive in the THz region. Because of high electrical conductivity, metals are highly reflective at THz frequencies. Nonpolar and nonmetallic materials, i.e., dielectrics such as paper, plastic, clothes, wood, and ceramics that are usually opaque at optical wavelengths, are transparent to THz radiation. A brief description of the optical properties of each material type is shown in Table 1.1.

Table 1.1. Optical Properties of Condensed Matter in the THz Band

Material Type	Optical Property
liquid water	high absorption ($\alpha \approx 250$ cm^{-1} at 1 THz)
metal	high reflectivity (>99.5% at 1 THz)
plastic	low absorption ($\alpha < 0.5$ cm^{-1} at 1 THz)
	low refractive index ($n \approx 1.5$)
semiconductor	low absorption ($\alpha < 1$ cm^{-1} at 1 THz)
	high refractive index ($n \sim$ 3-4)

These stark contrasts of THz properties are useful for many imaging applications. Since common packaging materials are dielectric, THz imaging is applied to nondestructive testing to inspect sealed packages. Because of the high absorption of water in the THz region, hydrated substances are easily differentiated from dried ones. Metal objects also can be easily identified due to their high reflectivity and complete opacity. The same concept is applied to security applications. THz imaging is used to identify weapons, explosives, and illegal drugs concealed underneath typical wrapping and packaging materials. The high sensitivity of THz radiation to water is useful for medical applications because, in a biological system, small changes in water content could indicate crucial defects emerging in the region.

In addition to the applications we briefly described in this section, THz spectroscopy and imaging techniques have been applied to many others ranging from basic scientific missions to commercial projects. We will look into them in following chapters.

2

Basic Theories of Terahertz Interaction with Matter

This chapter is concerned with basic concepts and theories that form the foundation for understanding the unique characteristics of THz radiation and its interaction with materials. Classical electromagnetic theory provides a general description of THz waves which propagate in and interact with macroscopically uniform media. The basic framework of quantum theory is utilized to describe elementary exciations at THz frequencies.

2.1 Electromagnetic Waves in Matter

We shall begin with Maxwell's equations to describe THz waves as we would do for any other spectral regions. The macroscopic form of Maxwell's equations have the form

$$\nabla \cdot \mathbf{D} = \rho_f, \tag{2.1}$$

$$\nabla \cdot \mathbf{B} = 0, \tag{2.2}$$

$$\nabla \times \mathbf{E} = -\frac{\partial \mathbf{B}}{\partial t}, \tag{2.3}$$

$$\nabla \times \mathbf{H} = \mathbf{J}_f + \frac{\partial \mathbf{D}}{\partial t}, \tag{2.4}$$

where ρ_f and $\mathbf{J}_f$ are the free charge density and the free current density. These deceptively simple equations, together with the Lorentz force law

$$\mathbf{F} = q\left(\mathbf{E} + \mathbf{v} \times \mathbf{B}\right), \tag{2.5}$$

constitute the entire theoretical basis of classical electrodynamics. The macroscopic fields $\mathbf{D}$ and $\mathbf{H}$ are related to the fundamental fields $\mathbf{E}$ and $\mathbf{B}$ as

$$\mathbf{D} \equiv \epsilon_0 \mathbf{E} + \mathbf{P} = \epsilon \mathbf{E}, \tag{2.6}$$

$$\mathbf{H} \equiv \frac{1}{\mu_0}\mathbf{B} - \mathbf{M} = \frac{1}{\mu}\mathbf{B}, \tag{2.7}$$

where ϵ_0 and μ_0 are the permittivity and the permeability of free space. The polarization **P** and the magnetization **M** contain the information about the macroscopic-scale electromagnetic properties of the matter. The last terms in Eqs. 2.6 and 2.7, where ϵ and μ denote the the electric permittivity and the magnetic permeability, are valid only if the media is isotropic and linear. Typical magnetic responses of matter are subtle, $|\mu - \mu_0| < 10^{-4}\mu_0$, comparing with their electric counter parts, largely because of the nonexistence of magnetic monopoles.

2.1.1 The Wave Equation

The coupled electric and magnetic fields in Maxwell's equations are disentangled by taking the curl of Eqs. 2.3 and 2.4 and using the linear relations of Eqs. 2.6 and 2.7:

$$\nabla \times (\nabla \times \mathbf{E}) + \epsilon\mu \frac{\partial^2 \mathbf{E}}{\partial t^2} = -\mu \frac{\partial \mathbf{J}_f}{\partial t}, \tag{2.8}$$

$$\nabla \times (\nabla \times \mathbf{H}) + \epsilon\mu \frac{\partial^2 \mathbf{H}}{\partial t^2} = \nabla \times \mathbf{J}_f. \tag{2.9}$$

These are the most general wave equations for **E** and **H**. Using the vector identity

$$\nabla \times (\nabla \times \mathbf{A}) = \nabla(\nabla \mathbf{A}) - \nabla^2 \mathbf{A} \tag{2.10}$$

with Eqs. 2.1 and 2.2 we can rewrite the wave equations as

$$\nabla^2 \mathbf{E} - \epsilon\mu \frac{\partial^2 \mathbf{E}}{\partial t^2} = \mu \frac{\partial \mathbf{J}_f}{\partial t} + \frac{1}{\epsilon} \nabla \rho_f, \tag{2.11}$$

$$\nabla^2 \mathbf{H} - \epsilon\mu \frac{\partial^2 \mathbf{H}}{\partial t^2} = -\nabla \times \mathbf{J}_f. \tag{2.12}$$

Assuming $\mathbf{J}_f$ is linear with **E**,

$$\mathbf{J}_f = \sigma \mathbf{E}, \tag{2.13}$$

where σ is the electric conductivity, and neglecting charge density fluctuations, i.e., $\nabla \rho_f = 0$, we simplify the wave equation for **E** as

$$\nabla^2 \mathbf{E} = \sigma\mu \frac{\partial \mathbf{E}}{\partial t} + \epsilon\mu \frac{\partial^2 \mathbf{E}}{\partial t^2}. \tag{2.14}$$

Here σ and ϵ are real and independent. The wave equation for **H** takes an identical form. These time-varying fields are closely intertwined by Maxwell's equations: if one is known, the other is fully determined. The coupled entity of the two fields is called an *electromagnetic wave*.

If the material is a dielectric or an insulator, the wave equation takes the universal form

$$\nabla^2 \mathbf{E} = \epsilon\mu \frac{\partial^2 \mathbf{E}}{\partial t^2} = \frac{1}{v^2}\frac{\partial^2 \mathbf{E}}{\partial t^2}, \tag{2.15}$$

which signifies that electromagnetic waves propagate in homogeneous media at a speed

$$v = \frac{1}{\sqrt{\epsilon\mu}} = \frac{c}{n}, \tag{2.16}$$

where $c\left(= 1/\sqrt{\epsilon_0\mu_0}\right)$ is the speed of light in free space and $n\left(= \sqrt{\epsilon/\epsilon_0}\right)$ is the refractive index, assuming $\mu = \mu_0$.

General solutions of the wave equation are linearly-polarized monochromatic plane waves:

$$\mathbf{E}(\mathbf{r}, t) = \mathbf{E}_0 e^{i(\mathbf{k}\cdot\mathbf{r}-\omega t)} \quad \text{and} \quad \mathbf{H}(\mathbf{r}, t) = \mathbf{H}_0 e^{i(\mathbf{k}\cdot\mathbf{r}-\omega t)}, \tag{2.17}$$

where $\mathbf{k}$ is the wave vector and ω is the angular frequency. From Maxwell's equations we can draw the relations between $\mathbf{E}$ and $\mathbf{H}$ associated with $\mathbf{k}$ and ω. Substituting the plane waves into $\nabla \cdot \mathbf{E} = 0$ and $\nabla \cdot \mathbf{B} = 0$ we obtain

$$\mathbf{k}\cdot\mathbf{E} = 0 \quad \text{and} \quad \mathbf{k}\cdot\mathbf{H} = 0. \tag{2.18}$$

This means that $\mathbf{E}$ and $\mathbf{H}$ are both perpendicular to the wave vector, that is, electromagnetic waves are transverse. The curl equations give the relation

$$\mathbf{k} \times \mathbf{E} = \omega\mu\mathbf{H}. \tag{2.19}$$

Inserting Eq. 2.17 into Eq. 2.15 we obtain the dispersion relation

$$k^2 = \epsilon\mu\omega^2. \tag{2.20}$$

As ϵ and μ quantify the electromagnetic properties of the material, the dispersion relation governs how the wave propagates in the medium. For a non-magnetic medium, the wavenumber k is related to the wavelength λ by the relation

$$k = \frac{2\pi}{\lambda} = n\frac{\omega}{c}. \tag{2.21}$$

The energy flux of an electromagnetic wave is the time-averaged Poynting vector

$$\langle \mathbf{S} \rangle = \frac{1}{2}\mathbf{E} \times \mathbf{H}^* = \frac{1}{2} v\,\epsilon |E_0|^2 \mathbf{e}_k, \tag{2.22}$$

where $\mathbf{e}_k(=\mathbf{k}/k)$ is a unit vector in the direction of the wave propagation. The magnitude of the energy flux,

$$I = |\langle \mathbf{S} \rangle| = \frac{1}{2} v\,\epsilon |E_0|^2, \tag{2.23}$$

is the radiation intensity, which is the measurement quantity of typical light detectors. A commonly used unit of light intensity is W/cm^2.

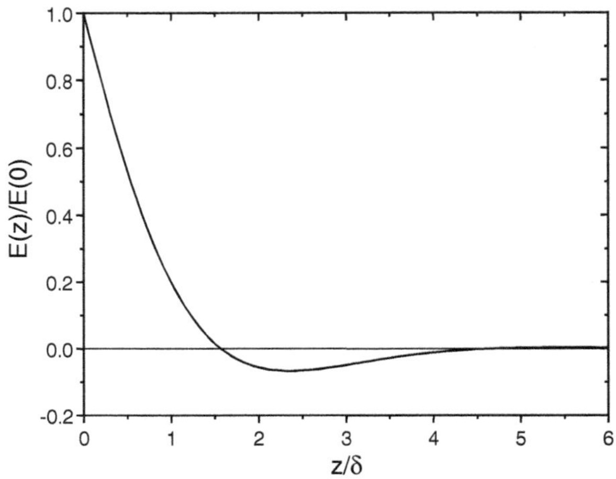

Fig. 2.1. Electric field decay in a conductor

In a conducting medium, general solutions of the wave equation, Eq. 2.14, also take the form of the plane waves in Eq. 2.17. Even Eqs. 2.18 and 2.19 are still valid. Wave propagation in a conductor is, however, quite different from that in a dielectric medium. If the medium has a very large conductivity such that $\sigma \ll \omega\epsilon$, the wave equation leads to the dispersion relation

$$k^2 \approx i\sigma\mu\omega. \tag{2.24}$$

Evidently the amplitude of the wave vector is a complex number,

$$k = k_R + ik_I \approx \sqrt{\frac{\omega\mu\sigma}{2}}(1+i). \tag{2.25}$$

This means that when an electromagnetic wave is incident on a conductor the field decays exponentially with an attenuation length δ, which is called the penetration depth or the skin depth:

$$\delta = \sqrt{\frac{2}{\omega\mu\sigma}}\ . \tag{2.26}$$

Typical metals behave like an ideal conductor for THz waves. For example, the skin depth of copper is $\delta \approx 0.07$ μm for the frequency $\nu(=\omega/2\pi) =$ 1 THz, which is almost negligible when compared to the free-space wavelength, 300 μm.

2.1.2 Reflection and Transmission

When an electromagnetic wave reflects from and transmits through an interface of two linear dielectric media, both **E** and **H** obey the boundary condition that the parallel components of the vector fields are continuous across the interface. An apparent ramification of the boundary condition is Snell's law

$$n_1 \sin\theta_1 = n_2 \sin\theta_2, \tag{2.27}$$

where $n_{1,2}$ are the refractive indices of the media and $\theta_{1,2}$ are the angle of incidence and the angle of refraction.

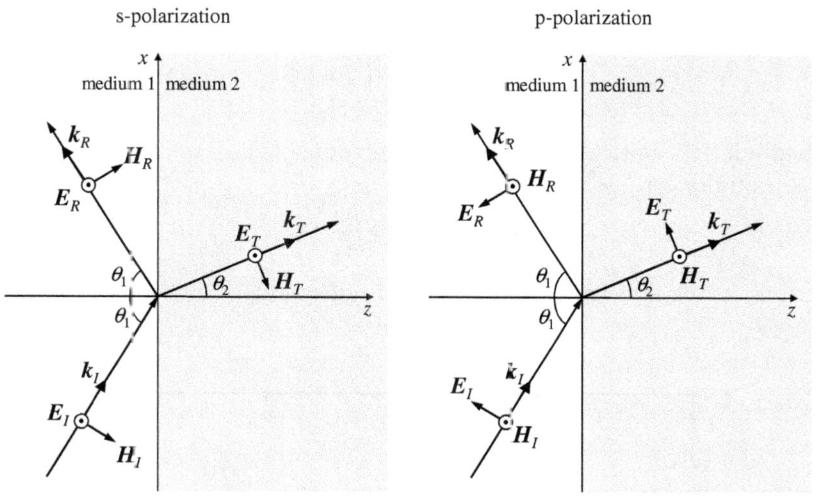

Fig. 2.2. Reflection and transmission of s- and p-polarized electromagnetic waves

Figure 2.2 illustrates the incident, reflected, and transmitted waves on a plane of incidence. S-polarization denotes the case when the polarization of the incident wave is perpendicular to the plane of incidence, and p-polarization is when the polarization of the incident wave is parallel to the plane of incidence. Eventually, the boundary conditions determine the ratios of the reflected and transmitted field amplitudes to the incident field amplitude. These relations are expressed in the Fresnel equations:

$$\frac{E_{R,s}}{E_{I,s}} = \frac{n_1\cos\theta_1 - n_2\cos\theta_2}{n_1\cos\theta_1 + n_2\cos\theta_2} \quad \text{and} \quad \frac{E_{T,s}}{E_{I,s}} = \frac{2n_1\cos\theta_1}{n_1\cos\theta_1 + n_2\cos\theta_2} \tag{2.28}$$

for s-polarization and

$$\frac{E_{R,p}}{E_{I,p}} = \frac{n_2\cos\theta_1 - n_1\cos\theta_2}{n_2\cos\theta_1 + n_1\cos\theta_2} \quad \text{and} \quad \frac{E_{T,p}}{E_{I,p}} = \frac{2n_1\cos\theta_1}{n_2\cos\theta_1 + n_1\cos\theta_2} \tag{2.29}$$

for p-polarization.

Reflectivity R is defined as the fraction of incident radiation power that reflects from the boundary, and transmittance T that transmits through. Since the radiation intensity striking the interface is the normal component of the Poynting vector $\langle \mathbf{S} \rangle \cdot \mathbf{e}_k$, the reflectivity and the transmittance are given as

$$R = \frac{|E_R|^2}{|E_I|^2} \quad \text{and} \quad T = \frac{n_2|E_T|^2}{n_1|E_I|^2}. \tag{2.30}$$

Figure 2.3 shows an example of the reflectivity of s- and p-polarized waves versus the angle of incidence for n_1=1 and n_2=2. The reflectivity of p-polarization is completely expunged at *Brewster's angle*,

$$\theta_B = \tan^{-1}\left(\frac{n_2}{n_1}\right). \tag{2.31}$$

If medium 1 is optically denser than medium 2, i.e., $n_1 > n_2$, reflectivity becomes unity for

$$\theta_1 > \sin^{-1}\left(\frac{n_2}{n_1}\right). \tag{2.32}$$

This phenomenon is called *total internal reflection.*

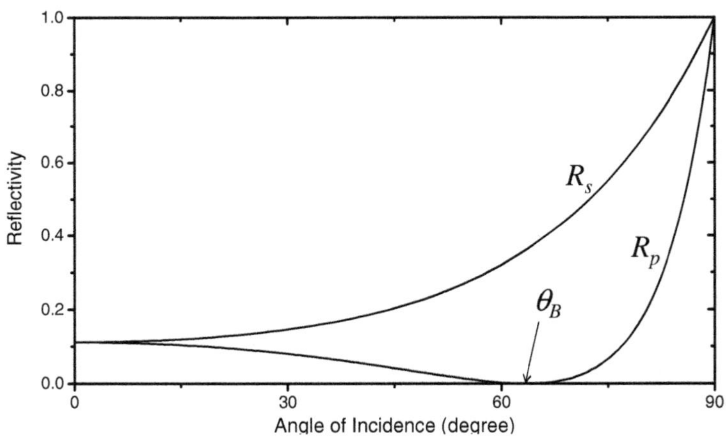

Fig. 2.3. Reflectivity of s- and p-polarization versus angle of incidence

The same boundary conditions are applied to a conducting surface. Consider an electromagnetic wave incident on an interface of air and a conductor with normal angle. Eq. 2.28 is still valid if we substitute n_2 with a complex index of refraction,

$$\tilde{n} = \frac{ck}{\omega} = c\sqrt{\frac{\mu\sigma}{2\omega}}(1+i), \tag{2.33}$$

using Eq. 2.25. Due to the large conductivity, $|\tilde{n}| \gg 1$,

$$E_R = \frac{1-\tilde{n}}{1+\tilde{n}} E_I \cong -E_I, \tag{2.34}$$

$$E_T = \frac{2}{1+\tilde{n}} E_I \cong 0. \tag{2.35}$$

The reflectivity is close to unity, and very little energy is dissipated into the conducting medium.

2.1.3 Coherent Transmission Spectroscopy

Coherent THz spectroscopy in a transmission geometry is a commonly used technique to measure the optical constants of materials. The coherent detection scheme measuring both the amplitude and phase of THz fields warrants simultaneous determination of the real and imaginary part of a dielectric function $\epsilon_r(\omega)$ or a conductivity $\sigma(\omega)$.

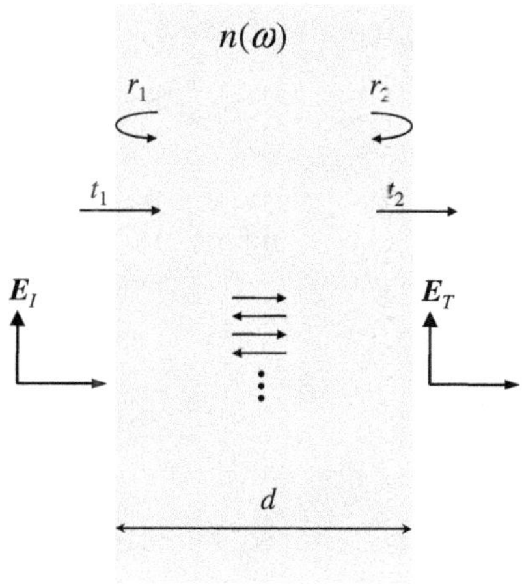

Fig. 2.4. Transmission of an electromagnetic wave through a flat single-layer of material with a complex index of refraction $\tilde{n}(\omega)$ and a thickness d. r_1 and r_2 are the reflection coefficients at the entrance and exit surfaces, respectively, and t_1 and t_2 are the transmission coefficients.

Figure 2.4 illustrates a normal-incident electromagnetic wave passing through a single-layer of material with a thickness d and a complex index of refraction

$$\tilde{n}(\omega) = n(\omega) + i\kappa(\omega). \tag{2.36}$$

Inside the layer some parts of the wave undergo multiple reflections at the interfaces before they transmit through. Ultimately the total transmission is the superposition of all the parts having endured multiple reflections. Thus, the transmitted field E_T can be expressed as

$$\begin{aligned} E_T &= E_I t_1 t_2 e^{i\phi_d} + E_I t_1 t_2 e^{i\phi_d} \cdot \left(r_1 r_2 e^{2i\phi_d}\right) + \cdots \\ &= E_I t_1 t_2 e^{i\phi_d} \sum_{m=0}^{\infty} \left(r_1 r_2 e^{2i\phi_d}\right)^m \\ &= \frac{E_I t_1 t_2 e^{i\phi_d}}{1 - r_1 r_2 e^{2i\phi_d}}, \end{aligned} \tag{2.37}$$

where

$$r_1(\omega) = r_2(\omega) = \frac{\tilde{n}(\omega) - 1}{\tilde{n}(\omega) + 1}, \tag{2.38}$$

$$t_1(\omega) = \frac{2}{\tilde{n}(\omega) + 1}, \tag{2.39}$$

$$t_2(\omega) = \frac{2\tilde{n}(\omega)}{\tilde{n}(\omega) + 1}, \tag{2.40}$$

obtained from Eq. 2.28 with the incident angle $\theta_1 = 0$, are the reflection and transmission coefficients at the entrance and exit surfaces, and

$$\phi_d(\omega) = \tilde{n}(\omega)\frac{\omega}{c}d \tag{2.41}$$

is the phase shift while the wave propagates a distance d within the material. What we actually measure is the complex transmission coefficient $t(\omega)$ with the amplitude $|t(\omega)|$ and the phase $\Phi(\omega)$, which is expressed as

$$\begin{aligned} t(\omega) &= |t(\omega)| e^{\Phi(\omega)} \\ &= \frac{E_T(\omega)}{E_I(\omega)} = \frac{4\tilde{n}(\omega) e^{i\phi_d(\omega)}}{\left[\tilde{n}(\omega) + 1\right]^2 - \left[\tilde{n}(\omega) - 1\right]^2 e^{2i\phi_d(\omega)}} \end{aligned} \tag{2.42}$$

in terms of $\tilde{n}(\omega)$. $\tilde{n}(\omega)$ is determined by fitting the transmission data to Eq. 2.42.

The complex dielectric function

$$\epsilon_r(\omega) = \left[\tilde{n}(\omega)\right]^2 = \epsilon_{r1}(\omega) + i\epsilon_{r2}(\omega) \tag{2.43}$$

and the complexity conductivity

$$\sigma(\omega) = \sigma_1(\omega) + i\sigma_2(\omega) \tag{2.44}$$

have the relation

$$\sigma_1(\omega) = -\epsilon_0\omega\, \epsilon_{r2}(\omega), \tag{2.45}$$

$$\sigma_2(\omega) = \epsilon_0\omega \left[\epsilon_{r1}(\omega) - \epsilon_{r1}(\infty)\right], \tag{2.46}$$

where $\epsilon_{r1}(\infty)$ is the dielectric constant in the high-frequency limit.

2.1.4 Absorption and Dispersion

Frequency dispersion refers to the phenomenon in which waves of different frequencies propagate at different speeds. Dispersion, together with absorption, characterizes how media respond to external electromagnetic fields. All electromagnetic phenomena involve the interaction of fields with charged particles, electrons and nuclei at a microscopic scale, in matter. Electromagnetic waves force charged particles to move; their accelerated motion induces radiation. The effects of magnetic fields on naturally occurring materials are mostly negligible and the amplitude of the electron motions are usually very small. Consequently, electromagnetic properties of a medium are dominated by electric dipoles induced by the applied electric fields. In the linear optical regime, the electric dipole moments are proportional to the amplitude of the applied electric fields.

The classical Lorentzian model provides a good qualitative description of this phenomenon. Assuming that a bound charge oscillates about its equilibrium position with a very small amplitude, we model the system as a simple harmonic oscillator. Figure 2.5 illustrates the Lorentzian harmonic oscillator model. The potential energy of the charged particle is quadratic for small displacements from equilibrium. The size of the oscillator is much smaller than the wavelength of the applied field, hence the electric field is constant near the equilibrium position for a given time t.

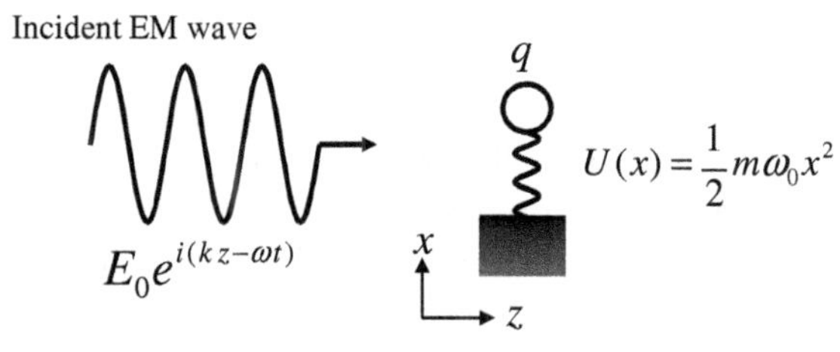

Fig. 2.5. The classical Lorentzian model accounts for the optical response of bound electrons in dielectric media.

When the monochromatic wave

$$E(t) = E_0 e^{-i\omega t} \tag{2.47}$$

of angular frequency ω, polarized along the x axis, interacts with the charge q, its equation of motion is expressed as

$$\frac{d^2x}{dt^2} + \gamma \frac{dx}{dt} + \omega_0^2 x = \frac{q}{m} E(t), \tag{2.48}$$

where q and m are the charge and the mass. γ and ω_0 are the damping constant and the resonant frequency, respectively. The solution of the equation of motion is

$$x(t) = x_0 e^{-i\omega t}, \quad \text{where} \quad x_0 = \frac{q}{m} \frac{E_0}{\omega_0^2 - \omega^2 - i\omega\gamma}. \tag{2.49}$$

The electric dipole moment of the harmonic oscillator is

$$p(t) = qx(t). \tag{2.50}$$

We suppose that a medium has N oscillators per unit volume, then the electric polarization is given by

$$P(t) = Nqx(t) = \frac{Nq^2}{m} \frac{E_0 e^{-i\omega t}}{\omega_0^2 - \omega^2 - i\omega\gamma} \equiv \epsilon_0 \chi_e(\omega) E_0 e^{-i\omega t}, \tag{2.51}$$

where $\chi_e(\omega)$ denotes the linear electric susceptibility of the medium. Inserting Eq. 2.51 into Eq. 2.6 we obtain the dielectric constant of the medium

$$\epsilon_r(\omega) \equiv \frac{\epsilon(\omega)}{\epsilon_0} = 1 + \chi_e(\omega) = 1 + \frac{Nq^2}{m\epsilon_0} \frac{1}{\omega_0^2 - \omega^2 - i\omega\gamma}. \tag{2.52}$$

The real and imaginary parts of the complex dielectric constant are written as

$$\Re\left[\epsilon_r\right] - 1 = \frac{Nq^2}{m\epsilon_0} \frac{\omega_0^2 - \omega^2}{\left(\omega_0^2 - \omega^2\right)^2 + \omega^2\gamma^2}, \tag{2.53}$$

$$\Im\left[\epsilon_r\right] = \frac{Nq^2}{m\epsilon_0} \frac{\omega\gamma}{\left(\omega_0^2 - \omega^2\right)^2 + \omega^2\gamma^2}. \tag{2.54}$$

Figure 2.6 shows characteristic dielectric dispersion in the vicinity of the resonant frequency. This medium is dispersive because its response to an external electromagnetic wave depends on frequency. The imaginary part of the dielectric constant indicates that absorption is maximized at the resonant frequency, and the bandwidth is $\sim \gamma$.

From the dispersion relation of Eq. 2.20 we get the complex amplitude of the wave vector

$$k(\omega) = k_R(\omega) + ik_I(\omega) = \sqrt{\epsilon_r(\omega)} \frac{\omega}{c}, \tag{2.55}$$

which governs how the wave propagates in the medium. The plane wave propagating along the z-axis is written as

$$\mathbf{E}(z,t) = \mathbf{E}_0 e^{i(kz - \omega t)} = \mathbf{E}_0 e^{-\frac{\alpha}{2}z} e^{-i\omega\left(t - \frac{n}{c}z\right)}. \tag{2.56}$$

The radiation intensity exponentially decays with an absorption coefficient

$$\alpha(\omega) = 2\Im[k(\omega)], \tag{2.57}$$

and the wave propagates with the phase velocity

$$v = \frac{n(\omega)}{c} = \frac{\Re[k(\omega)]}{\omega}. \tag{2.58}$$

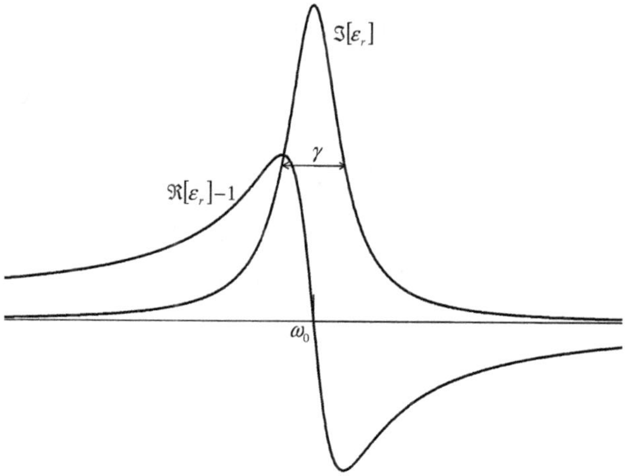

Fig. 2.6. Dielectric dispersion at a resonance.

2.1.5 Plasma Frequency

Consider the interaction of electromagnetic waves with a system in which charges move freely, and scattering between the particles is negligible. Doped semiconductors or plasmas may behave in such a way upon incidence of THz radiation. Substituting $\omega_0 = 0$ and $\gamma = 0$ into Eq. 2.52 we get the dielectric constant,

$$\epsilon_r(\omega) = 1 - \frac{\omega_p^2}{\omega^2}, \tag{2.59}$$

where

$$\omega_p = \sqrt{\frac{Nq^2}{m\epsilon_0}} \tag{2.60}$$

is the *plasma frequency*. The dispersion relation,

$$ck = \sqrt{\omega^2 - \omega_p^2}, \tag{2.61}$$

indicates that waves can propagate through the medium for $\omega > \omega_p$, while it decays with the absorption coefficient,

$$\alpha(\omega) = \frac{2}{c}\sqrt{\omega_p^2 - \omega^2}, \tag{2.62}$$

for $\omega < \omega_p$. A typical electron density, 10^{16} cm^{-3}, of laboratory plasmas and doped semiconductors corresponds to $\omega_p \cong 6$ THz.

2.1.6 Electric Dipole Radiation

Electromagnetic waves are generated by accelerating charges and time-varying currents. In the present section, we discuss the emission of radiating fields from an oscillating electric dipole. In most cases, it is the predominant source of electromagnetic radiation.

Before dealing with this specific case, we briefly overview the theoretical background. Here we introduce the potential formulation, which provides a relatively simple way to describe the radiation process. Scalar and vector potentials, Φ and $\mathbf{A}$, are defined as the relations

$$\mathbf{E} = -\nabla\Phi - \frac{\partial \mathbf{A}}{\partial t}, \tag{2.63}$$

$$\mathbf{B} = \nabla \times \mathbf{A}. \tag{2.64}$$

with electric and magnetic fields, $\mathbf{E}$ and $\mathbf{B}$. In the potential formalism, Maxwell's equations reduce to

$$\nabla^2\Phi - \frac{1}{c^2}\frac{\partial^2\Phi}{\partial t^2} = -\frac{1}{\epsilon_0}\rho \tag{2.65}$$

$$\nabla^2\mathbf{A} - \frac{1}{c^2}\frac{\partial^2\mathbf{A}}{\partial t^2} = -\mu_0\mathbf{J} \tag{2.66}$$

assuming the potentials are related by the Lorentz gauge

$$\nabla \cdot \mathbf{A} + \frac{\partial \Phi}{\partial t} = 0. \tag{2.67}$$

It is well known that the retarded potentials

$$\Phi(\mathbf{r}, t) = \frac{1}{4\pi\epsilon_0}\int \frac{\rho(\mathbf{r}', t')}{|\mathbf{r} - \mathbf{r}'|} d\mathbf{r}'^3 \tag{2.68}$$

and

$$\mathbf{A}(\mathbf{r}, t) = \frac{\mu_0}{4\pi}\int \frac{\mathbf{J}(\mathbf{r}', t')}{|\mathbf{r} - \mathbf{r}'|} d\mathbf{r}'^3, \tag{2.69}$$

where $\mathbf{r}'$ is the position of the sources and $t'(= t - |\mathbf{r} - \mathbf{r}'|/c)$ is the retarded time, are solutions of the wave equations for the potentials.

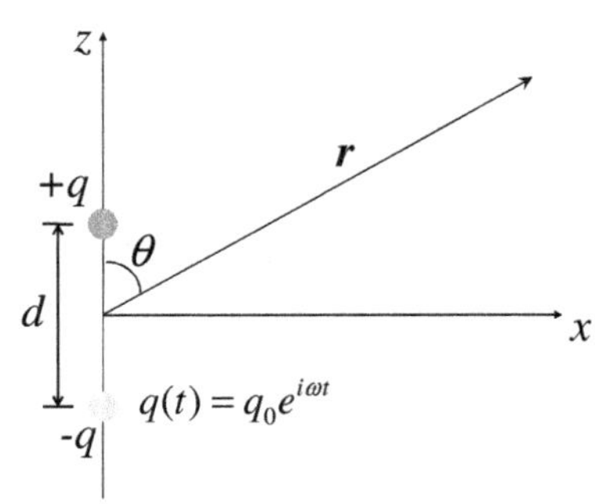

Fig. 2.7. Oscillating dipole

Now we apply this result to an oscillating dipole (Fig. 2.7). Two opposite charges, separated by a distance d, oscillate with an angular frequency ω: $q(t) = q_0 e^{-i\omega t}$. Then, the dipole moment is expressed as

$$\mathbf{p}(t) = \mathbf{p}_0 e^{-i\omega t} \tag{2.70}$$

with $p_0 = q_0 d$. The current of this system is

$$I(t) = \frac{dq(t)}{dt} = -i\omega q(t) = -\frac{ikc}{d}p(t). \tag{2.71}$$

Inserting these charge and current distributions into Eqs. 2.68 and 2.69, we can rewrite the scalar and vector potentials as

$$\Phi(\mathbf{r}, t) = \frac{q(t)}{4\pi\epsilon_0}\left(\frac{e^{ik|\mathbf{r}-\mathbf{d}/2|}}{|\mathbf{r}-\mathbf{d}/2|} - \frac{e^{ik|\mathbf{r}+\mathbf{d}/2|}}{|\mathbf{r}+\mathbf{d}/2|}\right) \tag{2.72}$$

and

$$\mathbf{A}(\mathbf{r}, t) = -\frac{i\mu_0 kc\mathbf{p}(t)}{4\pi d}\int_{-d/2}^{d/2}\frac{e^{ik|\mathbf{r}-z\mathbf{e}_z|}}{|\mathbf{r}-z\mathbf{e}_z|}dz. \tag{2.73}$$

Here we consider waves in the radiation or far zone, $d \ll \lambda \ll r$, where observations occur far from the source, the source size is much smaller than the radiation wavelength, and the wavelength is much shorter than the distance between the source and the observing position. In fact, most realistic circumstances render observations of electromagnetic radiation occurring in the far zone. In this limit we can approximate the potentials as

$$\Phi(\mathbf{r}, t) = -\frac{ik\mathbf{r}\cdot\mathbf{p}_0}{4\pi\epsilon_0 r^2}e^{i(kr-\omega t)} \tag{2.74}$$

and

$$\mathbf{A}(\mathbf{r}, t) = -\frac{ik\mathbf{p}_0}{4\pi c\epsilon_0 r}e^{i(kr-\omega t)}. \tag{2.75}$$

Using Eqs. 2.63 and 2.64 we obtain

$$\begin{aligned}\mathbf{E}(\mathbf{r}, t) &= \frac{k^2}{4\pi\epsilon_0 r}\left[\mathbf{p}_0 - \frac{(\mathbf{p}_0\cdot\mathbf{r})\mathbf{r}}{r^2}\right]e^{i(kr-\omega t)} \\ &= -\left[\frac{k^2 p(t-r/c)}{4\pi\epsilon_0 r}\sin\theta\right]\mathbf{e}_\theta\end{aligned} \tag{2.76}$$

and

$$\begin{aligned}\mathbf{B}(\mathbf{r}, t) &= \frac{1}{c}\frac{k^2}{4\pi\epsilon_0 r}\frac{\mathbf{p}_0\times\mathbf{r}}{r}e^{i(kr-\omega t)} \\ &= -\left[\frac{1}{c}\frac{k^2 p(t-r/c)}{4\pi\epsilon_0 r}\sin\theta\right]\mathbf{e}_\phi.\end{aligned} \tag{2.77}$$

These transverse fields are related by

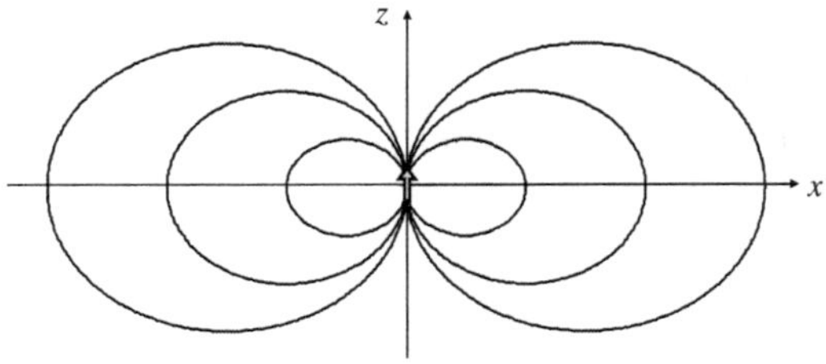

Fig. 2.8. Electric dipole radiation pattern

$$\nabla \times \mathbf{B} = -\frac{ik}{c}\mathbf{E}. \tag{2.78}$$

The time-averaged energy flux is given by

$$\langle \mathbf{S} \rangle = \frac{1}{2\mu_0}\mathbf{E}_0 \times \mathbf{B}_0 = \left[\frac{\mu_0 p_0^2}{32\pi^2 c}\frac{\omega^4}{r^2}\sin^2\theta\right]\mathbf{e}_r \propto \frac{\sin^2\theta}{r^2}\mathbf{e}_r \tag{2.79}$$

The radiation propagates radially outward with an anisotropic power distribution proportional to $\sin^2\theta$. Figure 2.8 shows the angular distribution of the radiated power. The total radiation power,

$$P = \oint \langle \mathbf{S} \rangle \cdot d\mathbf{a} = \frac{\mu_0}{4\pi c}\frac{p_0^2\omega^4}{3}, \tag{2.80}$$

goes up as the square of the dipole moment and the forth power of the frequency.

2.1.7 Quasi-Optical Propagation in Free Space

Many THz optical systems require high-level control of lateral extent and mode quality of a THz beam. Propagating in such a system, the beam is focused on some components with a spot size comparable to its wavelength. In this case, diffraction is of great importance to describe the beam propagation. Quasi optics deals with such cases where a beam of long wavelength propagates in free space under the influence of strong diffraction.

Given that an electromagnetic wave in free space is monochromatic and linearly polarized,

$$\mathbf{E}(\mathbf{r}, t) = \mathbf{e}_x E(x, y, z)e^{-i\omega t}, \tag{2.81}$$

the wave equation 2.15 takes the form of the Helmholtz equation

$$\nabla^2 E(x, y, z) + k^2 E(x, y, z) = 0 \tag{2.82}$$

with the wavenumber $k = \omega/c = 2\pi/\lambda$. For the beam propagating along the z-axis, the electric field can be written as

$$E(x, y, z) = \psi(x, y, z)e^{ikz}, \tag{2.83}$$

where $\psi(x, y, z)$ varies slowly along the z-axis. With the approximation

$$\left|\frac{\partial^2 \psi}{\partial z^2}\right| \ll 2k\left|\frac{\partial \psi}{\partial z}\right|, \tag{2.84}$$

substitution of Eq. 2.83 into Eq. 2.82 leads to

$$\frac{\partial^2 \psi}{\partial x^2} + \frac{\partial^2 \psi}{\partial y^2} + 2ik\frac{\partial \psi}{\partial z} = 0. \tag{2.85}$$

When the paraxial approximation, in which the beam is confined to the vicinity of the propagation axis, is applied, the solutions to this wave equation are Hermite polynomials in rectangular coordinates and Laguerre polynomials in cylindrical coordinates with a Gaussian distribution as an envelope function.

In general, the lower order modes are dominant in free-space beam propagation. Thus we consider only the lowest order solution, the fundamental Gaussian mode

$$\psi(x, y, z) = \psi(r, z), \tag{2.86}$$

where $r = \sqrt{x^2 + y^2}$. As the spot size is being minimized at $z = 0$, the lateral beam profile in the plane of the beam waist takes the Gaussian distribution

$$\psi(r, 0) = e^{-r^2/w_0^2}, \tag{2.87}$$

where w_0 is called the beam waist radius. Conventionally, the beam diameter $2w_0$ refers to the spot size at the focal plane. Eq.2.87 also implies that the beam has a flat phase front at the beam waist. The beam expands away from the focal plane, while its profile remains a Gaussian shape. Figure 2.9 illustrates the Gaussian beam expansion near the beam waist.

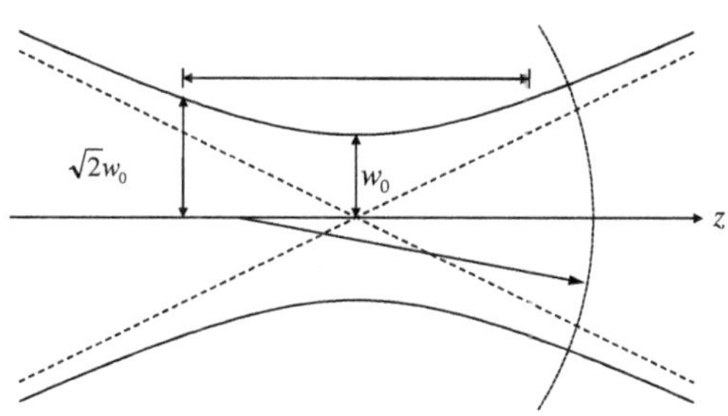

Fig. 2.9. Gaussian beam near the beam waist

The field amplitude at a distance z from the beam waist is expressed as

$$\psi(r, z) = \frac{w(z)}{w_0} e^{-r^2/w(z)^2} e^{ikr^2/2R(z)+i\phi(z)}, \tag{2.88}$$

where the beam parameters, $w(z)$, $R(z)$, and $\phi(z)$, are defined as

$$w(z) = w_0 \left[1 + \left(\frac{z}{z_0}\right)^2\right]^{\frac{1}{2}}, \tag{2.89}$$

$$R(z) = z \left[1 + \left(\frac{z}{z_0}\right)^2\right], \tag{2.90}$$

$$\phi(z) = \tan^{-1}\left(\frac{z}{z_0}\right) \tag{2.91}$$

with the Rayleigh length

$$z_0 = \frac{1}{2}kw_0^2 = \frac{\pi w_0^2}{\lambda}. \tag{2.92}$$

The beam divergence is insignificant within the confocal range, $-z_0 < z < z_0$: $w(\pm z_0) = \sqrt{2}w_0$. The field amplitude reduces to $1/e$ of its on-axis maximum at $z = \pm z_0$. The equiphase surface forms the spherical phase front, having the radius of curvature $R(z)$. The phase delay $\phi(z)$ is called the Gouy phase of the beam. The beam diverges with the angle

$$\theta_0 = \lim_{z\to\infty} \tan^{-1}\left[\frac{w(z)}{z}\right] = \tan^{-1}\left(\frac{w_0}{z_0}\right). \tag{2.93}$$

If the beam diameter is much larger than the wavelength, i.e., $w_0 \gg \lambda$, the divergence angle approaches

$$\theta_0 \simeq \frac{\lambda}{\pi w_0}. \tag{2.94}$$

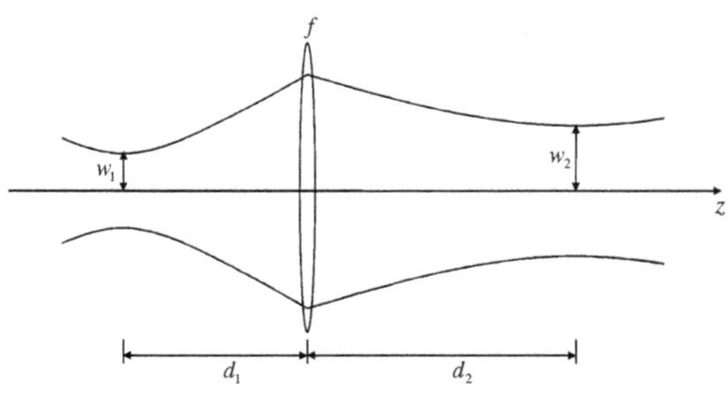

Fig. 2.10. Focusing of a Gaussian beam with a thin lens of focal length f

A simple and practical application of wave optical analysis is the transformation of a Gaussian beam by a focusing element. Figure 2.10 illustrates the focusing of a Gaussian beam, where w_1 and w_2 are the radii of the beam waists, and d_1 and d_2 are the distances of the beam waists from the lens. When w_1 and d_1 are given, w_2 and d_2 have the following relations:

$$w_2 = w_1 \cdot \frac{f}{[(d_1 - f)^2 + z_1^2]^{1/2}}, \tag{2.95}$$

$$d_2 = f \cdot \left[\frac{d_1(d_1 - f) + z_1^2}{(d_1 - f)^2 + z_1^2} \right], \tag{2.96}$$

where $z_1 = \pi w_1^2/\lambda$ is the Rayleigh length of beam waist 1. Figure 2.11 shows the ratio of w_2 to w_1 and the distance d_2 scaled with f as a function of d_1/f for several Rayleigh lengths.

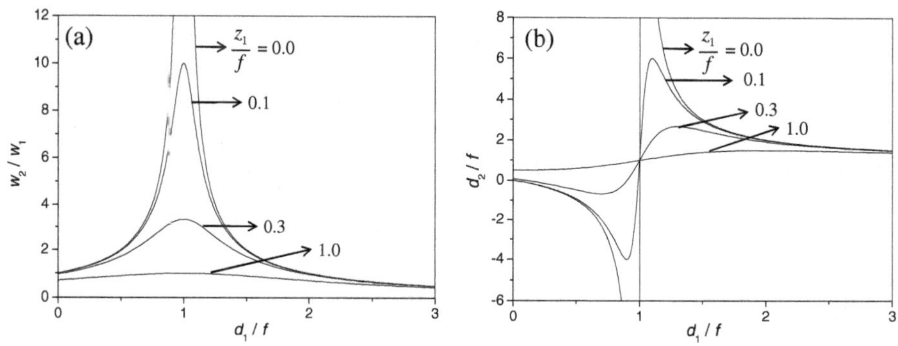

Fig. 2.11. Beam parameters of a Gaussian beam focused by a thin lens: (a) ratio of the beam waist radius w_2 to w_1 and (b) the distance of the focal plane from the lens d_2 scaled with f as function of the initial beam-waist location d_1 for several Rayleigh lengths, $z_1/f = 0.0$, 0.1, 0.3, and 1.0.

The equations describing the beam parameters tell us some important aspects of quasi-optical systems. First of all, the location of the focal plane is different from the prediction of geometrical optics, in which the lens equation

$$\frac{1}{d_1} + \frac{1}{d_2} = \frac{1}{f} \tag{2.97}$$

leads to

$$d_2 = f \cdot \left[\frac{d_1(d_1 - f)}{(d_1 - f)^2} \right]. \tag{2.98}$$

Comparing Eqs. 2.96 and 2.98, we see that the discrepancy becomes evident when the focal length of the lens f is comparable to or less than the Rayleigh length z_1. Second, as the lens has a clear aperture of a_L, the beam waist should be kept within a distance $\pi a_L^2/2\lambda$ from the lens for efficiently collimating the beam. Third, Fig. 2.11 indicates that $d_1 = f$ is a unique arrangement: d_2 is also equal to f and the beam waist has the maximum value, $w_2 = w_1(f/z_1)$. Fourth, for a telescope with a pair of lenses separated by the sum of their focal lengths, $f + f'$, the magnification and the focal-plane location are same as those of geometrical optics:

$$\frac{w_2}{w_1} = \frac{f'}{f}, \tag{2.99}$$

$$d_2 = \left(\frac{f'}{f}\right)^2 d_1 - \frac{f'}{f}(f + f'). \tag{2.100}$$

Note that these parameters are independent of wavelength. Because of this unique property, the telescope is a very useful device for broadband applications.

2.2 Terahertz Radiation and Elementary Excitations

Interactions of THz waves with matter involve low-energy excitations corresponding to THz frequencies. Some elementary excitations of cardinal interest include Rydberg transitions in atoms [1, 2, 3], transitions among impurity states in semiconductors [4], intraband transitions in semiconductor nanostructures [5, 6], many-body interactions in strongly correlated electron systems [7], phonon modes in organic and inorganic crystals [8], rotation-vibration transitions in molecules [9], and collective large-amplitude motions in biological molecules [10]. In the present section, we briefly review the basic quantum theories of the elementary excitations to gain insights into their fundamental properties on the microscopic scale.

2.2.1 Quantum Theory of Electric Dipole Interaction

Stationary quantum states of a single electron are governed by the time-independent Schödinger equation

$$H_0(\mathbf{r})\Psi_n(\mathbf{r}) = E_n\Psi_n(\mathbf{r}), \tag{2.101}$$

where the Hamiltonian H_0 is the summation of the kinetic and potential energies,

$$H_0(\mathbf{r}) = -\frac{\hbar^2}{2m_e}\nabla^2 + V(\mathbf{r}), \tag{2.102}$$

and Ψ_n and E_n are the energy eigenstates and eigenvalues. If the electron is confined to a finite space by the potential, the energy levels become discrete.

Consider an electromagnetic wave $\mathbf{E}(t)$ interacting with the electron. Here we assume the dipole approximation, in which the electron is confined to a region whose size is much smaller than the radiation wavelength. When the dipole approximation is made, the Hamiltonian of the electric dipole interaction is expressed as

$$H_I(t) = -\mathbf{p} \cdot \mathbf{E}(t), \tag{2.103}$$

where $\mathbf{p} = -e\mathbf{r}$ is the electric dipole moment. The time-dependent Schrödinger equation of the electron has the form

$$i\hbar \frac{\partial}{\partial t} \Psi(\mathbf{r}, t) = [H_0 + H_I(t)] \Psi(\mathbf{r}, t). \tag{2.104}$$

The electron response to the applied field becomes prominent when the radiation frequency is resonant with the energy difference between two energy eigenstates, $\hbar\omega = E_1 - E_2$. Absorption, stimulated emission, and spontaneous emission are the fundamental phenomena near the resonance, as shown in Fig. 2.12. In the quantum mechanical point of view, absorption is the process in which the electron in the lower energy state is excited to the higher energy state by absorbing an incident photon. Stimulated emission is the opposite process to absorption where the energy transition is from the higher to the lower state. It is essential to note that the emitted photon is in phase with the incident photon, which is the underlying process driving the coherent radiation from lasers. Spontaneous emission is a quantum optical effect in which a photon is emitted due to the relaxation of the electron from the higher energy state to the lower energy state without any external perturbation.

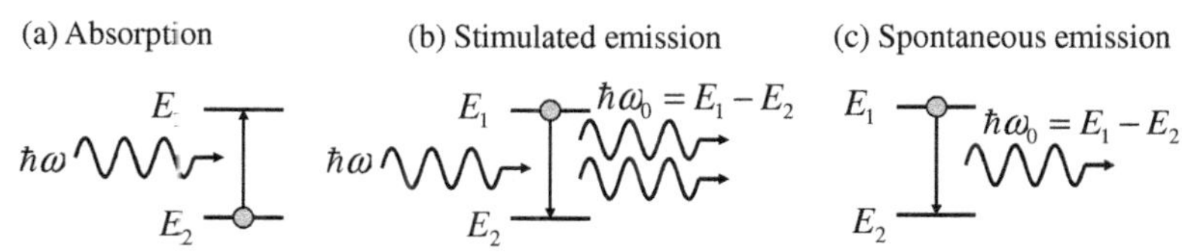

Fig. 2.12. Absorption, stimulated emission, and spontaneous emission at a resonance

Now we consider that a linearly-polarized electromagnetic wave of angular frequency ω interacts with a quantum system having N two-level atoms per unit volume. The energy eigenfunctions are given as $\phi_1(\mathbf{r})$ and $\phi_2(\mathbf{r})$. We define the resonance angular frequency as

$$\omega_0 = \frac{E_1 - E_2}{\hbar}. \tag{2.105}$$

The applied electric field is expressed as

$$E(t) = E_\omega e^{-i\omega t} + c.c., \tag{2.106}$$

where *c.c.* is complex conjugate. Time-independent perturbation theory is applied to obtain the linear electric susceptibility $\chi_e(\omega)$ of the system. We assume that initially the electrons are in the ground state,

$$\Psi(\mathbf{r}, t = -\infty) = \phi_2(\mathbf{r}), \tag{2.107}$$

and the incident light is weak so that only a small number of electrons are excited to the higher energy level by the dipole interaction.

The first-order approximation of perturbation theory yields the time-varying wavefunction

$$\Psi(\mathbf{r},t) = \phi_2(\mathbf{r}) + c_1^{(1)}(t)\phi_1(\mathbf{r}) \tag{2.108}$$

with the coefficient

$$\begin{aligned} c_1^{(1)}(t) &= -\frac{e^{-i\omega_0 t}}{i\hbar}\int_{-\infty}^{t} dt' \left[-\frac{d_{12}}{\sqrt{2}}E(t')\right] e^{i\omega_0 t'} \\ &= \frac{d_{12}}{\hbar\sqrt{2}}\left[\frac{E_\omega e^{-i\omega t}}{\omega_0 - \omega - i\gamma/2} + \frac{E_\omega^* e^{i\omega t}}{\omega_0 + \omega + i\gamma/2}\right]. \end{aligned} \tag{2.109}$$

Here we introduce the decay rate γ, which reflects the finite lifetime of the excited state. If the two level atoms are independent of each other and isolated from their surroundings, γ represents the transition rate of spontaneous emission. The matrix elements of the dipole moment operator, $p_x = -ex$, are given as

$$d_{12} = d_{21}^* = \sqrt{2}\,\langle\phi_1|p_x|\phi_2\rangle = -\sqrt{2}e\int \phi_1^*(\mathbf{r})x\phi_2(\mathbf{r})d^3\mathbf{r}. \tag{2.110}$$

The field induced polarization is obtained by integrating over the expectation value of the dipole moment,

$$P(t) = N\langle p\rangle = -Ne\int \Psi^*(\mathbf{r},t)\,x\,\Psi(\mathbf{r},t)d^3\mathbf{r}. \tag{2.111}$$

Inserting Eq. 2.108 into the Eq. 2.111, we obtain

$$\begin{aligned} P(t) &= \frac{1}{\sqrt{2}}c_1^{(1)}(t)\,d_{21} + \frac{1}{\sqrt{2}}\left[c_1^{(1)}(t)\right]^* d_{12} \\ &= \frac{N|d_{12}|^2}{2\hbar}\left[\frac{1}{\omega_0 - \omega - i\gamma/2} + \frac{1}{\omega_0 + \omega + i\gamma/2}\right] E_\omega e^{-i\omega t} + c.c. \end{aligned} \tag{2.112}$$

Using a complex number formalism in which we keep the first term proportional to $e^{-i\omega t}$, we get the linear susceptibility,

$$\chi_e(\omega) = \frac{N|d_{12}|^2}{2\hbar\epsilon_0}\left[\frac{1}{\omega_0 - \omega - i\gamma/2} + \frac{1}{\omega_0 + \omega + i\gamma/2}\right], \tag{2.113}$$

from the relation

$$P(t) = \epsilon_0\chi_e(\omega)E_\omega e^{-i\omega t}. \tag{2.114}$$

It is noteworthy to compare the quantum mechanical analysis with the classical Lorentz model discussed in section 2.1.4. Near the resonance $\omega \approx \omega_0$, Eq. 2.113 becomes

$$\chi_e(\omega) \cong \frac{N|d_{12}|^2}{2\hbar\epsilon_0}\left[\frac{1}{\omega_0 - \omega - i\gamma/2}\right] \tag{2.115}$$

assuming $\gamma \ll \omega_0$. This is consistent with the classical susceptibility in Eq. 2.51,

$$\chi_e(\omega) = \frac{Ne^2}{m\epsilon_0}\left[\frac{1}{\omega_0^2 - \omega^2 - i\omega\gamma}\right]$$
$$\cong \frac{Ne^2}{2m\omega_0\epsilon_0}\left[\frac{1}{\omega_0 - \omega - i\gamma/2}\right]. \tag{2.116}$$

You can see that the simple classical model is a good approximation of the quantum mechanical system in the linear response regime.

No classical picture, however, accounts for the case when the applied field is sufficiently strong so that a significant portion of the atoms are excited into the higher energy level. In order to describe the nonperturbative regime we introduce the density matrix formalism. As a representative case, we deal with the simplest quantum system: a two-level system. By studying this simple example in detail, we can appreciate many of the fundamental properties of nonperturbative transitions. Furthermore, when it comes to a resonant transition in a quantum system, the effects of other levels are usually negligible.

When the two-level system is a statistical mixture of states $\{\,|\Psi_\alpha\rangle\}$ with corresponding fractional population P_α, it is described by the density operator

$$\rho = \sum_\alpha P_\alpha |\Psi_\alpha\rangle\langle\Psi_\alpha|. \tag{2.117}$$

For a pure quantum state, $|\Psi\rangle = c_1|\phi_1\rangle + c_2|\phi_2\rangle$, the density matrix is given as

$$\rho = |\Psi\rangle\langle\Psi| = \begin{pmatrix} |c_1|^2 & c_1c_2^* \\ c_1^*c_2 & |c_2|^2 \end{pmatrix}. \tag{2.118}$$

In general the density matrix is expressed as

$$\rho = \begin{pmatrix} \rho_{11} & \rho_{12} \\ \rho_{21} & \rho_{22} \end{pmatrix}. \tag{2.119}$$

Comparing Eq. 2.119 with Eq. 2.118 we can see the physical meaning of the density matrix. The diagonal elements are the population of the levels 1 and 2, satisfying $\rho_{11} - \rho_{22} = 1$, and the off-diagonal elements represent the coherence, or the relative phase between the two levels, of the system.

The density matrix obeys the Liouville equation,

$$i\hbar\frac{\partial\rho}{\partial t} = [H_0 + H_I(t), \rho] + i\hbar\left(\frac{\partial\rho}{\partial t}\right)_{\text{damping}}. \tag{2.120}$$

Neglecting the damping term we can easily derive it from the Schrödinger equation. In matrix formalism, the unperturbed Hamiltonian is written as

$$H_0 = \begin{pmatrix} E_1 & 0 \\ 0 & E_2 \end{pmatrix} = \frac{1}{2}\hbar\omega_0\begin{pmatrix} 1 & 0 \\ 0 & -1 \end{pmatrix} \tag{2.121}$$

and the dipole interaction Hamiltonian is expressed as

$$\begin{aligned} H_I(t) &= -\mathbf{p} \cdot \mathbf{E}(t) \\ &= -\left[p_x E_x(t) + p_y E_y(t)\right] = -\left[p_+ E_-(t) + p_- E_+(t)\right] \end{aligned} \quad (2.122)$$

with the dipole matrices,

$$p_+ = d_{12} \begin{pmatrix} 0 & 1 \\ 0 & 0 \end{pmatrix} \quad \text{and} \quad p_- = d_{12} \begin{pmatrix} 0 & 0 \\ 1 & 0 \end{pmatrix}, \quad (2.123)$$

where d_{12} is real, and the circularly polarized fields and dipoles,

$$E_\pm = \frac{1}{\sqrt{2}}(E_x \pm iE_y) \quad (2.124)$$

$$p_\pm = \frac{1}{\sqrt{2}}(p_x \pm ip_y). \quad (2.125)$$

We introduce a phenomenological damping of population and polarization. The population decay and the dephasing are described by

$$-\frac{1}{T_1}\left\{\left[\rho_{11}(t) - \rho_{22}(t)\right] - \left[\rho_{11}(0) - \rho_{22}(0)\right]\right\} \quad (2.126)$$

and

$$-\frac{\rho_{12}(t)}{T_2} \quad \text{and} \quad -\frac{\rho_{21}(t)}{T_2}, \quad (2.127)$$

respectively. The longitudinal relaxation time T_1 is the population decay time. The transverse relaxation time T_2 (decoherence or dephasing time) is the inverse of the resonance linewidth. In this phenomenological model, the Liouville equation leads to the set of three equations,

$$i\hbar\frac{\partial \rho_{12}}{\partial t} = \hbar\omega_0 \rho_{12} + d_{12}E_-(\rho_{11} - \rho_{22}) - \frac{i\hbar}{T_2}\rho_{12}, \quad (2.128)$$

$$i\hbar\frac{\partial \rho_{21}}{\partial t} = -\hbar\omega_0 \rho_{21} - d_{12}E_+(\rho_{11} - \rho_{22}) - \frac{i\hbar}{T_2}\rho_{21}, \quad (2.129)$$

$$i\hbar\frac{\partial}{\partial t}(\rho_{11} - \rho_{22}) = d_{12}(E_-\rho_{21} - E_+\rho_{12}) - \frac{1}{T_1}\left[(\rho_{11} - \rho_{22}) + 1\right]. \quad (2.130)$$

The ensemble average of the electric dipole moment is obtained from $\mathbf{d} = [\mathbf{p}] = Tr(\rho\mathbf{p})$, then

$$d_x = [p_x] = \frac{d_{12}}{\sqrt{2}}(\rho_{21} + \rho_{12}) \quad (2.131)$$

$$d_y = [p_y] = \frac{d_{12}}{i\sqrt{2}}(\rho_{21} - \rho_{12}). \quad (2.132)$$

Here we define the pseudo-dipole

$$d_z \equiv d_{12}(\rho_{11} - \rho_{22}). \quad (2.133)$$

For a near-resonant excitation by a circularly-polarized field

$$\mathbf{E}(t) = \frac{1}{\sqrt{2}}(\mathbf{e}_x + i\mathbf{e}_y)E_\omega e^{-i\omega t} \tag{2.134}$$

we can simplify Eqs. 2.128, 2.129, and 2.130 in terms of the dipole moment $\mathbf{d}$ as

$$\frac{\partial \mathbf{d}}{\partial t} = \boldsymbol{\Omega}_R \times \mathbf{d} - \frac{1}{T_2}(d_x\mathbf{e}'_x + d_y\mathbf{e}'_y) - \frac{1}{T_1}(d_z + 1) \tag{2.135}$$

in a rotating frame with an angular frequency ω around the z-axis, where the Rabi frequency is given as

$$\boldsymbol{\Omega}_R = (\omega_0 - \omega)\mathbf{e}_z - \frac{2d_{12}}{\hbar}E(\omega)\mathbf{e}'_x. \tag{2.136}$$

Equation 2.135 is called the optical Bloch equation.

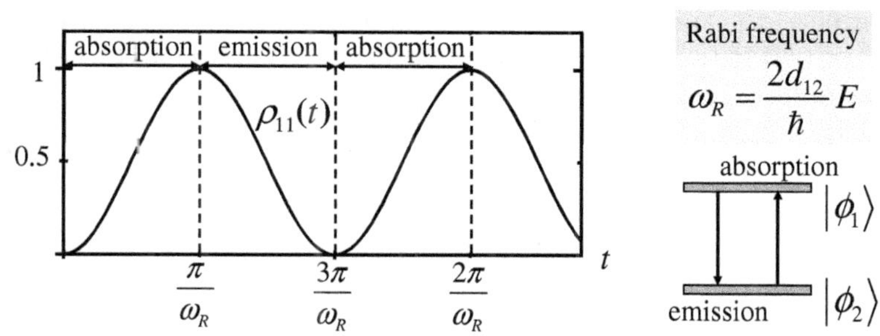

Fig. 2.13. Rabi oscillation

Neglecting the relaxation terms we get

$$\frac{\partial \mathbf{d}}{\partial t} = \boldsymbol{\Omega}_R \times \mathbf{d} \tag{2.137}$$

with the Rabi frequency

$$\boldsymbol{\Omega}_R = -\frac{2d_{12}}{\hbar}E(\omega)\mathbf{e}'_x = -\omega_R\mathbf{e}'_x \tag{2.138}$$

at the resonance $\omega = \omega_0$. The solution of this equation is that the vector $\mathbf{d}$ precesses around the x-axis in the rotating frame with an angular frequency ω_R. If the system is initially in the ground state, $d_z(0) = \rho_{11}(0) - \rho_{22}(0) = -1$ and $d_z(t) = \rho_{11}(t) - \rho_{22}(t) = -\cos\omega_R t$, then the time evolution of the excited state population is expressed as

$$\rho_{11}(t) = \frac{1 - \cos\omega_R t}{2}. \tag{2.139}$$

Figure 2.13 illustrates the cyclic evolution of the state population, called Rabi oscillation. Rabi oscillations are a fundamental phenomenon in quantum optics. Demonstration of Rabi oscillations for a quantum system is considered to be a very important step towards arbitrary quantum manipulation of the system.

2.2.2 Energy Levels of Hydrogen-like Atoms

An atom consists of a nucleus and bound electrons. Because of the spatial confinement of the electrons, atomic energy levels are discrete. Transitions between these energy eigenstates coincide with the absorption or emission of photons. Analyzing the spectra, we disentangle the threads of atomic structure. Hydrogen atoms, with only one electron in the system, have a relatively simple level structure which is not complicated by mutual interactions among electrons occurring in multi-electron atoms. Nevertheless, it provides an exemplary picture of how atomic energy levels are configured in general. Furthermore, the theoretical framework is pertinent to understanding the interactions of THz waves with Rydberg atoms (section 6.1) and with electrons in semiconductor nanostructures (section 8.1).

The time-independent Schrödinger equation for a hydrogen atom is written as

$$\left[-\frac{\hbar^2}{2m_e}\nabla^2 - \frac{e^2}{4\pi\epsilon_0 r}\right]\Psi(\mathbf{r}) = E\Psi(\mathbf{r}) \tag{2.140}$$

with the Coulomb potential

$$V(r) = -\frac{e^2}{4\pi\epsilon_0 r}. \tag{2.141}$$

Solving this equation, we obtain the energy eigenvalues of a hydrogen atom,

$$E_n = -\frac{\mathcal{R}}{n^2}, \tag{2.142}$$

where the positive integer n is the principle quantum number and

$$\mathcal{R} = -\frac{\hbar^2}{2m_e a_0^2} = -13.6 \text{ eV} \tag{2.143}$$

is the Rydberg constant. The Bohr radius,

$$a_0 = \frac{4\pi\epsilon_0\hbar^2}{m_e e^2} = 0.529 \text{ Å}, \tag{2.144}$$

estimates the size of a hydrogen atom.

The corresponding eigenfunctions have the form

$$\psi_{nlm}(r,\theta,\phi) = R_{nl}(r)Y_{lm}(\theta,\phi), \tag{2.145}$$

where $l(=0,1,2,\ldots,n-1)$ and $m(=-l,-l+1,\ldots,l-1,l)$ are the angular momentum quantum number and the magnetic quantum number, respectively.

The angular functions are spherical harmonics

$$Y_{lm}(\theta,\phi)=\sqrt{\frac{(2l+1)}{4\pi}\frac{(l-m)!}{(l+m)!}}P_l^m(\cos\theta)e^{im\phi}, \tag{2.146}$$

where $P_l^m(\cos\theta)e^{im\phi}$ are the associated Legendre polynomials. The first few spherical harmonics are

$$\begin{aligned}
Y_{00}(\theta,\phi)&=\frac{1}{\sqrt{4\pi}}\\
Y_{1\pm1}(\theta,\phi)&=\mp\sqrt{\frac{3}{8\pi}}\sin\theta e^{\pm i\phi}\\
Y_{10}(\theta,\phi)&=\sqrt{\frac{3}{4\pi}}\cos\theta\\
Y_{2\pm2}(\theta,\phi)&=\sqrt{\frac{15}{32\pi}}\sin^2\theta e^{\pm 2i\phi}\\
Y_{2\pm1}(\theta,\phi)&=\mp\sqrt{\frac{15}{8\pi}}\sin\theta\cos\theta e^{\pm i\phi}\\
Y_{20}(\theta,\phi)&=\sqrt{\frac{15}{32\pi}}(3\cos^2\theta-1).
\end{aligned}$$

The radial wave functions are

$$R_{nl}(r)=\left[\left(\frac{2Z}{na_0}^3\frac{(n-l-1)!}{2n\{(n+l)!\}^3}\right)\right]^{\frac{1}{2}}e^{-\frac{1}{2}\rho}\rho^l L_{n+l}^{2l+1}(\rho), \tag{2.147}$$

where $L_{n+l}^{2l+1}(\rho)$ are the associated Laguerre polynomials with

$$\rho=\frac{2}{na_0}r. \tag{2.148}$$

The first few $R_{nl}(r)$s are

$$\begin{aligned}
R_{10}(r)&=\left(\frac{Z}{a_0}\right)^{\frac{3}{2}}2e^{-\frac{Zr}{a_0}}\\
R_{20}(r)&=\left(\frac{Z}{2a_0}\right)^{\frac{3}{2}}\left(2-\frac{Zr}{a_0}\right)e^{-\frac{Zr}{2a_0}}\\
R_{21}(r)&=\left(\frac{Z}{2a_0}\right)^{\frac{3}{2}}\frac{Zr}{a_0\sqrt{3}}e^{-\frac{Zr}{2a_0}}.
\end{aligned}$$

Figure 2.14 shows these radial functions. When an electron is in a highly excited state having a large angular momentum, i.e., $l\gg1$, the radial wave function is pushed far out from the center.

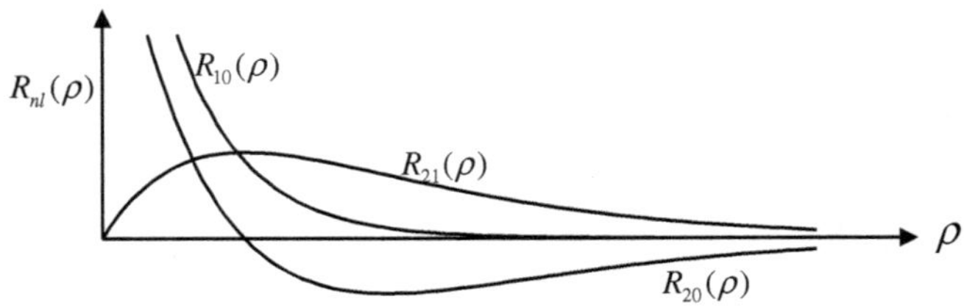

Fig. 2.14. Radial wave functions of a hydrogen atom

We will briefly overview how THz waves interact with atoms. THz radiation has much smaller photon energy ($h\nu$=4.1 meV at ν=1 THz) than the Rydberg constant, therefore atomic transitions involving THz photons can occur only between highly excited states. Such an atom, having an electron in a state with a very high principle quantum number n, is called a Rydberg atom. Rydberg atoms have several interesting properties. Their size, $2n^2a_0$, reaches up to ~1 μm with $n \approx 100$. Consequently, dipole moments are extremely large: $n^2ea_0 \sim 10^4$ D, which is several thousand times larger than typical polar molecules such as CO and H_2O. Their long lifetimes, ~1 ms, facilitate the control of the delicate quantum states. Taming these exotic states in a controlled manner is of great interest for several areas from wavepacket dynamics and quantum chaos to studies on interstellar materials. We will look into relevant studies in section 6.1.

2.2.3 Rotational and Vibrational Modes of Molecules

In order to understand the fundamental properties of molecules at the microscopic level, we start with describing the molecular energy levels and wave functions in the frame of quantum chemistry. While establishing the Hamiltonian of a molecule is straightforward, it is virtually impossible to get exact solutions of the time-independent Schrödinger equation. The total Hamiltonian includes five components:

$$\begin{aligned} H &= T_n + T_e + V_{en} + V_{ee} + V_{nn} \\ &= \sum_{\alpha} \frac{P_\alpha^2}{2M_\alpha} + \sum_i \frac{p_i^2}{2m_e} \\ &\quad - \frac{1}{4\pi\epsilon_0} \sum_{i,\alpha} \frac{Z_\alpha e^2}{r_{i\alpha}} + \frac{1}{4\pi\epsilon_0} \sum_{i,j} \frac{e^2}{r_{ij}} + \frac{1}{4\pi\epsilon_0} \sum_{\alpha,\beta} \frac{Z_\alpha Z_\beta e^2}{r_{\alpha\beta}}, \end{aligned} \tag{2.149}$$

where T_n and T_e are the nuclear and electron kinetic energies, and V_{en}, V_{ee}, and V_{nn} denote the electron-nuclear, electron-electron, and nuclear-nuclear Coulomb interaction energies.

It is practically impossible to solve multi-dimensional differential equations of this complexity without any approximations. The Born-Oppenheimer approximation is the most common way to make this problem less formidable. This method is based on the fact that nuclear masses are much greater than the electron mass—the mass ratio is in the order of 10^4. Because of their greater inertia, nuclei move exceedingly slow compared with electrons. For the same reason, the nuclear kinetic energy T_n is significantly smaller compared to other terms. Consequently, electron dynamics can be decoupled from the motion of nuclei.

The first step is to compute the electron energy eigenvalues and wave functions under the assumption that the nuclei are frozen in space and the nuclear kinetic energy is negligible. They are determined by solving the eigenvalue equation for the electrons

$$H_e \Psi_e(\mathbf{r};\mathbf{R}) = E_e(\mathbf{R})\Psi_e(\mathbf{r};\mathbf{R}), \tag{2.150}$$

where $H_e = T_e + V_{en} + V_{ee}$, and $\mathbf{R} \equiv \{\mathbf{R}_\alpha\}$ and $\mathbf{r} \equiv \{\mathbf{r}_i\}$ are the nuclear and electron position vectors. Then, the total wave function can be written as

$$\Psi(\mathbf{r},\mathbf{R}) = \Psi_e(\mathbf{r};\mathbf{R})\Psi_n(\mathbf{R}). \tag{2.151}$$

The nuclear wave function $\Psi_n(\mathbf{R})$ is determined in the next step of the approximation by solving the eigenvalue equation

$$[T_n(\mathbf{P}) + V_n(\mathbf{R})]\,\Psi_n(\mathbf{R}) = E\,\Psi_n(\mathbf{R}), \tag{2.152}$$

where the nuclear potential energy $V_n(\mathbf{R}) = T_n(\mathbf{R}) - E_e(\mathbf{R})$ is the combined effect of the nuclear-nuclear Coulomb interaction and electron shielding. The motion of the nuclei is of capital interest because their relatively low energies are relevant to the excitations in the THz regime.

For simplicity, we shall devote our attention to the dynamics of a diatomic molecule. Figure 2.15 illustrates the coordinate system of the diatomic molecule.

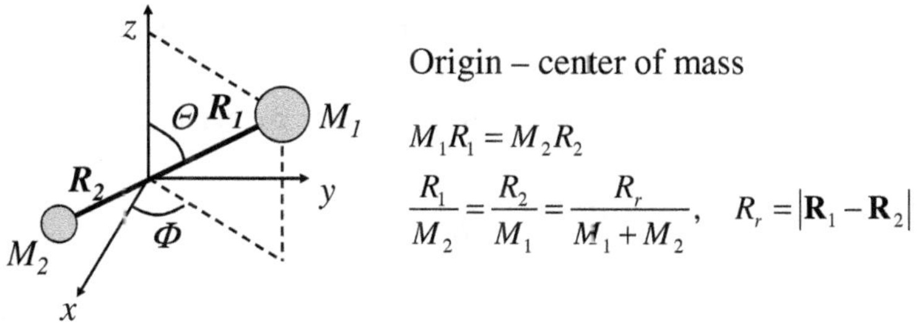

Fig. 2.15. Center-of-mass and relative coordinates of a diatomic molecule

Introducing the center of mass coordinates,

$$\mathbf{R}_c = \frac{M_1\mathbf{R}_1 + M_2\mathbf{R}_2}{M_1 + M_2} \quad \text{and} \quad \mathbf{P}_c = \mathbf{P}_1 + \mathbf{P}_2, \tag{2.153}$$

and the relative coordinates,

$$\mathbf{R}_r = \mathbf{R}_1 - \mathbf{R}_2, \quad \text{and} \quad \mathbf{P}_r = \frac{M_2\mathbf{P}_1 - M_1\mathbf{P}_2}{M_1 + M_2}, \tag{2.154}$$

we can rewrite the nuclear kinetic energy as

$$T_n(P_1, P_2) = \frac{P_1^2}{2M_1} + \frac{P_2^2}{2M_2} = \frac{P_c^2}{2\mu_c} + \frac{P_r^2}{2\mu_r}, \tag{2.155}$$

where $\mathbf{P}_1$ and $\mathbf{P}_2$ are the momenta of particle 1 and 2, $\mathbf{P}_c$ is the total momentum, $\mathbf{P}_r$ is the relative momentum, $\mu_c = M_1 + M_2$ is the total mass, and $\mu_r = M_1M_2/(M_1 + M_2)$ is the reduced mass. Since the nuclear potential $V_n(\mathbf{R}_1, \mathbf{R}_2)$ depends only on the relative distance $R_r(= |\mathbf{R}_r|)$, the center-of-mass motion obeys free particle dynamics. Neglecting its constant energy, the eigenvalue equation for nuclear motion is reduced to

$$\left[\frac{P_r^2}{2\mu_c} + V_n(\mathbf{R}_r)\right] \Psi_n(\mathbf{R}_r) = E\,\Psi_n(\mathbf{R}_r). \tag{2.156}$$

From now on we will drop the subscripts r and n for simplicity. Further separation of variables in spherical coordinate parts the equation

$$\left[-\frac{\hbar^2}{2\mu R^2}\frac{\partial}{\partial R}\left(R^2\frac{\partial}{\partial R}\right) + \frac{L^2}{2\mu R^2} + V(R)\right] \Psi(R, \Theta, \Phi) = E\,\Psi(R, \Theta, \Phi) \tag{2.157}$$

into the angular equation

$$L^2 Y_{lm}(\Theta, \Phi) = l(l+1)\hbar^2\; Y_{lm}(\Theta, \Phi) \tag{2.158}$$

and the radial equation

$$-\frac{\hbar^2}{2\mu}\frac{d^2U(R)}{dR^2} + \left[\frac{l(l+1)\hbar^2}{2\mu R^2} + V(R)\right] U(R) = E\; U(R), \tag{2.159}$$

where L is the angular momentum and the wave function is expressed as

$$\Psi(R, \Theta, \Phi) = \frac{1}{R}U(R)Y_{lm}(\Theta, \Phi). \tag{2.160}$$

Rotation

First we analyze the rotational modes of the molecule. If the relative distance between the two atoms is constant, we can treat the molecule as a rigid rotor. From the eigenvalue equation

$$H\Psi(\mathbf{R}) = \frac{L^2}{2I}\Psi(\mathbf{R}) = E\Psi(\mathbf{R}), \quad \mathbf{R} = (X, Y, Z), \tag{2.161}$$

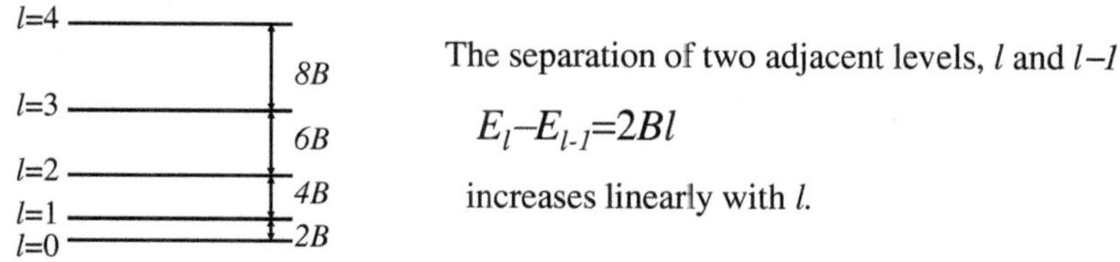

Fig. 2.16. Rotational energy levels of a diatomic molecule

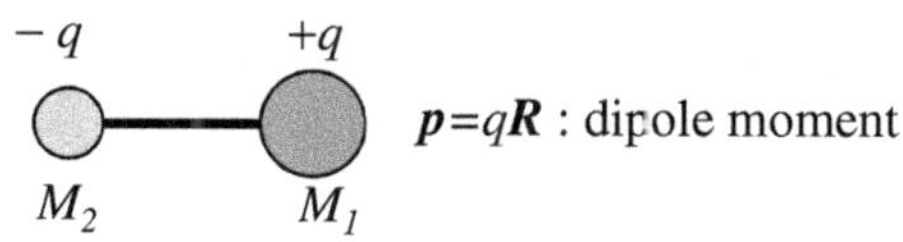

Fig. 2.17. Permanent dipole moment of a heteropolar diatomic molecule

with moment of inertia $I = \mu R^2$, we obtain the rotational energy levels of the diatomic molecule

$$E_{rot}(l) = Bl(l+1), \tag{2.162}$$

where $B = \hbar^2/2I$ is the rotational constant. Figure 2.16 shows the rotational energy levels for the first few states.

Rotational transitions absorb or emit photons only when the molecule has a permanent dipole moment (Fig. 2.17). When an electromagnetic wave interacts with the molecule, the interaction Hamiltonian is written as

$$H_I = -\mathbf{p} \cdot \mathbf{E}(t) = -qE(t)Z, \tag{2.163}$$

where the electric field is aligned along the z-axis. The expectation value of Z is expressed as

$$\langle Z \rangle(t) = \langle \Psi(t)|Z|\Psi(t)\rangle = \sum_{ll'} \sum_{mm'} c^*_{l'm'} c_{lm} \langle l'm'|Z|lm\rangle. \tag{2.164}$$

We get the matrix element

$$\begin{aligned}\langle l'm'|Z|lm\rangle &= \int d\Omega\, Y^*_{l'm'}(\Theta,\Phi) R \cos\Theta Y_{lm}(\Theta,\Phi) \\ &= R\delta_{mm'} \left[\delta_{l',l-1}\sqrt{\frac{l^2-m^2}{4l^2-1}} + \delta_{l',l+1}\sqrt{\frac{(l+1)^2-m^2}{4(l+1)^2-1}}\right]\end{aligned} \tag{2.165}$$

by using the spherical harmonics recurrence relation

$$\begin{aligned}\cos\Theta Y_{lm}(\Theta,\Phi) &= \sqrt{\frac{(l+m+1)(l-m+1)}{(2l+1)(2l+3)}} Y_{l+1,m}(\Theta,\Phi) \\ &+ \sqrt{\frac{(l+m)(l-m)}{(2l+1)(2l-1)}} Y_{l-1,m}(\Theta,\Phi).\end{aligned} \tag{2.166}$$

Inserting Eq. 2.165 into Eq. 2.164 we obtain the time-dependent dipole moment

$$\begin{aligned}\langle p\rangle(t) &= q\langle Z\rangle(t)\\ &= qR\sum_l\sum_m\left[\sqrt{\frac{l^2-m^2}{4l^2-1}}c^*_{l-1,m}(t)c_{l,m}(t)\right.\\ &\qquad\left.+\sqrt{\frac{(l+1)^2-m^2}{4(l+1)^2-1}}c^*_{l+1,m}(t)c_{l,m}(t)\right].\end{aligned} \tag{2.167}$$

This quantum mechanical quantity corresponds to the classical dipole moment of Eq. 2.50 and obeys the classical equation of motion.

Vibration

Now we go back to the equation for radial motion, Eq. 2.159. A typical molecular potential energy function $V(R)$ is shown in Fig. 2.18. The potential near the equilibrium distance R_0 can be approximated as a quadratic function of the displacement from R_0,

$$V(R) \cong -V_0 + \frac{1}{2}\mu\omega_0^2(R-R_0)^2. \tag{2.168}$$

If rotation is neglected, the radial equation becomes an eigenvalue equation for a simple harmonic oscillator,

$$\left[-\frac{\hbar^2}{2\mu}\frac{d^2}{dx^2}+\frac{1}{2}\mu\omega_0^2x^2\right]U(x) = E\,U(x), \tag{2.169}$$

where $x = R - R_0$, and the potential energy minimum $-V_0$ is neglected.

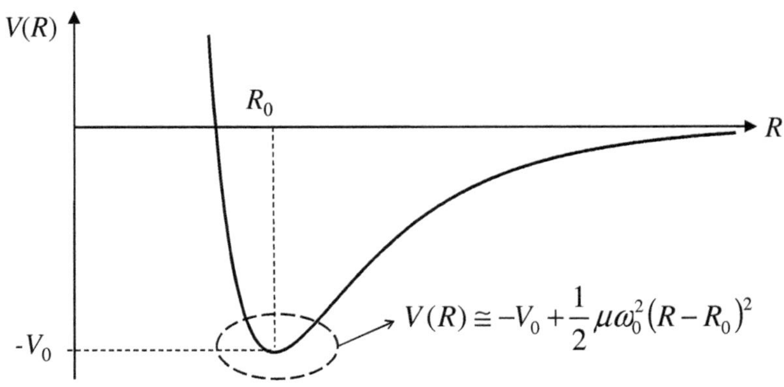

Fig. 2.18. Typical molecular potential energy function

The energy eigenvalues for the harmonic oscillator are

$$E_v = \left(v + \frac{1}{2}\right)\hbar\omega_0, \tag{2.170}$$

where $v = 0, 1, 2, \ldots$, and the eigenfunctions are

$$U_v(x) = \left(\frac{1}{\pi^{1/4}\sqrt{2^v v!}}\right)\left(\frac{1}{x_0^{v+1/2}}\right)\left(x - x_0^2\frac{d}{dx}\right)^v e^{-\frac{x^2}{2x_0^2}}, \tag{2.171}$$

where $x_0 = \sqrt{\hbar/\mu\omega_0}$.

Vibrational-Rotational Energy Levels

If the rotational and the vibrational motions are independent, the total energy is simply the sum of the two energies,

$$E^{tot} = E^{vib} + E^{rot} = \left(v + \frac{1}{2}\right)\hbar\omega_0 + Bl(l+1). \tag{2.172}$$

In reality, rotational and vibrational modes are coupled to each other, and there are two dominant effects of the coupling.

Figure 2.18 implies that the asymmetric potential gives rise to bond stretching as the molecule vibrates more. Centrifugal distortion induced by rotational motion also contributes to bond stretching. The vibrational and rotational bond stretching results in a reduction of the rotational constant:

$$B \longrightarrow B - a\left(v + \frac{1}{2}\right) - Dl(l+1), \tag{2.173}$$

where a and D are the coefficients of vibrational and rotational stretching, respectively. Therefore, the energy of a rotational-vibrational state is given as

$$E^{tot} = \left(v + \frac{1}{2}\right)\hbar\omega_0 + Bl(l+1) - a\left(v + \frac{1}{2}\right)[l(l+1)] - D[l(l+1)]^2. \tag{2.174}$$

Usually rotational energy is much smaller than vibrational energy. Figure 2.19 illustrates the rotational level splitting for a given vibrational state. The dotted vertical lines indicate uncoupled rotational states.

For a harmonic oscillator, transitions among the vibrational-rotational states must obey the selection rules $\Delta v = \pm 1$ and $\Delta l = \pm 1$. $\Delta l = 1$ is a forbidden transition for the diatomic molecule. Because of the anharmonic characteristics of real molecules, overtones resulting from the transitions of $|\Delta v| > 1$ are present, but they are usually very weak.

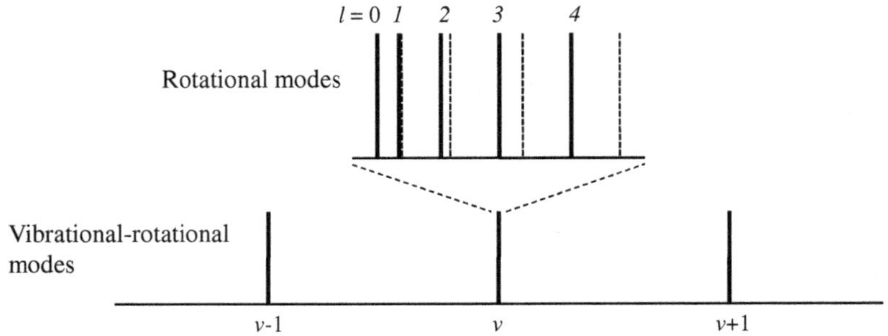

Fig. 2.19. Vibrational-rotational energy levels. l and v are the rotational and vibrational quantum numbers, respectively.

Polyatomic Molecules

It is straightforward to extend this theoretical analysis to the dynamics of polyatomic molecules, yet they have more complicated vibrational-rotational modes than diatomic molecules. For example, a water molecule (H_2O) has three degrees of vibrational and rotational freedom as shown in Fig. 2.20. The water molecule is an asymmetric top—the least symmetric form of a rigid rotor—having three different principle moments of inertia. The three normal modes of vibration involve the stretching and bending of the OH bonds. For a large molecule the bending modes generally have lower energies than the stretching modes, which often fall into the THz region.

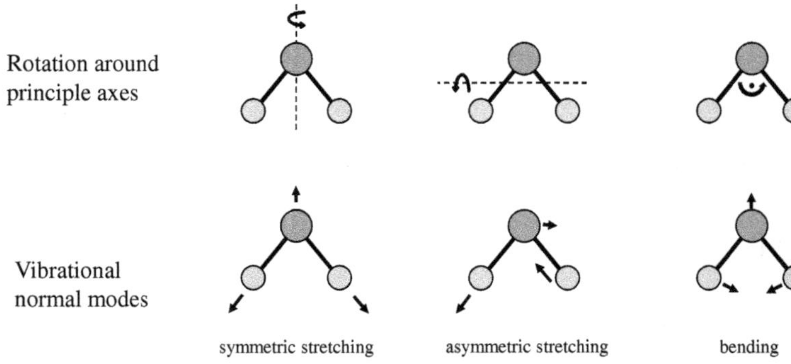

Fig. 2.20. Rotation and vibration of water molecules

If we treat a polyatomic molecule as a rigid rotor, the Hamiltonian of the rotational motion has the general form

$$H = \frac{J_a^2}{2I_a} + \frac{J_b^2}{2I_b} + \frac{J_c^2}{2I_c}, \tag{2.175}$$

where J_a, J_b, and J_c are the angular momentum operators along the three principle axes. Because the total angular momentum, $J_{tot}^2 = J_a^2 + J_b^2 + J_c^2$, commutes with the Hamiltonian, each energy level contains only one total angular momentum quantum number J. It is, however, nontrivial to express the eigenfunctions and energy levels of an asymmetric top ($I_a \neq I_b \neq I_c$) in terms of the angular momentum eigenstates and quantum numbers. The rotational energy levels of an asymmetric top are characterized by three quantum numbers $J_{K_{-1},K_{+1}}$ in the King-Hainer-Cross notation. The $K_{\pm 1}$ subscripts correspond to the K values in the limiting cases of prolate and oblate symmetric-tops, respectively.

Given that the rotator is a symmetric top characterized by $I_a = I_b = I$, the Hamiltonian is reduced to

$$H = \frac{J_{tot}^2}{2I} + \left(\frac{1}{2I_c} - \frac{1}{2I}\right) J_c^2, \tag{2.176}$$

and the energy eigenvalues are given as

$$E_{rot}(J, K) = BJ(J+1) + CK^2 \tag{2.177}$$

with the rotational constants $B = \hbar^2/2I$ and $C = \hbar^2(1/2I_c - 1/2I)$. The quantum number K can take the values $K = 0, 1, 2, \ldots, J$, where all levels are doubly degenerate except $K = 0$. The selection rules for symmetric tops are $\Delta J = \pm 1$ and $\Delta K = 0$. Because of the $\Delta K = 0$ rule, the transition energy has the simple form of $2B(J+1)$, independent of J.

The total number of vibrational normal modes depends on the number of atoms, N, and the shape of the molecule. A linear molecule such as CO_2 has $3N - 5$ normal modes while a nonlinear molecule such as H_2O has $3N - 6$ normal modes. In the harmonic oscillator approximation, each normal mode has the eigenvalues

$$E(v_i) = \left(v_i + \frac{1}{2}\right) \hbar\omega_i \tag{2.178}$$

with its resonant frequency ω_i. The normal modes are independent, not interacting with each other.

2.2.4 Lattice Vibrations

A crystalline solid has ions arranged in a periodic structure on the microscopic level. The ions ,however, are not completely static: a closer look may find that each ion wiggles in the vicinity of its lattice site, while the average positions retain the periodic arrangement. A collective oscillation of the ions with well-defined frequency and wavelength is called a normal mode of lattice vibration, and its quantization is called a *phonon*. The phonon resonances are our great interest because, in general, the normal mode frequencies fall into the THz region. Not all normal modes, however, interact with electromagnetic radiation. Only the long wavelength optical modes in ionic crystals can be involved in such interactions.

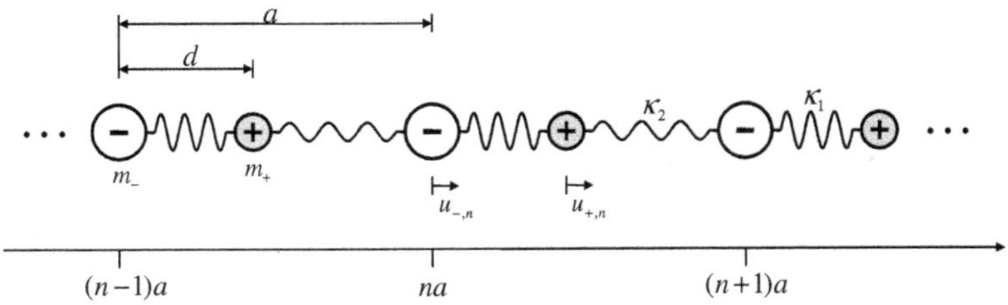

Fig. 2.21. Harmonic model for a one-dimensional lattice with two ions in a unit cell. a: lattice constant, $m_{\pm}$: masses of the ions, $u_{\pm,n}$: displacements of the ions from equilibrium positions at nth unit cell, $\kappa_{1,2}$: spring constants

We consider a one-dimensional lattice with two species of ions to describe the normal modes, and discuss how they interact with external electromagnetic waves. Figure 2.21 illustrates the harmonic model for a one-dimensional ionic crystal. It is convenient to assume that (i) the number of unit cells in the system, N, is large, yet finite, and (ii) the displacements of the ions satisfy the periodic boundary condition, $u_1 = u_N$. The equations of motion at the nth unit cell are written as

$$\begin{aligned} m_+\ddot{u}_{+,n} &= -\kappa_1(u_{+,n} - u_{-,n}) - \kappa_2(u_{+,n} - u_{-,n-1}), \\ m_-\ddot{u}_{-,n} &= -\kappa_1(u_{-,n} - u_{+,n}) - \kappa_2(u_{-,n} - u_{+,n+1}), \end{aligned} \quad (2.179)$$

where $m_{\pm}$ are the masses of the ions, and $u_{\pm,n}$ are the displacements of the ions from equilibrium positions at nth unit cell. A normal mode solution with angular frequency ω and wavenumber k has the form

$$\begin{aligned} u_{+,n} &= u_+ e^{i(kna-\omega t)}, \\ u_{-,n} &= u_- e^{i(kna-\omega t)}. \end{aligned} \quad (2.180)$$

Inserting Eq. 2.180 into Eq. 2.179 we obtain two coupled equations,

$$\begin{aligned} \left[m_+\omega^2 - (\kappa_1 + \kappa_2)\right] u_+ + \left[\kappa_1 + \kappa_2 e^{-ika}\right] u_- &= 0, \\ \left[\kappa_1 + \kappa_2 e^{ika}\right] u_+ + \left[m_-\omega^2 - (\kappa_1 + \kappa_2)\right] u_- &= 0. \end{aligned} \quad (2.181)$$

The solution of these equations requires

$$\begin{vmatrix} m_+\omega^2 - (\kappa_1 + \kappa_2) & \kappa_1 + \kappa_2 e^{-ika} \\ \kappa_1 + \kappa_2 e^{ika} & m_-\omega^2 - (\kappa_1 + \kappa_2) \end{vmatrix} = 0, \quad (2.182)$$

which leads to the dispersion relation,

$$\omega^2 = \frac{\kappa_1 + \kappa_2}{2\mu_r}\left[1 \pm \sqrt{1 - \frac{8\mu_r\kappa_1\kappa_2}{\mu_c(\kappa_1 + \kappa_2)^2}(1 - \cos ka)}\right] \quad (2.183)$$

with the reduced mass, $\mu_r = m_+m_-/(m_+ + m_-)$, and the total mass, $\mu_c = m_+ + m_-$. The dispersion relation is defined in a reciprocal unit cell, $-\pi/a <$

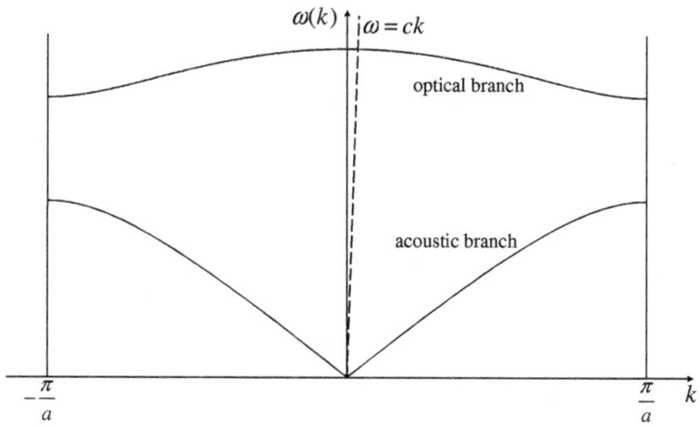

Fig. 2.22. Dispersion relation for a one-dimensional ionic crystal. The upper and lower curves are optical and acoustic branches, respectively.

$k < \pi/a$ as shown in Fig. 2.22. The two curves are referred to as the *optical branch* and the *acoustic branch.*

Analyzing the dynamics of lattice vibrations in the long wavelength limit helps us to understand how normal modes interact with external electromagnetic fields and where the branch names come from. In the limit of $k \ll \pi/a$, the dispersion relation is approximated as

$$\omega \cong \sqrt{\frac{\kappa_1 + \kappa_2}{\mu_r}} \text{ : optical branch,} \tag{2.184}$$

$$\omega \cong \sqrt{\frac{\kappa_1 \kappa_2}{\mu_c(\kappa_1 + \kappa_2)}}(ka) \text{ : acoustic branch.} \tag{2.185}$$

The frequency of an optical mode depends on the reduced mass, but not the total mass; the opposite is true for the acoustic mode. This is because optical modes are associated with relative motions between two ions, while acoustic modes are associated with center-of-mass motions (Fig. 2.23). Acoustic modes of long wavelength are responsible for sound propagation in media. The linear dispersion relation is a characteristic of sound waves, of which the velocity is constant. Now imagine that an electromagnetic wave with polarization parallel to the lattice axis is applied to the ionic crystal. Because the ions have opposite charges, the field exerts only relative motion. Therefore, the electromagnetic wave can only interact with optical modes. The light-matter

Fig. 2.23. Optical and acoustic modes are associated with relative and center-of-mass motions of the ions, respectively.

interaction requires energy conservation as well as momentum conservation. As shown in Fig. 2.22, the two conservations are simultaneously satisfied when the dispersion curve of light intersects with the dispersion curve of the optical branch. Due to the high speed of light, the coupling happens only in the long wavelength limit.

Table 2.1. Transverse Optical Phonon Frequency for Ionic Crystals[a]

Crystal	$\omega_T/2\pi$ in THz
LiF	9.19
NaF	7.36
KF	5.70
CsF	2.60
LiCl	5.74
NaCl	5.10
KCl	4.47
CsCl	3.14
LiBr	4.76
NaBr	4.06
KBr	3.45
CsBr	2.37

[a]Reference [11]

The one-dimensional model can be easily extended to three dimensions. Typical dispersion curves along a general direction in momentum space for an ionic crystal consist of three acoustic branches and three optical branches. The optical branches are composed of one branch of longitudinal optical-modes and two branches of transverse optical-modes. Ionic displacements of longitudinal and transverse optical-modes are parallel and perpendicular to **k**, respectively. In the long wavelength limit, the longitudinal and transverse optical-mode frequencies have a simple relation,

$$\omega_L^2 = \frac{\epsilon(0)}{\epsilon(\infty)}\omega_T^2, \tag{2.186}$$

known as the *Lyddane-Sachs-Teller relation.* Normally, crystals are more polarizable at lower frequencies, thus ω_L is larger than ω_T. Transverse optical-phonon frequencies for several ionic crystals are summarized in Table 2.1. Naturally, crystals composed of heavier ions have lower resonance frequencies.

2.3 Laser Basics

Several types of lasers have been developed for the THz region of the electromagnetic spectrum. They are characterized by underlying quantum mechanical transitions between different energy levels. For instance, traditional molecular gas lasers are based on the transitions between rotational modes of molecules, and solid state lasers such as quantum cascade lasers and p-type germanium lasers rely on intraband transitions in semiconductors.

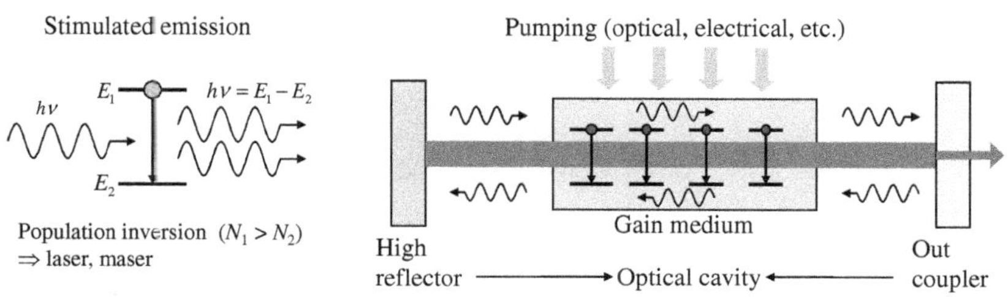

Fig. 2.24. Schematics of a typical laser operation

The key process governing radiation from these different types of lasers is stimulated emission. Figure 2.24 illustrates the basic concepts of how a typical laser works. The elemental parts of a laser system include a gain medium, a laser cavity, and a pump. The gain medium is a material system in which stimulated emission takes place. Population inversion, which means that more atoms or molecules in the gain medium are in the excited states than the lower energy states, is a prerequisite of stimulated emission. Therefore, a pump source is necessary to maintain the system in the high energy state. Light is confined within the laser cavity by high reflectors. One of the reflectors is partially transmissive so that its transmitted radiation can be used as the laser output. The light confinement leads to amplification of the radiation intensity within the cavity, which encourages stimulated emission because, as Eq. 2.109 indicates, the transition rate of stimulated emission W_{12} increases linearly with the incident light intensity $I(\omega)$,

$$W_{1\to 2} \propto |c_2^{(1)}|^2 \propto |E(\omega)|^2 \propto I(\omega). \tag{2.187}$$

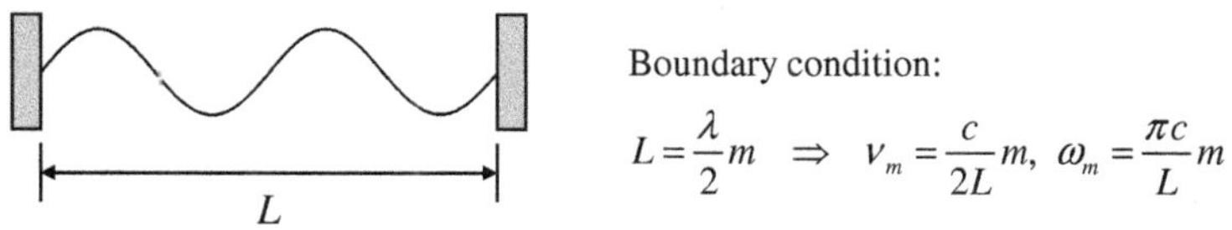

Fig. 2.25. Cavity modes with discrete frequencies

Satisfying the boundary condition at the reflecting surfaces, the laser cavity supports modes with discrete frequencies

$$\nu_m = \frac{c}{2L}m, \tag{2.188}$$

where m is a positive integer, as shown in Fig. 2.25. Because the radiation loses energy when interacting with cavity components, the cavity confinement has a finite lifetime, τ_c. The radiation energy loss is expressed as an exponential decay

$$E(t) = E_0 e^{-t/\tau_c} e^{-i\omega_0 t}, \tag{2.189}$$

and the corresponding emission spectrum

$$|E(\omega)|^2 = \frac{E_0^2}{(\omega - \omega_0)^2 + 1/\tau_c^2} \tag{2.190}$$

is centered at a cavity mode frequency ω_0 and has a bandwidth of $1/\tau_c$. The Q factor, or quality factor,

$$Q_c = \frac{\omega_0}{\Delta\omega} = \omega\tau_c, \tag{2.191}$$

is a commonly used measure for the decay of cavity modes. Roughly, Q_c is equal to the number of round trips that light makes in the cavity before it fades away.

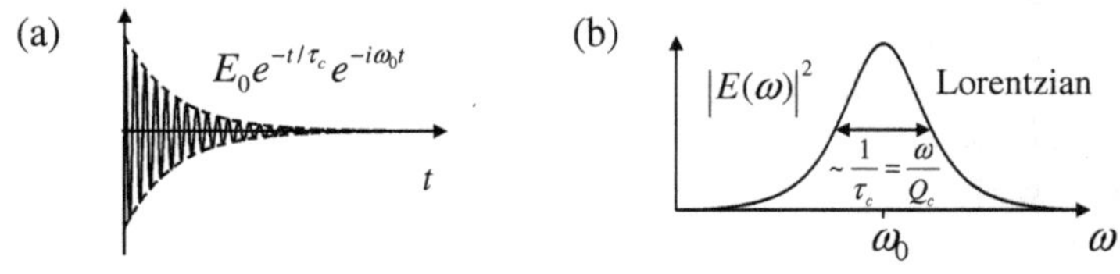

Fig. 2.26. Cavity decay and emission spectrum

As stated previously, a population inversion is imperative for laser operation. It, however, cannot be attained in a two-level system by any pumping schemes because, for an external perturbation, the rate of the transition from the lower to the higher level is exactly the same as the one from the higher to the lower level. Additionally, spontaneous emission is an extra pathway for the higher-to-lower transition. Therefore, however hard the system is pumped, an equal population of the two levels is as good as it gets. The only exception is Rabi flopping attained by a coherent excitation (see section 2.2.1), but, as the pump source has the same frequency as the laser output, there is no point in building such a laser.

In order to achieve a population inversion, most lasers adopt either a three-level or a four-level system. A four-level system is generally much more efficient than a three-level one, because its lower level used for laser operation keeps a very low population throughout the process. Figure 2.27 illustrates the transitions in a four-level system. A pump excites the system from the lowest level 1 to the highest level 4. Level 4 is then depleted quickly through a non-radiative relaxation to level 3. Level 3, the upper laser level, has a long lifetime, and the dominant relaxation process for this level is the radiative transition to level 2, the lower laser level, by spontaneous or stimulated emission. Having a short lifetime, level 2 undergoes a fast relaxation to the ground level. Given steady-state pumping, a substantial population remains in level 3 while level 2 is almost empty. Consequently, a population inversion is obtained between levels 2 and 3.

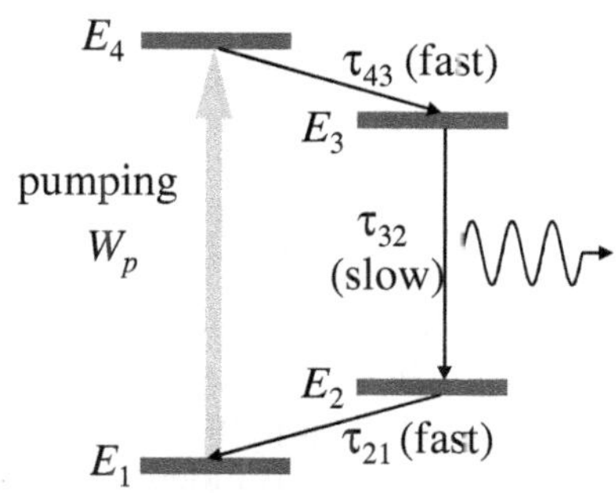

Fig. 2.27. Four-level laser system

Quantitative analysis of the transitions is carried out using rate equations. Here N_i is the population of level-i, and $\gamma_{ij}(= 1/\tau_{ij})$ is the relaxation rate for the transition from level-i to level-j. The rate equation for level 4 has the form

$$\begin{aligned} \frac{dN_4}{dt} &= W_p(N_1 - N_4) - (\gamma_{43} + \gamma_{42} + \gamma_{41})N_4 \\ &= W_p(N_1 - N_4) - \frac{N_4}{\tau_4}, \end{aligned} \tag{2.192}$$

where $W_p = W_{14} = W_{41}$ is the pump transition probability, and

$$\frac{1}{\tau_4} = \gamma_4 = \gamma_{43} + \gamma_{42} + \gamma_{41} \tag{2.193}$$

is the total relaxation rate of level 4. The steady state population for $dN_4/dt = 0$ obeys the relation

$$N_4 = \frac{W_p\tau_4}{1 + W_p\tau_4}N_1 \approx W_p\tau_4 N_1, \text{ for } W_p\tau_4 \ll 1, \tag{2.194}$$

where $W_p\tau_4$ is the normalized pumping rate. If level 4 has an extremely short lifetime, the pumping cannot catch up to the fast depletion of the level 4, and level 4 will have a substantially smaller population than level 1. The rate equations for levels 2 and 3 are written as

$$\frac{dN_3}{dt} = \gamma_{43}N_4 - (\gamma_{32} + \gamma_{31})N_3 = \frac{N_4}{\tau_{43}} - \frac{N_3}{\tau_3}, \tag{2.195}$$

$$\frac{dN_2}{dt} = \gamma_{42}N_4 + \gamma_{32}N_3 - \gamma_{21}N_2 = \frac{N_4}{\tau_{42}} + \frac{N_3}{\tau_{32}} - \frac{N_2}{\tau_{21}}. \tag{2.196}$$

The steady-state solutions of these equations lead to the following relations between N_2, N_3, and N_4:

$$N_3 = \frac{\tau_3}{\tau_{43}}N_4, \tag{2.197}$$

$$N_2 = \left(\frac{\tau_{21}}{\tau_{32}} + \frac{\tau_{43}\tau_{21}}{\tau_{42}\tau_3}\right)N_3 = \beta N_3, \tag{2.198}$$

where

$$\beta \equiv \frac{\tau_{21}}{\tau_{32}} + \frac{\tau_{43}\tau_{21}}{\tau_{42}\tau_3}. \tag{2.199}$$

In a good laser system, level 3 has a long lifetime, while the $4 \longrightarrow 3$ relaxation is extremely short: $\tau_3 \gg \tau_{43}$, therefore Eq. 2.197 assures that N_3 is much larger than N_4. Eq. 2.198 shows that a population inversion is attained when $\beta < 1$. The $4 \longrightarrow 2$ transition is nearly forbidden ($\gamma_{42} \approx 0$) for an efficient laser system, and level 4 relaxes primarily into level 3. In this case,

$$\beta \approx \frac{\tau_{21}}{\tau_{32}}. \tag{2.200}$$

Fast depletion of level 2 and slow relaxation of level 3 ($\tau_{21} \ll \tau_{32}$) warrants the efficient population inversion, $\beta \ll 1$.

3

Generation and Detection of Broadband Terahertz Pulses

The time representation of typical broadband THz radiation takes the shape of single-cycle pulse, the ultimate transform-limited waveform. These extremely short THz pulses are obtained by using either nonlinear optical responses of bound electrons in nonlinear crystals or transient photocurrents in semiconductors induced by ultrashort optical pulses from femtosecond lasers. The electron responses are not exactly transient, yet fast enough that the femtosecond optical pulses can induce polarizations or currents on a subpicosecond time scale. In the present chapter, we shall discuss the ultrafast and nonlinear optical methods to generate and detect broadband THz pulses. The ultrashort waveforms in the THz region have found a wide range of applications from imaging to communications, some of which will be presented in the later chapters.

3.1 Ultrafast Optics

3.1.1 Optical Pulse Propagation in Linear and Dispersive Media

The Gaussian waveform is a useful representation of optical pulses because of its straightforward mathematical description. It is also effective in characterizing the output pulse shape of actively mode-locked lasers. The time representation of the electric field of a Gaussian pulse has the form,

$$E(t) = E_0 e^{-a_0 t^2} e^{-i\omega_0 t}, \tag{3.1}$$

and the instantaneous intensity in a linear medium is expressed as

$$I(t) = \frac{1}{2}\epsilon_0 cn|E(t)|^2 = \frac{1}{2}\epsilon_0 cnE_0^2 e^{-2a_0 t^2}, \tag{3.2}$$

where ϵ_0 is the permittivity of free space, c is the speed of light in free space, and n is the refractive index of the medium. The full-width at half-maximum (FWHM) pulse duration τ_p is given as $\sqrt{2\ln 2/a_0}$. Having the form

$$E(\omega) = \sqrt{\frac{\pi}{a_0}} E_0 e^{-(\omega-\omega_0)^2/4a_0}, \tag{3.3}$$

the Fourier spectrum is also a Gaussian function. The bandwidth of its power spectrum $|E(\omega)|^2$ is

$$\Delta\nu = \frac{\Delta\omega}{2\pi} = \frac{\sqrt{(2\ln 2)\cdot a_0}}{\pi}. \tag{3.4}$$

The time-bandwidth product of transform-limited Gaussian pulse, independent of the parameter a_0, is the constant

$$\Delta\nu \cdot \tau_p = \frac{2\ln 2}{\pi} \cong 0.44. \tag{3.5}$$

Transform limit means that the time-bandwidth is at its minimum. In other words, a transform-limited pulse has the shortest pulse duration for a given bandwidth. Equivalently, its spectral phase is independent of frequency. The time evolution of the electric field in a Gaussian pulse and its Fourier spectrum are shown in Fig. 3.1.

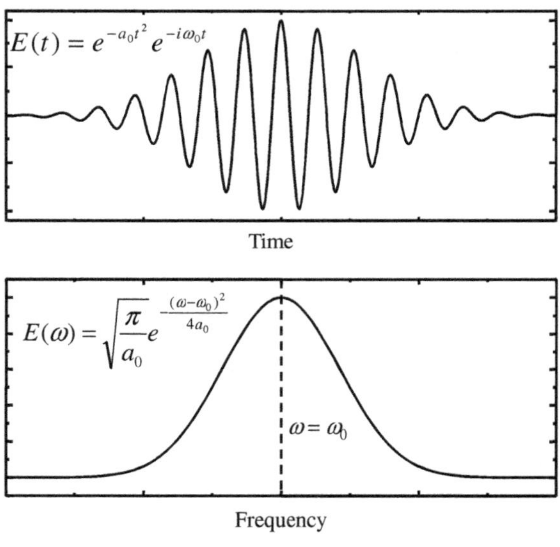

Fig. 3.1. The Gaussian waveform and its Fourier spectrum showing the dependence of electric field on time and on frequency.

When the Gaussian pulse propagates through a uniform, lossless, and dispersive medium, the waveform deforms gradually throughout the propagation. The change of pulse shape is governed by the dispersion relation

$$k(\omega) = \frac{\omega}{c} n(\omega), \tag{3.6}$$

where k is the wavenumber. Suppose that the wavenumber varies slowly in the vicinity of the central frequency ω_0. Then, the second order approximation

$$k(\omega) \cong k(\omega_0) + k'(\omega_0)(\omega - \omega_0) + \frac{1}{2}k''(\omega_0)(\omega - \omega_0)^2 \tag{3.7}$$

is valid to describe the dispersion.

Since a monochromatic electric field picks up the phase factor $e^{ik(\omega)z}$ after propagating a distance z, the output Gaussian pulse spectrum after the propagation becomes

$$\begin{aligned} E(z,\omega) &= E(\omega)e^{ik(\omega)z} \\ &= E_0 \exp\{ik(\omega_0)z + ik'(\omega_0)z(\omega - \omega_0) \\ &\qquad + \left[\frac{ik''(\omega_0)z}{2} - \frac{1}{4a_0}\right](\omega - \omega_0)^2\Big\}. \end{aligned} \tag{3.8}$$

The time evolution of the pulse is obtained by the inverse Fourier transform of $E(z,\omega)$,

$$E(z,t) = E_0 e^{-a(z)(t-z/v_{gr})^2} e^{-i\omega_0(t-z/v_{ph})}, \tag{3.9}$$

where the complex parameter $a(z)$ has the relation

$$\frac{1}{a(z)} = \frac{1}{a_0} - 2ik''(\omega_0)z, \tag{3.10}$$

and the group and the phase velocities are defined as

$$v_{gr}(\omega_0) = \frac{1}{k'(\omega_0)} = \left[\frac{\partial\omega}{\partial k}\right]_{\omega=\omega_0}, \tag{3.11}$$

$$v_{ph}(\omega_0) = \frac{\omega_0}{k(\omega_0)}. \tag{3.12}$$

The consequent pulse shape is determined by the real and imaginary parts of $a(z)$,

$$a_R(z) = \frac{a_0}{1 + 4a_0^2 k''(\omega_0)^2 z^2}, \tag{3.13}$$

$$a_I(z) = \frac{2a_0^2 k''(\omega_0)z}{1 + 4a_0^2 k''(\omega_0)^2 z^2}. \tag{3.14}$$

The pulse duration depends on the real part, having the form

$$\tau_p(z) = \tau_p(0)\sqrt{1 + 4a_0^2 k''(\omega_0)^2 z^2}. \tag{3.15}$$

This means that the transform-limited pulse gradually broadens while propagating in the dispersive medium. The imaginary part introduces a quadratic temporal phase,

$$\phi(t') = \phi_0 + \omega_0 t' + a_I t'^2, \tag{3.16}$$

where $t' = t - z/v_{gr}$, and the instantaneous frequency,

$$\omega \equiv \frac{d\phi(t')}{dt'} = \omega_0 + 2a_I t', \tag{3.17}$$

varies in time monotonically. Such a waveform is called a chirped pulse. Figure 3.2 illustrates the effects of pulse propagation in a dispersive medium. The pulse broadens while the field amplitude decreases. The instantaneous frequency increases gradually in the positively-chirped pulse.

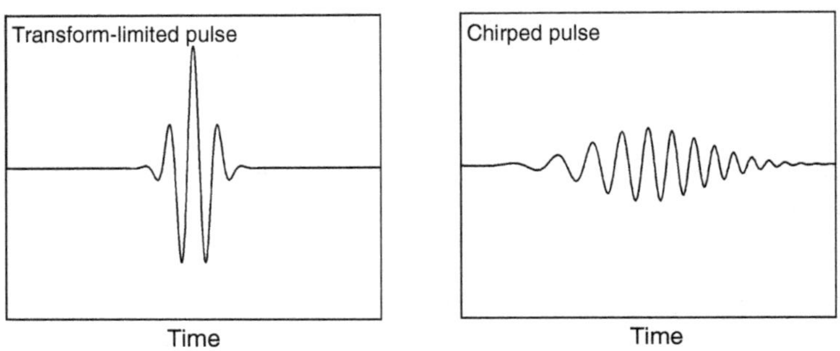

Fig. 3.2. A transform-limited Gaussian pulse becomes broadened and chirped after propagation in a dispersive medium.

3.1.2 Femtosecond Lasers

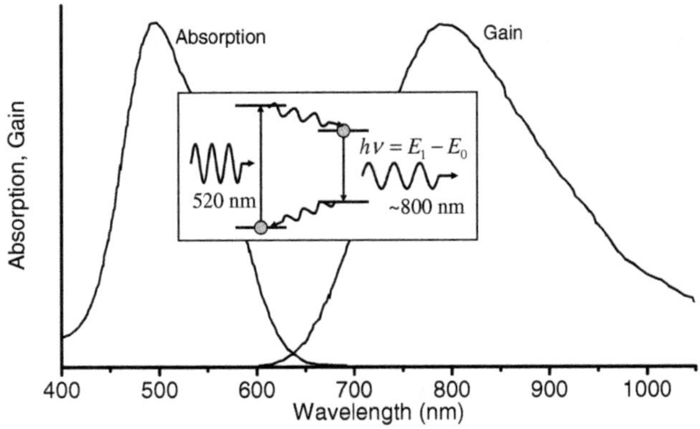

Fig. 3.3. Normalized absorption and gain spectra of a Ti-doped aluminium oxide crystal. Its simplified lasing mechanism is described by the four-level scheme.

Ultrashort optical pulses are produced by femtosecond lasers. The predominant gain medium of femtosecond laser systems is titanium-doped aluminium oxide (Ti:sapphire), in which Ti^{3+} ions (~0.1%) substitute Al^{3+} ions in a sapphire matrix. Ti:sapphire has several outstanding properties to produce ultrashort laser pulses: the gain spectrum is extremely broad ranging from 650 to 1100 nm, the crystal can take high optical pumping power (~20 W) due to its high thermal conductivity, and the carrier lifetime (~3.2 ms) is relatively short. Figure 3.3 shows the normalized absorption and gain spectra of a Ti:sapphire crystal. The lasing mechanism of a Ti:sapphire laser is simplified as a four-level scheme. Electrons excited by a green pump laser relax into lower and more stable states building up a population inversion, which gives rise to stimulated emission near 800 nm. A few other materials have been used as gain media for femtosecond laser systems. Their spectral ranges are listed in Table 3.1.

Table 3.1. Gain Materials for Femtosecond Lasers

Material	Gain Spectrum
Ti:sapphire	650-1100 nm
Rhodamine 6G (dye)	600-650 nm
Cr:LiSAF, Cr:LiCAF, Cr:LiSCAF	800-1000 nm
Nd:glass, Nd-doped fiber	1040-1070 nm
Yb-doped fiber	1030-1080 nm
Er-doped fiber	1520-1580 nm

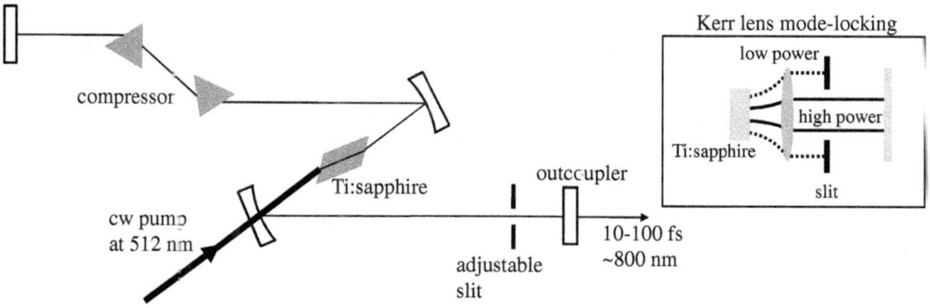

Fig. 3.4. Schematic of a Ti:sapphire laser. Kerr lens mode-locking (inset) is accomplished by adjusting the slit.

Figure 3.4 illustrates a schematic diagram of a mode-locked Ti:sapphire oscillator. Diode-pumped solid-state lasers and argon-ion lasers, whose wavelengths are near the peak of the absorption spectrum at 500 nm, are used as optical pump lasers. A prism pair is introduced inside the laser cavity in order to compensate the dispersion induced by the gain medium and other optical components. The adjustable slit forces the Kerr lens mode-locking by suppressing the continuous-wave modes of low power while short pulses of high power are allowed to pass through the hard aperture. A typical mode-locked Ti:sapphire oscillator is characterized by ~10 nJ pulse energy, ~80 MHz repetition rate, and 10-100 fs pulse duration.

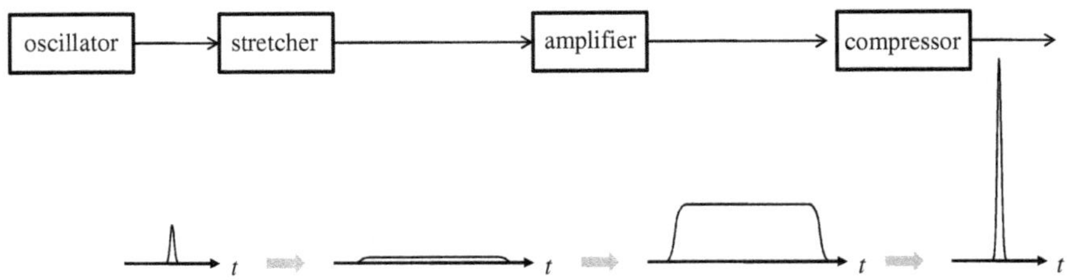

Fig. 3.5. Chirped pulse amplification

Many scientific applications concerning nonlinear and non-equilibrium phenomena require high peak powers of broadband THz pulses. Femtosecond laser amplifiers are indispensable to produce high-power THz pulses. Taking the output of a femtosecond oscillator as a seed, a femtosecond laser amplifier system inflates pulse energy, while reducing repetition rate. In order to avoid optical damage of components by amplified pulses, chirped pulse amplification is employed. The basic scheme is illustrated in Fig. 3.5. The pulse stretcher reduces the peak intensity of the seed pulse from the oscillator while temporally expanding the pulse to several hundreds of picoseconds. Pulse stretching is accomplished by a pair of gratings, exploiting their extremely large dispersion. After amplification, the compressor, another pair of gratings, squeezes the pulse to near the initial pulse duration and heightens the peak intensity.

Two schemes, regenerative and multipass amplification, are used for amplifying femtosecond pulses. Regenerative amplification, shown in Fig. 3.6(a), employs a resonant laser cavity to compel the pulse to pass through the gain medium multiple times. Using the different reflectivities of s- and p-polarization, injection of seed pulses and ejection of amplified pulses are accomplished by Pockel cells being switched electronically between quarter- and half-wave plates. A trapped pulse makes 10-20 round trips between injection and ejection. Optimal ejection timing is adjusted to dump the pulse right after the amplification is saturated. The operational scheme of multipass amplification, shown in Fig. 3.6(b), is straightforward. A seed pulse is sequentially amplified by passing through a gain medium several times with a slightly different angle of incidence at each time. Some high power systems employ multiple stages of amplification by combining the two schemes in series.

Q-switched Nd:YLF lasers or Nd:YAG lasers are commonly used for optical pumping. An amplifier works at repetition rates from a few hertz to several hundred kilohertz depending on the pump laser. Pulse energy is greatly en-

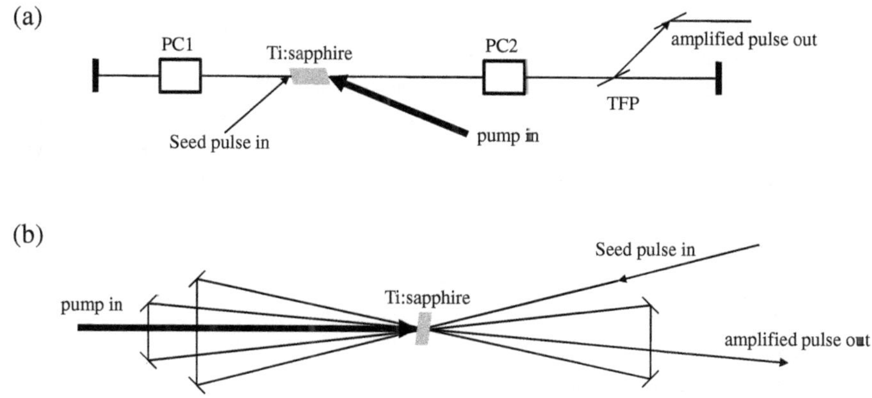

Fig. 3.6. (a) Regenerative amplification. Electronically controlled Pockel cells (PCs) rotate polarization to inject seed pulses and to eject amplified pulses. TFP: thin film polarizer. (b) Multipass amplification

Table 3.2. Repetition Rate and Pulse Energy of Femtosecond Lasers

Repetition rate	80 MHz	250 kHz	5 kHz	1 kHz	10 Hz
Pulse energy	12 nJ	4 μJ	0.2 mJ	1 mJ	100 mJ

hanced by the amplification processes, yet average output power, ~1 W, does not change much. Typical pulse energies and repetition rates of femtosecond lasers are listed in Table 3.2.

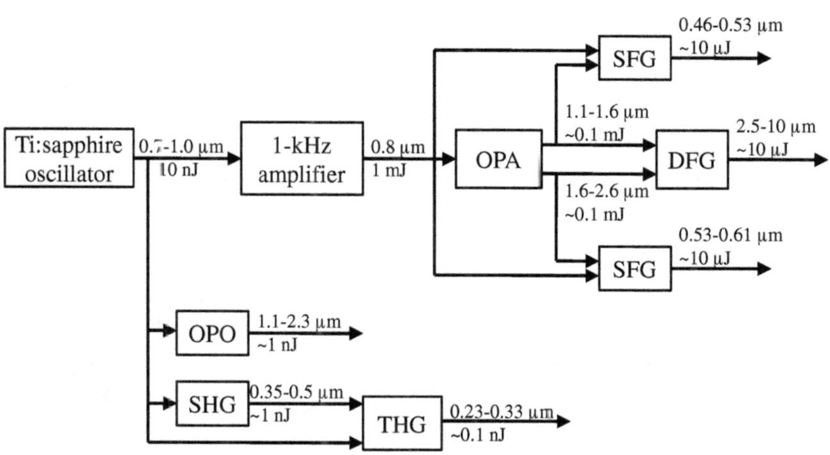

Fig. 3.7. Wavelength tuning via nonlinear optical processes

Ti:sapphire femtosecond laser systems have limited tunability in the spectral region near 800 nm. Femtosecond pulses tunable from UV to IR can bring out new THz generation and detection techniques. They can also extend the scope of optical and THz ultrafast studies. In order to get broadband tunability, nonlinear optical processes such as second harmonic generation (SHG), third harmonic generation (THG), optical parametric oscillator (OPO), optical parametric amplification (OPA), difference frequency generation (DFG), and sum frequency generation (SFG) are employed. Figure 3.7 sketches a femtosecond laser system for broadband tuning with the spectral ranges and pulse energies noted.

3.1.3 Time-resolved Pump-Probe Technique

Femtosecond lasers are an essential tool to study ultrafast phenomena on a subpicosecond time scale. Temporal resolution of such studies is primarily determined by the optical pulse duration. In addition to short pulse duration, another crucial property of ultrashort optical pulses is that the peak intensity can be extremely high because all of the optical energy is concentrated in such a short time period. Short and intense pulses interacting with a material system induce changes in optical properties of the material. The origin of the changes could be any optical excitations such as ionizations, transient currents, induced polarizations, lattice vibrations, etc.

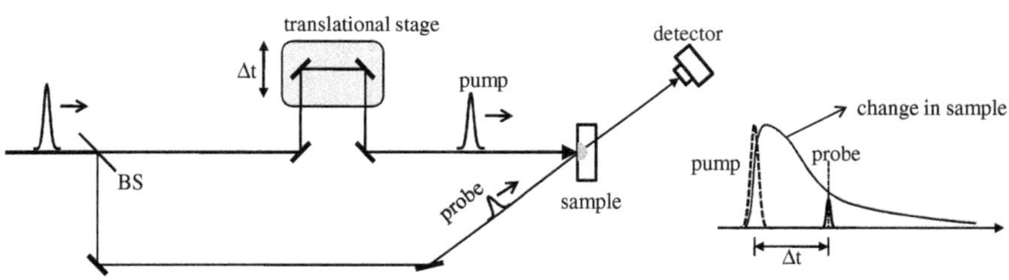

Fig. 3.8. Typical setup of a time-resolved pump-probe experiment. BS: beam splitter, Δt: relative time delay between pump and probe.

In order to investigate transient events, pump-probe techniques are employed. Figure 3.8 illustrates the general scheme of such methods. No absolute reference of time exists, therefore the relative time delay between pump and probe pulses is used as a temporal reference frame. A laser beam is split into two for use as pump and probe beams. The relative time delay is adjusted by a translational stage which changes the path length of the pump line. The temporal accuracy of this method is usually on the order of 0.1 fs. The pump-induced transients in the sample are analyzed by measuring probe transmission, reflectivity, and/or scattering.

3.1.4 Terahertz Time-Domain Spectroscopy

The basic experimental scheme for generation and detection of THz pulses using a femtosecond laser is similar to the pump-probe technique. A schematic of the typical setup is shown in Fig. 3.9. The optical beam is split into two parts, one of which goes through a translational stage to provide a relative time delay. The optical pump pulse illuminates the emitter and generates the THz pulse, which travels through a distance in free space, and focuses on the detector. The THz-induced transients in the detector are measured by the probe pulses. THz pulses are generated by either transient currents in a photoconductive antenna or optical rectification in a nonlinear optical crystal. Detection of THz fields is also done with either a photoconductive antenna or nonlinear crystal.

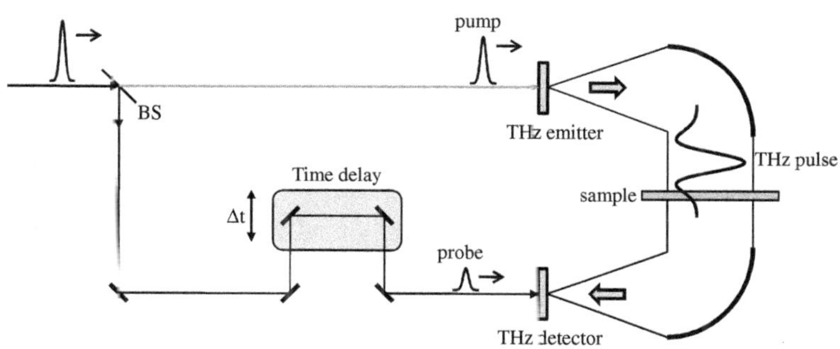

Fig. 3.9. Schematic of a typical setup for generation and detection of THz pulses using femtosecond optical pulses

In THz time-domain spectroscopy (THz-TDS), THz pulses are measured with and without a sample. Since THz-TDS determines both the amplitude and the phase of the THz radiation, not only the absorption but also the dispersion of the sample can be obtained by analyzing the Fourier transforms of the waveforms.

3.2 Terahertz Emitters and Detectors Based on Photoconductive Antennas

3.2.1 Photoconductive Antenna

A photoconductive (PC) antenna is an electrical switch exploiting the increase in electrical conductivity of semiconductors and insulators when they are exposed to light. The photoconductivity results from an increase in the number of free carriers—electrons and holes—generated by photons. The photon energy must be sufficiently large to overcome the bandgap of the material.

Figure 3.10 illustrates a PC switch, in which a bias voltage and a load resistor are connected in series with the semiconductor. The photocurrent flows through the circuit when light generates free electrons and holes.

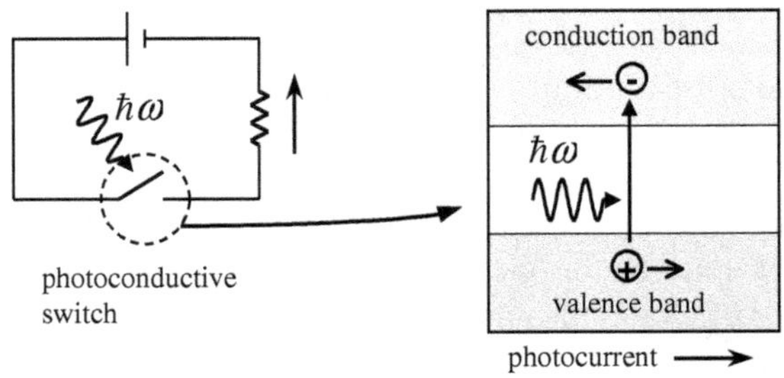

Fig. 3.10. Photoconductive switch

In order to either emit or detect THz radiation, the switching action in the PC antenna should occur in the subpicosecond time range. The switch-on time is a function of the laser pulse duration, and the switch-off time is mainly determined by the photoexcited carrier lifetime in the semiconductor substrate of the antenna; therefore, in addition to a short laser pulse duration, a short carrier lifetime is a vital property for ultrafast photoconductive switching. High carrier mobility and high breakdown voltage are also desirable for photoconductive materials of high quality. Several photoconductive materials have been tested for PC switches: low-temperature grown gallium arsenide (LT-GaAs), radiation-damaged silicon-on-sapphire (RD-SOS), chromium-doped gallium arsenide (Cr-GaAs), indium phosphide (InP), and amorphous silicon.

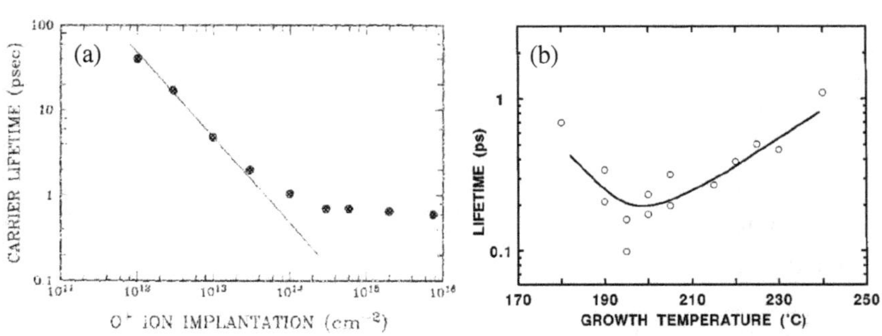

Fig. 3.11. Carrier lifetime vs (a) ion-implantation dose in RD-SOS (Reprinted with permission from [12]. ©1987, American Institute of Physics.) and (b) growth temperature for LT-GaAs (Reprinted with permission from [13]. ©1997, American Institute of Physics.)

The most commonly used materials for THz emitters and detectors are RD-SOS and LT-GaAs. Their carrier lifetimes are in the subpicosecond range: the ultrashort carrier lifetimes result from a high concentration of defects, at which carriers are trapped and recombined. The defects in RD-SOS are O^+ ions implanted by ion bombardment. Its defect density can be controlled by the amount of ion implantation. Figure 3.11(a) shows that the carrier lifetime of RD-SOS declines with increasing ion implantation [12]. It lowers to ~0.6 ps for the highly implanted, amorphous samples. LT-GaAs is grown by molecular beam epitaxy at low substrate temperatures (~200 °C). The growth is followed by rapid thermal annealing. This material contains a high density ($>10^{18}$ cm^3) of point defects such as As antisites, As interstitials, and Ga-related vacancies [14]. Figure 3.11(b) shows the carrier lifetime of LT-GaAs epilayers annealed for 10 min at 600 °C as a function of growth temperature [13]. The LT-GaAs samples grown between 180 and 240 °C have subpicosecond carrier lifetimes. The shortest lifetime of ~0.2 ps is reached for growth temperatures near 200 °C. The effective carrier mobilities of RD-SOS and LT-GaAs are reported as 10-100 cm^2/V·s [15] and 200-400 cm^2/V·s, respectively [16]. Since the hole mobility in LT-GaAs is one order of magnitude lower than the electron mobility, carrier transport in the THz frequency range is dominated by electrons.

3.2.2 Generation of Terahertz Pulses from Biased Photoconductive Antennas

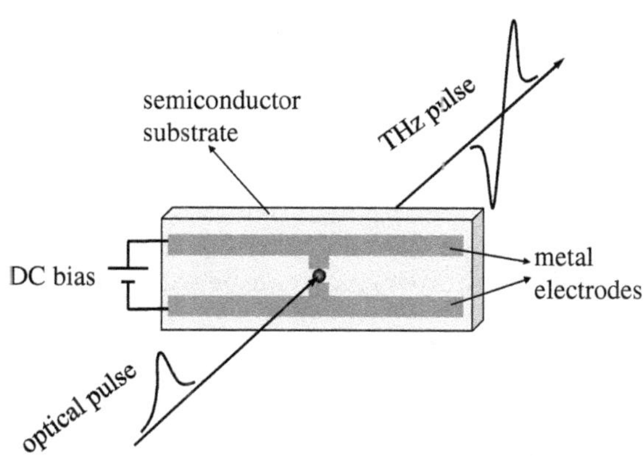

Fig. 3.12. Schematic diagram of THz pulse emission from a PC antenna excited by a femtosecond laser pulse

Subpicosecond THz pulses can be generated from a biased PC antenna excited by femtosecond laser pulses. Figure 3.12 illustrates THz pulse emission from a commonly used THz emitter structure. The THz emitter has two metal electrodes deposited on a semiconductor substrate. A DC bias is applied between the electrodes. Femtosecond optical pulses with photon energy larger than the bandgap of the semiconductor generate free electron and hole pairs in the gap between the electrodes. The static bias field accelerates the free carriers and, simultaneously, the charge density declines primarily by trapping of carriers in defect sites on the time scale of carrier lifetimes. The impulse current arising from the acceleration and decay of free carriers is the source of the subpicosecond pulses of electromagnetic radiation.

The radiation source of a PC emitter can be modelled as a Hertzian dipole antenna whose size is much smaller than the wavelength of the emitted radiation. Figure 3.13 illustrates the electric dipole radiation from a PC antenna. The dipole approximation is valid since the size of the source, which is comparable to the spot size of the optical beam w_0 ($\sim 10\mu$m), is usually much smaller than the wavelength of the THz radiation λ_{THz} (300 μm at 1 THz). We are interested in the radiation fields that survive at large distances from the source, in the so-called far-field range: $r \gg \lambda_{THz}$.

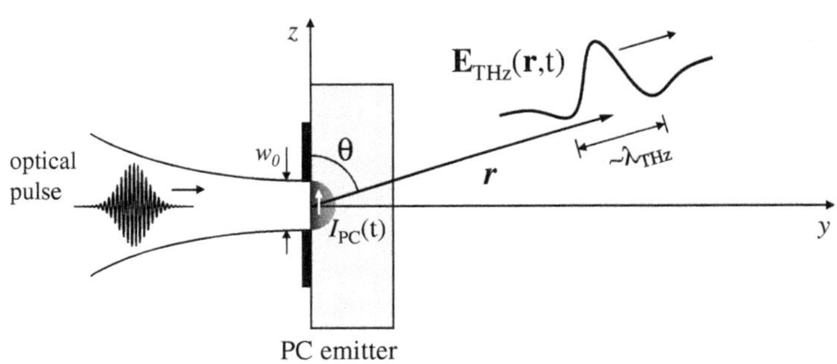

Fig. 3.13. Electric dipole radiation from a PC antenna

Here we assume dipole radiation in free space for the sake of simplicity. We will discuss the effects of the dielectric interface in the following section. The THz dipole radiation in free space can be expressed as

$$\mathbf{E}_{THz}(t) = \frac{\mu_0}{4\pi} \frac{\sin\theta}{r} \frac{d^2}{dt_r^2} [p(t_r)]\, \hat{\theta}, \tag{3.18}$$

where $p(t_r)$ is the dipole moment of the source at the retarded time $t_r = t - r/c$. The time derivative of the dipole moment can be written as

$$\frac{d\mathbf{p}(t)}{dt} = \frac{d}{dt} \int \rho(\mathbf{r}', t)\mathbf{r}' d^3\mathbf{r}' = \int \mathbf{r}' \frac{\partial \rho(\mathbf{r}', t)}{\partial t} d^3\mathbf{r}', \tag{3.19}$$

where $\rho(\mathbf{r}, t)$ is the charge carrier density and $\mathbf{J}(\mathbf{r}, t)$ is the photocurrent density. We simplify the integration using the continuity equation

$$\nabla \cdot \mathbf{J} + \frac{\partial \rho}{\partial t} = 0 \tag{3.20}$$

and integration by parts:

$$\frac{d\mathbf{p}(t)}{dt} = -\int \mathbf{r}'\nabla \cdot \mathbf{J}(\mathbf{r}', t)d^3\mathbf{r}' = \int \mathbf{J}(\mathbf{r}', t)d^3\mathbf{r}'. \tag{3.21}$$

We assume that the carrier transport is one dimensional. Under this condition

$$\frac{dp(t)}{dt} = \int J(z', t)d^3\mathbf{r}' = \int_{-w_0/2}^{w_0/2} I_{PC}(z', t)\, dz' = w_0 I_{PC}(t), \tag{3.22}$$

where w_0 is the spot size of the optical beam, and I_{PC} is the photocurrent. Subsequently, the THz electric field can be written as

$$\mathbf{E}_{THz}(t) = \frac{\mu_0 w_0}{4\pi} \frac{\sin\theta}{r} \frac{d}{dt_r} [I_{PC}(t_r)]\, \hat{\theta} \propto \frac{dI_{PC}(t)}{dt}, \tag{3.23}$$

which is proportional to the time derivative of the photocurrent in the photoconductive gap of the antenna.

The Drude-Lorentz model is effective in describing the carrier transport of the photoexcited electron-hole pairs. Since the electron mobility is much higher than the hole mobility in commonly used photoconductive materials such as LT-GaAs, we consider electrons as the dominant charge carriers. The time-dependent photocurrent is expressed as the convolution of the optical pulse envelope and the impulse response of the photocurrent [17]:

$$I_{PC}(t) = \int I_{opt}(t - t')\, [e\, n(t')v(t')]\, dt', \tag{3.24}$$

where $I_{opt}(t)$ is the intensity profile of the optical pulses, e is the electron charge, $n(t)$ is the carrier density, and $v(t)$ is the average electron velocity.

The dynamics of the carrier density under an impulsive excitation obeys

$$\frac{dn_c(t)}{dt} = -\frac{n_c(t)}{\tau_c} + \delta(t), \text{ thus } n_c(t) = \begin{cases} e^{-t/\tau_c} & \text{for } t > 0 \\ 0 & \text{for } t < 0 \end{cases}, \tag{3.25}$$

where τ_c is the carrier lifetime and $\delta(t)$ is a Dirac delta function representing the impulsive optical excitation.

In the Drude-Lorentz model, the equation of motion describing the average velocity is given as

$$\frac{dv(t)}{dt} = -\frac{v(t)}{\tau_s} + \frac{e}{m} E_{DC}, \tag{3.26}$$

where τ_s is the momentum relaxation time, m is the effective mass of the carriers, and E_{DC} is the DC bias field. Therefore, the time dependence of the average velocity has the form

$$v(t) = \begin{cases} \mu_e E_{DC}\left[1 - e^{-t/\tau_s}\right] & \text{for } t > 0 \\ 0 & \text{for } t < 0 \end{cases}, \tag{3.27}$$

where the electron mobility, $\mu_e = e\tau_s/m$. Here we neglect the screening effect of the space-charge field generated by the separation of electron-hole pairs. However, the dynamics of the screening field from the accelerated charge carriers becomes a crucial factor to account for the characteristics of the THz pulses from a PC emitter when the carrier density is high enough to satisfy $\omega_p \tau_s > 0$ with the plasma frequency $\omega_p^2 = n_c e^2/\epsilon_0 m$ [18].

We suppose that the optical pulse is Gaussian with a pulse duration of $2\sqrt{\ln 2} \cdot \tau_p$. Then we can integrate Eq. 3.24 analytically by inserting Eqs. 3.25 and 3.27 into it:

$$\begin{aligned} I_{PC}(t) &= \int_0^\infty I_{opt}^0 e^{-(t-t')^2/\tau_p^2} \cdot e^{-t'/\tau_c} \cdot \mu_e E_{DC}\left[1 - e^{-t'/\tau_s}\right] dt' \\ &= \mu_e E_{DC} I_{opt}^0 \int_0^\infty e^{-(t-t')^2/\tau_p^2 - t'/\tau_c}\left[1 - e^{-t'/\tau_s}\right] dt' \end{aligned} \tag{3.28}$$

which leads to

$$\begin{aligned} I_{PC}(t) = \frac{\sqrt{\pi}}{2}\mu_e E_{DC} I_{opt}^0 &\left[\exp\left(\frac{\tau_p^2}{4\tau_c^2} - \frac{t}{\tau_c}\right) \cdot \operatorname{erfc}\left(\frac{\tau_p}{2\tau_c} - \frac{t}{\tau_p}\right)\right. \\ &\left. - \exp\left(\frac{\tau_p^2}{4\tau_{cs}^2} - \frac{t}{\tau_{cs}}\right) \cdot \operatorname{erfc}\left(\frac{\tau_p}{2\tau_{cs}} - \frac{t}{\tau_p}\right)\right], \end{aligned} \tag{3.29}$$

where $1/\tau_{cs} = 1/\tau_c + 1/\tau_s$ and $\operatorname{erfc}(x) = 1 - \operatorname{erf}(x) = \frac{2}{\sqrt{\pi}}\int_x^\infty e^{-t^2} dt$ [17].

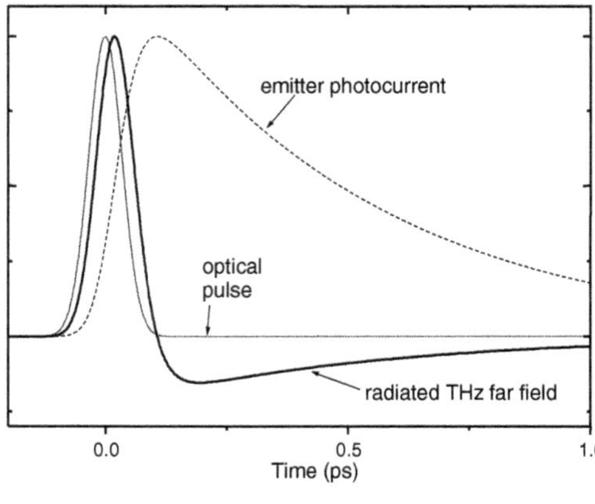

Fig. 3.14. Calculated photocurrent (dashed line) in the emitter and electric field amplitude of the THz radiation (solid line) versus time. The dotted line indicates temporal shape of the laser pulses. The curves are calculated with τ_s=0.03 ps, τ_c=0.5 ps, and τ_p=0.048 ps.

Figure 3.14 shows a photocurrent and a corresponding THz far-field with typical material and physical parameters, calculated from Eq. 3.29 and Eq. 3.23, respectively. For the calculation, we set the carrier lifetime τ_c=0.5 ps, the momentum relaxation time τ_s=0.03 ps, and the optical pulse duration $2\sqrt{\ln 2} \cdot \tau_p$=0.08 ps [17].

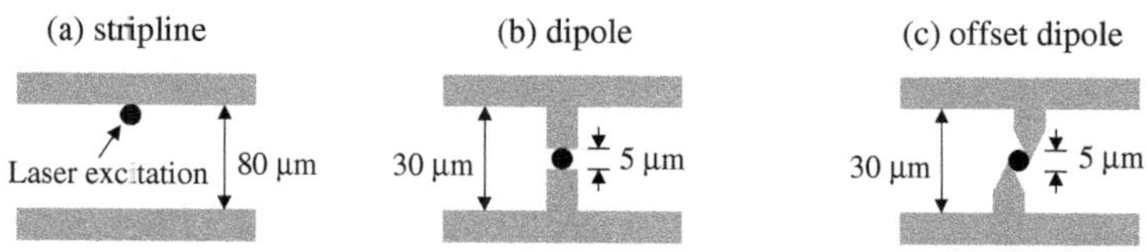

Fig. 3.15. Schematic diagram of stripline, dipole, and offset dipole metal electrodes of PC switches.

The power and bandwidth of THz emission from a PC switch vary widely depending on its metal electrode structure. Figure 3.15 illustrates the commonly used electrode structures of PC switches: stripline, dipole, and offset dipole electrodes. Typical values of the dipole gap and the stripline separation are 5-10 μm and 30-100 μm, respectively.

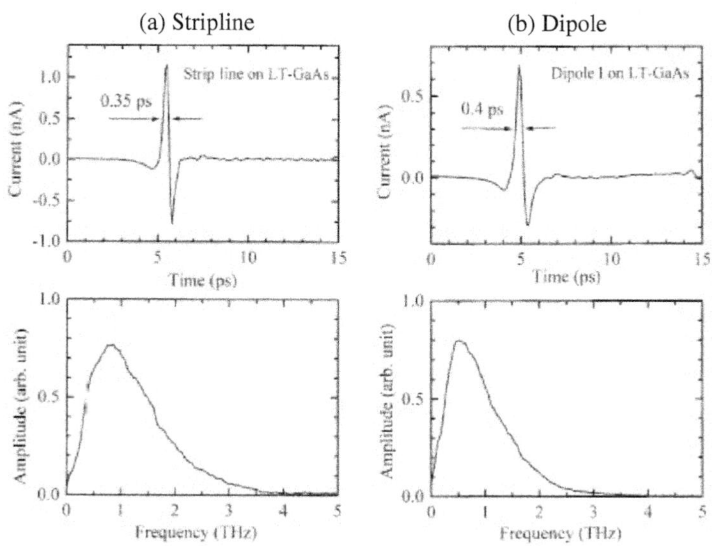

Fig. 3.16. THz radiation pulse shapes and amplitude spectra from PC emitters with (a) stripline and (b) dipole electrodes. The dimensions of the structures are the same as those in Fig. 3.15. (Reprinted from [19])

Figure 3.16 shows temporal waveforms and amplitude spectra of THz radiation from PC emitters with stripline and dipole electrodes on LT-GaAs substrates [19]. The pulse duration is shorter for the stripline structure than for the dipole structure. The amplitude spectrum extends up to 4 THz, which is partly limited by the detection response time. THz radiation from the dipole structure has a narrower bandwidth and a lower peak frequency, but the radiation power is much higher under the same optical pump power and DC bias. The THz generation efficiency can be improved by optimizing the electrode structure. For example, the laterally offset sharp triangular electrodes shown in Fig. 3.15(c) are more efficient than the simple dipole electrodes [20]. An average THz radiation power of 2-3 μW from the structure has been obtained under a 60-V bias voltage and a 20-mW optical excitation with a Ti:sapphire femtosecond oscillator.

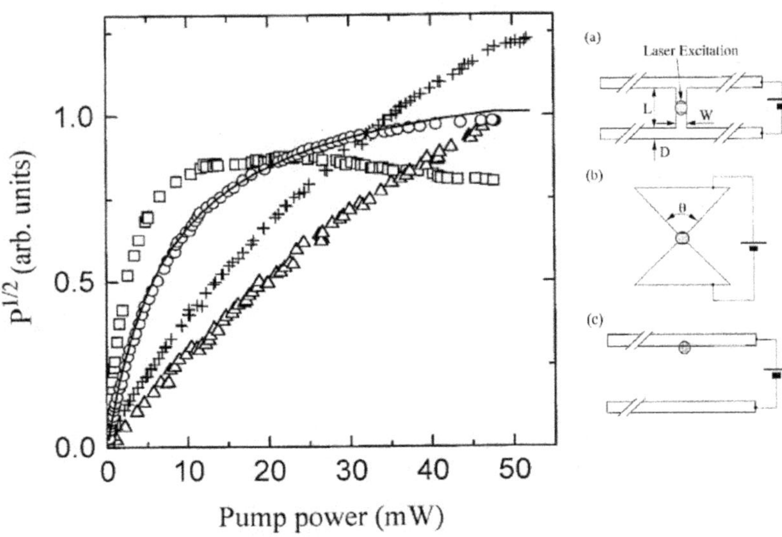

Fig. 3.17. The THz radiation field amplitude versus optical pump power for Hertzian dipole (open circle), bow tie (cross), and stripline (open triangle) antennas on LT-GaAs, and Hertzian dipole on SI-GaAs (open square). The vertical axis represents the square root of the radiation power P, and the data are arbitrarily normalized. The solid curve is the theoretical curve fitted to the data for the dipole antenna on LT-GaAs. The schematics of the PC antennas are shown on the right-hand side: (a) the Hertzian dipole antenna with a 5-μm gap, (b) the bow-tie antenna with a 10×10-μm^2 gap, and (c) the stripline antenna with a 80-μm gap. (Reprinted from [19])

The output power of a PC emitter depends on the bias voltage and the optical pump power. The amplitude of the radiation field increases linearly with both parameters when the optical pump power is low and the bias field is weak [19], which is consistent with Eq. 3.29. The maximum radiation power is limited by the breakdown voltage of its substrate material. The breakdown field of LT-GaAs is reported as ~300 kV/cm [19], which corresponds to a 150-V bias voltage for a 5 μm gap. The THz output power saturates at high optical pump powers due to the screening of the bias field by photocarriers. Figure 3.17 shows the radiation field amplitudes versus optical pump power for several PC antennas on LT-GaAs and on semi-insulating (SI) GaAs [19]. The field amplitude increases linearly with the radiation power only at low pump powers and becomes saturated as the pump power is further increased. Saturation is more notable in dipole antennas than in others due to the relatively small size of their mid-gap area.

3.2.3 Substrate Lenses: Collimating Lens and Hyper-Hemispherical Lens

The actual radiation pattern of a PC emitter is more complicated than the free-space dipole radiation described in Eq. 3.18 due to the dielectric interface between air and substrate. The photocurrent of a THz emitter is located just below the interface on which the antenna structure of metal electrodes is deposited. The power radiated directly into the substrate is stronger than into free space by a factor of the substrate dielectric constant ϵ_r [21]; since $\epsilon_r \approx 12$ for typical photoconductive materials such as GaAs and sapphire, a major portion of the generated THz pulse propagates into the substrate.

The radiation is highly divergent because the size of the source is much smaller than the THz wavelength. Therefore, a substrate lens, attached to the backside of a PC emitter to collimate the THz radiation, is a critical component for an efficient THz emitter. The substrate lenses are made of high resistivity silicon. This material is best suited for the component because the refractive index matches well with typical substrate materials, the linear absorption is very low at the THz frequencies, dispersion over the whole THz spectrum is almost negligible, and the fabrication of high-quality components is reasonably simple.

A collimating lens and a hyper-hemispherical lens, which are commonly used substrate-lens designs, along with their associated ray-tracing diagrams, are depicted in Fig. 3.18. The emission angle ϕ and the internal-incidence angle θ have a simple relation

$$(d - R)\sin\phi = R\sin\theta, \tag{3.30}$$

where R is the radius of the lens. Because the wavelength of the THz radiation is not negligible compared with the size of the lenses (~5 mm), diffraction can strongly affect the THz beam propagation. Ray-tracing analysis may overly simplify the realistic THz emission pattern emerging from the substrate lenses. Nevertheless, this simple picture provides some salient features of different lens designs.

The distance between the focal point and the tip of a collimating lens has the relation

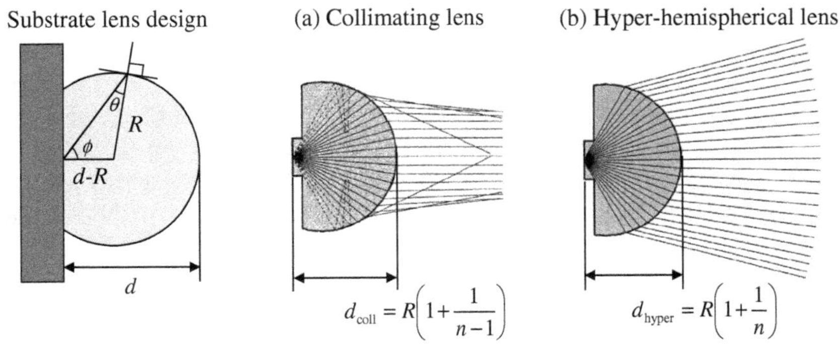

Fig. 3.18. Designs and ray-tracing diagrams of a collimating lens and a hyper-hemispherical lens. R is the radius of the lens, d is the distance between the focal point and the tip of the lens, ϕ is the emission angle, θ is the internal incidence angle, and n is the refractive index of the lens. The dashed lines inside the collimating lens indicate the rays trapped by total internal reflection. (Reprinted from [22].)

$$d_{coll} = R\left(1 + \frac{1}{n-1}\right), \tag{3.31}$$

where n is the refractive index of the lens material. The refractive index of high resistivity silicon is 3.418 over the broad THz spectral range [23], which leads to $d_{coll} = 1.414R$ for the silicon collimating lens. The dipole source of the emitter is located at the focus of the collimating lens, thus the ray tracing indicates a nearly collimated output beam near the optical axis. When θ approaches the critical angle $\theta_c = \sin^{-1}(1/n)$ of total internal reflection, output rays undergo strong refraction, which leads to severe astigmatism. The ray tracing diagram shows the wavefront aberration at large emission angles. At larger angles, the rays are trapped inside the lens by internal reflection. From Eq. 3.30 we can obtain the critical emission angle

$$\phi_c = \sin^{-1}\left(\frac{R}{d-R}\sin\theta_c\right) = \sin^{-1}\left(\frac{n-1}{n}\right), \tag{3.32}$$

which is 45° for a silicon lens. The critical angle of total internal reflection determines an effective aperture size for the lens.

For the hyper-hemispherical lens, the distance between the focal point and the tip of the lens is

$$d_{hyper} = R\left(1 + \frac{1}{n}\right). \tag{3.33}$$

For silicon, $d_{hyper} = 1.293R$. This design does not lead to spherical aberration. Because the critical emission angle $\phi_c = \sin^{-1}\left(n \cdot \frac{1}{n}\right) = 90°$, there is no loss by internal reflection in the hyper-hemispherical lens. At $\phi = 90°$ the output rays emerge with the angle $\theta = 17°$ from the optical axis, thus the output beam diverges with a 34° cone angle.

THz beam propagation can be described more accurately by a model based on wave optics [18]. Figure 3.19 illustrates the fundamental steps in the model to calculate the radiation pattern from the THz emitter. First, the source of radiation is a point dipole, thus Eq. 3.23 is still valid to express the THz electric field. Inside the emitter, the speed of light should be adjusted by the refractive index n. Second, internal reflection from the substrate interface is included. Third, at a point **P** just below the lens surface the reflected part of the field from the interface interferes with the part of the field emitted directly into the substrate. Fourth, the electric field inside the lens is transmitted into free space. The transmission and refraction at the lens-air interface is governed by Fresnel's and Snell's laws [24]. Fifth, the radiation propagates through free space to a point **Q** of detection. The electric field at **Q** is calculated by the Fresnel-Kirchhoff diffraction integral [18]:

$$E_Q = -\frac{ik}{4\pi}\int_A E_{surf}\frac{e^{ikr}}{r}(\cos\theta_o - n\cos\theta_i)dA, \qquad (3.34)$$

where k is the wavenumber and E_{surf} is the electric field on the lens surface. θ_o and θ_i are the refraction angle and the internal incidence angle at **P**, respectively. The integration extends over the total lens surface.

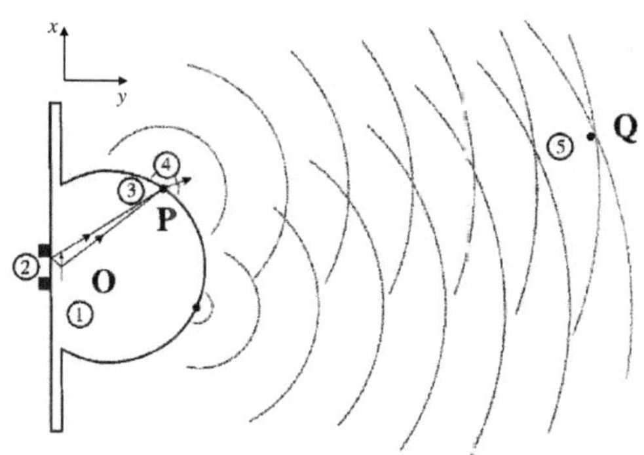

Fig. 3.19. Schematic illustration of the fundamental steps in the model for calculation of the radiation pattern from the THz emitter. (Reprinted from [18])

Figure 3.20 shows the calculated radiation patterns for the horizontal components of the electric fields with the frequencies of 0.5 and 1.0 THz. The radius of the collimating lens is set to 5 mm. The x axis of the plots is the horizontal position perpendicular to the emitter axis; the y axis is the distance from the lens tip. It is evident that the radiation patterns depend on frequency: the angular divergence of radiation is larger for lower frequency. The fringes in the radiation patterns are caused by diffraction from the aperture, which is determined by total internal-reflection inside the lens. In the far-field range the radiation pattern of the beam matches well with a Gaussian-beam profile.

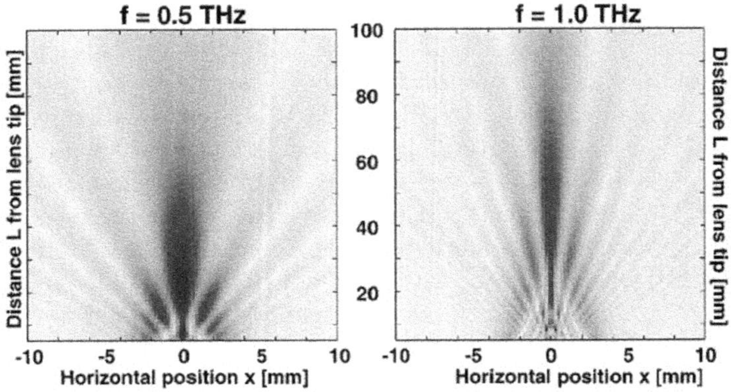

Fig. 3.20. Radiation patterns at the frequencies of 0.5 and 1.0 THz from a PC emitter with a collimating lens of R=5 mm. The density plots represent the amplitudes of the horizontal component of the THz electric fields. Dark shading corresponds to large field amplitudes. (Reprinted from [25])

3.2.4 Terahertz Radiation from Large-Aperture Photoconductive Emitters

Large-aperture PC antennas have been used for generating high-power THz pulses. The principal mechanism of THz generation in a large-aperture antenna is similar to that in a dipole antenna: the radiation source is the surge current of photoexcited carriers induced by a bias field. What makes a large-aperture PC antenna different from a dipole antenna is that the size of optically excited area between the electrodes is much greater than the radiated wavelength. Due to the large excitation area, large-aperture emitters are capable of producing high-power THz pulses. This requires a high DC bias voltage and amplified femtosecond laser pulses.

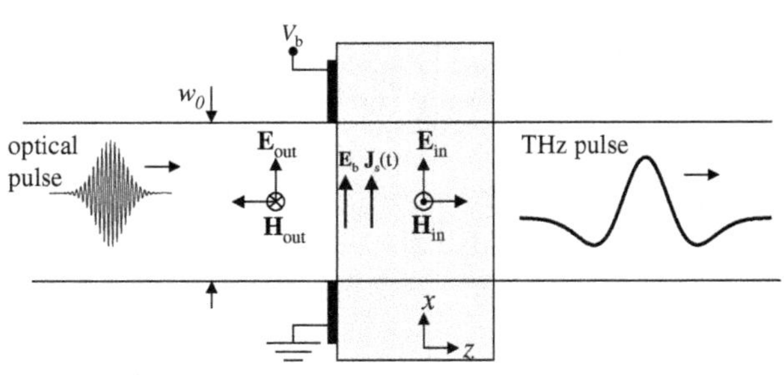

Fig. 3.21. Schematic of a large-aperture PC antenna excited at normal incidence [26]. $\mathbf{E}_b$ is the bias electric field and $\mathbf{J}_s(t)$ is an idealized surface current density. $\mathbf{E}_{in}(t)$ and $\mathbf{H}_{in}(t)$ are the THz electric and magnetic fields directly radiated into the substrate; $\mathbf{E}_{out}(t)$ and $\mathbf{H}_{out}(t)$ are the fields radiated into free space.

Figure 3.21 illustrates THz generation in a large-aperture PC antenna [26]. Since the photoexcited carriers are confined to a thin layer near the interface, we can idealize the photocurrent as a surface current flowing on the boundary surface. The corresponding surface current density is defined as

$$\mathbf{J}_s(t) = \int_0^{\infty} \mathbf{J}(t)\, dz, \tag{3.35}$$

where $\mathbf{J}(t)$ is the volume current density. Using Eqs. 3.18 and 3.22, we describe the on-axis THz electric field radiated from the surface current in the far field region ($z \gg w_0$),

$$\mathbf{E}_{THz}(z,t) = \frac{\mu_0}{4\pi z}\left[-\int \frac{\partial}{\partial t'}\mathbf{J}(\mathbf{r}',t')d^3\mathbf{r}'\right] = -\frac{\mu_0 A}{4\pi z}\frac{d\mathbf{J}_s(t_r)}{dt_r}, \tag{3.36}$$

where A is the optically excited area. We assume that the surface current is uniform.

$\mathbf{E}_{in}(t)$ and $\mathbf{H}_{in}(t)$ are the THz electric and magnetic fields radiated directly into the substrates; $\mathbf{E}_{out}(t)$ and $\mathbf{H}_{out}(t)$ are the fields radiated into free space. The boundary conditions at the air-substrate interface are expressed as [27]:

$$\mathbf{E}_{in}(t) = \mathbf{E}_{out}(t) \tag{3.37}$$

$$\mathbf{e}_z \times [\mathbf{H}_{in}(t) - \mathbf{H}_{out}(t)] = \mathbf{J}_s(t), \tag{3.38}$$

where $\mathbf{e}_z$ is the z-axis unit vector. The magnetic and electric fields are related by

$$\mathbf{e}_z \times \mathbf{H}_{in}(t) = -\frac{n}{Z_0}\mathbf{E}_{in}(t) \tag{3.39}$$

$$\mathbf{e}_z \times \mathbf{H}_{out}(t) = \frac{1}{Z_0}\mathbf{E}_{out}(t), \tag{3.40}$$

where $Z_0 = \sqrt{\frac{\mu_0}{\epsilon_0}} = 377\ \Omega$ is the impedance of free space. From Eqs. 3.37, 3.38, 3.39, and 3.40, we can derive a relation between the surface current density $\mathbf{J}_s(t)$ and the radiated electric field in the substrate $\mathbf{E}_{in}(t)$ [26],

$$\mathbf{J}_s(t) = -\left(\frac{1+n}{Z_0}\right)\mathbf{E}_{in}(t). \tag{3.41}$$

Consequently, the pulse shape of the THz radiation in the near field region replicates the time varying surface current.

Unless the bias field is very strong, the Drude-Lorentz model is valid to describe the carrier transport in the PC switch. Applying Ohm's law, we can express the surface current density as

$$\mathbf{J}_s(t) = \sigma_s(t)\left[\mathbf{E}_b + \mathbf{E}_{in}(t)\right], \tag{3.42}$$

where $\sigma_s(t)$ is the surface conductivity [26]. Note that the equation based on Ohm's law should be modified when the applied bias field is so strong that the carrier velocity saturates and is not linearly proportional to the mobility. The velocity saturation becomes effective when the bias field strength E_{th} exceeds 3 kV/cm in Si and 5 kV/cm in GaAs [28].

Using Eq. 3.41 and 3.42, we derive

$$\mathbf{J}_s(t) = \frac{\sigma_s(t)}{\left[\frac{\sigma_s(t)Z_0}{1+n} + 1\right]}\mathbf{E}_b. \tag{3.43}$$

From Eq. 3.36 and 3.43, the THz field in the far field region is obtained as [29]

$$\mathbf{E}_{THz}(z,t) = -\frac{\mu_0 A\mathbf{E}_b}{4\pi z}\frac{\frac{d\sigma_s(t_r)}{dt_r}}{\left[\frac{\sigma_s(t_r)Z_0}{1+n} + 1\right]^2}. \tag{3.44}$$

We use the following definition of surface conductivity:

$$\sigma_s(t) = \frac{e(1-R_{opt})}{\hbar\omega}\int_{-\infty}^{t} I_{opt}(t')\mu(t,t')n(t,t')\,dt', \tag{3.45}$$

where e is the electron charge, R_{opt} is the optical reflectivity of the photoconductor, $\hbar\omega$ is the photon energy, $\mu(t,t')$ is the mobility at time t of a photoexcited electron created at t', and $n(t,t')$ is the population of electrons at t that were created at t' [26]. The electron population undergoes an exponential decay with the carrier lifetime τ_c, $n(t,t') = e^{-(t-t')/\tau_c}$.

For a Gaussian optical pulse $I_{opt}(t) = I_0 e^{-t^2/\tau_p^2}$, Eq. 3.44 with a constant mobility μ_e results in

$$\mathbf{E}_{THz}(x) = -C\mathbf{E}_b\frac{S\left[e^{-x^2} - \frac{\tau_p}{\tau_c}\int_{-\infty}^{x} e^{-x'^2}dx'\right]}{e^{\frac{\tau_p}{\tau_c}x} + S\int_{-\infty}^{x} e^{-x'^2}dx'}, \tag{3.46}$$

where

$$C = \frac{\mu_0(1+n)A}{4\pi Z_0\tau_p z} \text{ and } S = \frac{e\,(1-R_{opt})\,\mu_e Z_0 \tau_p I_0}{(1+n)\hbar\omega}e^{-\frac{\tau_p^2}{4\tau_c^2}}. \tag{3.47}$$

S is the normalized optical pump intensity, and $x = t/\tau_p - \tau_p/2\tau_c$ is the normalized time.

Using Eq. 3.46, we calculate the THz electric fields at low (S=0.1, 0.2, and 0.5) and high (S=100, 300, and 1000) optical excitation regimes. Fig. 3.22 shows the THz waveforms with the optical pulse duration $2\sqrt{\ln 2}\cdot\tau_p$=0.08 ps and carrier lifetime τ_c=0.5 ps. At low excitations, the THz field amplitude increases linearly with the optical intensity S, while it shows a saturation behavior for the high intensity cases. This can be easily explained by the

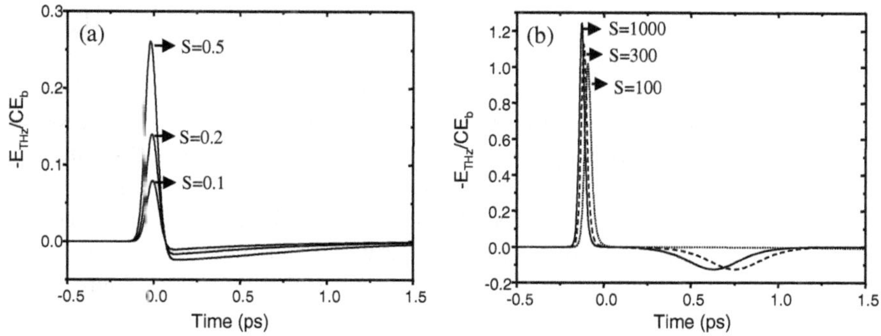

Fig. 3.22. Normalized THz electric field $-E_{THz}(t)/CE_b$ versus time for (a) low optical excitation intensities (S=0.1, 0.2, and 0.5) and (b) high optical excitation intensities, S=100 (dotted), 300 (dashed), and 1000 (solid). The optical pulse duration $2\sqrt{\ln 2} \cdot \tau_p$ is 0.08 ps and the carrier lifetime τ_c is 0.5 ps.

asymptotic behavior of Eq. 3.46 for the two extreme cases of low and high intensities:

$$\begin{aligned} \mathbf{E}_{THz}(x) &\approx -C\mathbf{E}_b S\, e^{-\frac{\tau_p}{\tau_c}x} \left[e^{-x^2} - \frac{\tau_p}{\tau_c} \int_{-\infty}^{x} e^{-x'^2} dx' \right] \propto S \quad \text{for } S \ll 1 \\ &\approx -C\mathbf{E}_b \frac{\left[e^{-x^2} - \frac{\tau_p}{\tau_c} \int_{-\infty}^{x} e^{-x'^2} dx' \right]}{\int_{-\infty}^{x} e^{-x'^2} dx'} \quad \text{for } S \gg 1. \end{aligned} \tag{3.48}$$

The intensity dependence of the THz field is summed up as $E_{THz} \propto \frac{S}{S+S_0}$, where S_0 is the threshold intensity. This saturation behavior is universal for any PC switch. The maximum THz pulse energy is limited by the capacitance

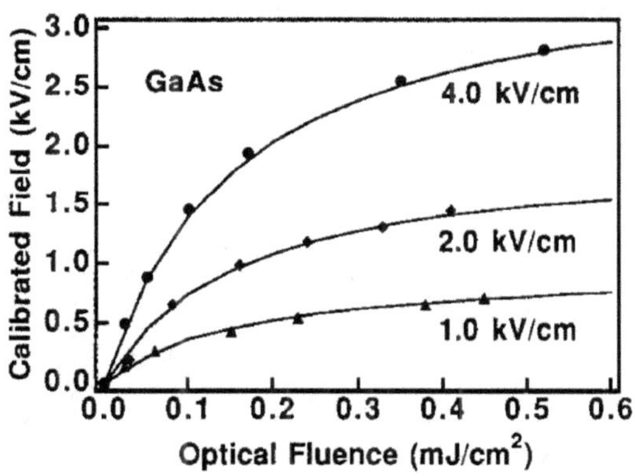

Fig. 3.23. Radiated THz field amplitude as a function of optical fluence of the excitation pulse. The THz radiation is generated by a 0.5-mm gap GaAs antenna at bias fields of 4.0 kV/cm, 2.0 kV/cm, and 1.0 kV/cm. (Reprinted from [26])

and the bias voltage of the PC switch. As we extract energy from the switch by optical excitation in the form of THz pulses, the THz pulse energy cannot exceed the amount stored in the closed gap of electrodes.

Some experimental measurements are shown in Fig. 3.23: the radiated field amplitude as a function of optical fluence from a 0.5-mm gap GaAs antenna at bias fields of 1.0 kV/cm, 2.0 kV/cm, and 4.0 kV/cm [26]. The maximum field amplitude is in the range of a few kV/cm. The curves show strong saturation at high optical excitation. In a separate study, the THz field amplitude reached up to 150 kV/cm by using a pulsed bias of ~10 kV/cm, which increases the electrical breakdown voltage [30].

3.2.5 Time-Resolved Terahertz Field Measurements with Photoconductive Antennas

The underlying mechanism of THz field detection in a PC antenna is almost identical with that of THz emission in a PC emitter. The carrier dynamics discussed in the previous sections are applicable to THz field detection in a PC antenna. Figure 3.24 shows a schematic diagram of a time-resolved measurement of THz electric fields with a PC receiver. In the absence of a bias field, the THz electric field induces a current in the photoconductive gap when the photocarriers are injected by the optical probe pulse. The photocurrent lasts for the carrier lifetime, which should be much shorter than the THz pulse duration for a time-resolved waveform measurement. The induced photocurrent is proportional to the field amplitude of the THz radiation focused on the photoconductive gap. The THz pulse shape is mapped out in the time domain by measuring the current while varying the time delay between the THz pulse and the optical probe. A typical photocurrent is in the sub-nanoamp range, thus a current amplifier is necessary to convert the weak current signals into measurable voltages. In order to enhance the signal-to-noise ratio, the signal is processed by a lock-in amplifier synchronized with an optical intensity modulator such as an optical chopper.

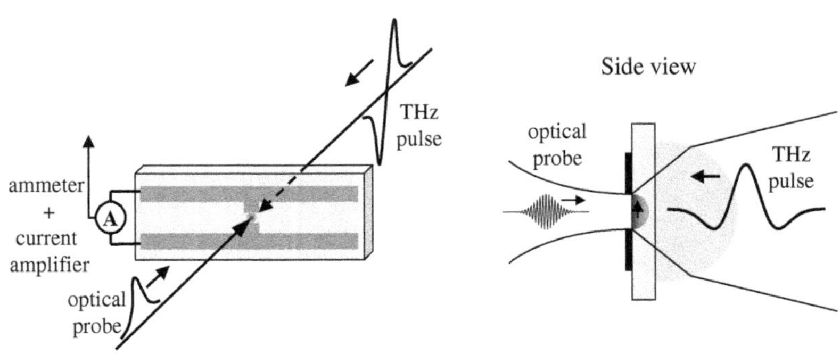

Fig. 3.24. Schematic representation of THz pulse detection with a PC antenna

The photocurrent depends not only on the incident THz electric field, but also on the transient surface conductivity $\sigma_s(t)$ defined in Eq. 3.45:

$$J(t) = \int_{-\infty}^{t} \sigma_s(t-t')E_{THz}(t')\,dt'. \tag{3.49}$$

The time-dependent conductivity implies that the current cannot flow instantaneously in response to the THz electric field. The photocurrent is a convolution of the THz field at previous times with the conductivity. By use of the convolution theorem, the Fourier transform of Eq. 3.49 results in

$$\tilde{J}(\nu) = \tilde{\sigma}_s(\nu)\tilde{E}_{THz}(\nu), \tag{3.50}$$

where $\tilde{J}(\nu)$, $\tilde{\sigma}_s(\nu)$, and $\tilde{E}_{THz}(\nu)$ are the Fourier transform of $J(t)$, $\sigma_s(t)$, and $E_{THz}(t)$, respectively. This equation shows that the detection bandwidth of a PC receiver is limited by the carrier dynamics in the PC material. The photocurrent signal is not an exact replica of the THz waveform, but exhibits frequency filtering through the conductivity.

The surface conductivity is determined by the optical intensity, carrier drift velocity, and carrier population. The ratio of carrier-drift velocity (Eq. 3.27) to a bias field is expressed as

$$\mu(t,t') = \mu_e\left[1 - e^{-(t-t')/\tau_s}\right], \tag{3.51}$$

where μ_e is the electron mobility for stationary charge transport and τ_s is the momentum relaxation time. The electron population decays exponentially with carrier lifetime τ_c as $n(t,t') = e^{-(t-t')/\tau_c}$. Assuming a Gaussian optical pulse $I_{opt}(t) = I_0 e^{-t^2/\tau_p^2}$, we calculate the surface conductivity:

$$\sigma_s(t) = \frac{2\sigma_0}{\sqrt{\pi}\tau_p}\int_{-\infty}^{t} e^{-t'^2/\tau_p^2}\left[1 - e^{-(t-t')/\tau_s}\right]e^{-(t-t')/\tau_c}\,dt', \tag{3.52}$$

where

$$\sigma_0 = \frac{\sqrt{\pi}e\mu_e(1-R_{opt})I_0\tau_p}{2\hbar\omega}. \tag{3.53}$$

The integration leads to

$$\begin{aligned}\sigma_s(\eta) = \sigma_0 \Big\{ & e^{-a\eta + a^2/4}\left[1 + \text{erf}(\eta - a/2)\right] \\ & - e^{-(a+b)\eta + (a+b)^2/4}\left[1 + \text{erf}(\eta - (a+b)/2)\right]\Big\},\end{aligned} \tag{3.54}$$

where $\eta = t/\tau_c$, $a = \tau_p/\tau_c$, and $b = \tau_p/\tau_s$.

Figure 3.25 shows the time-dependent surface conductivity and its Fourier transform. The curve in Fig. 3.25(a) is for an optical pulse duration of $2\sqrt{\ln 2}\cdot\tau_p$=0.08 ps, a carrier lifetime of τ_c=0.5 ps, and a momentum relaxation time of τ_s=0.03 ps. The Fourier transform of the conductivity indicates

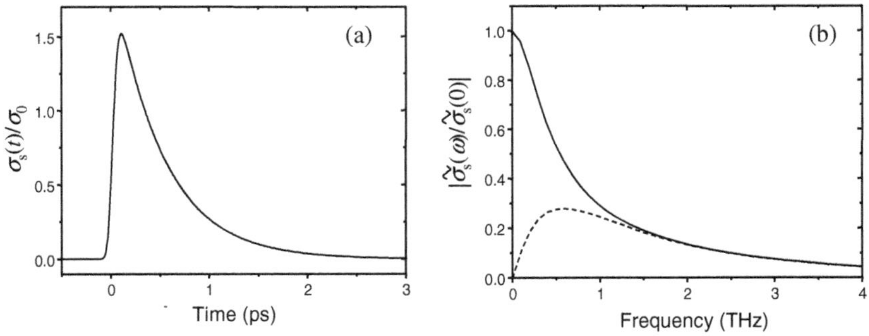

Fig. 3.25. (a) Surface conductivity as a function of time and (b) the amplitude of its Fourier transform. The dotted line indicates the response function of the detector when the diffraction effect is included. The optical pulse duration $2\sqrt{\ln 2} \cdot \tau_p$ is 0.08 ps, the carrier lifetime τ_c is 0.5 ps, and the momentum relaxation time τ_s is 0.03 ps.

that the PC receiver, having a finite bandwidth of ~1 THz, can measure frequencies up to ~4 THz.

There is another frequency-dependent limitation to the responsivity of the PC receiver. The spot size of the THz beam focused on the detector depends on frequency due to diffraction. Consequently, the lower the frequency, the weaker the field strength is at the detector. The response function relevant to the diffraction is proportional to the frequency at low frequencies [18, 31]. Including the diffraction effect, we modify Eq. 3.50 as $\tilde{J}(\nu) = \tilde{H}(\nu)\tilde{E}_{THz}(\nu)$, where $\tilde{H}(\nu)$ is the effective response function of the PC receiver. The dotted line in Fig. 3.25(b) indicates the amplitude of the response function.

3.3 Optical Rectification

In this section we turn our attention to optical phenomena in nonlinear optical crystals. In particular, we will focus on THz generation schemes exploiting optical rectification, a second-order nonlinear optical effect.

The response of electrons in matter to external electromagnetic waves is the primary source of most optical phenomena. Electromagnetic waves force electrons to move, and the accelerated motion of electrons induces radiation. Usually, the amplitude of the electrons motion is very small, and the influence of magnetic fields on naturally occurring material is almost negligible. Consequently, the optical response of a medium is dominated by the electric dipole oscillations of electrons.

In the linear optical regime, the electric dipole moments are proportional to the amplitude of the applied optical field. As we discussed in section 2.1.4, the classical Lorentzian model provides a good qualitative description of this phenomenon. We assume that electrons bound into atoms oscillate about their equilibrium positions with very small amplitude, then each bound electron behaves as a simple harmonic oscillator. Figure 3.26 illustrates the harmonic oscillator model. As given in Eq. 2.51, the bulk polarization for N electrons per unit volume is proportional to the applied electric field,

$$P(t) = \frac{Ne^2}{m} \frac{E_0 e^{-i\omega t}}{\omega_0^2 - \omega^2 - i\omega\gamma} = \epsilon_0 \chi(\omega) E_0 e^{-i\omega t} \propto E(t). \qquad (3.55)$$

In this linear regime the optical response of the medium oscillates with the same frequency of the external field.

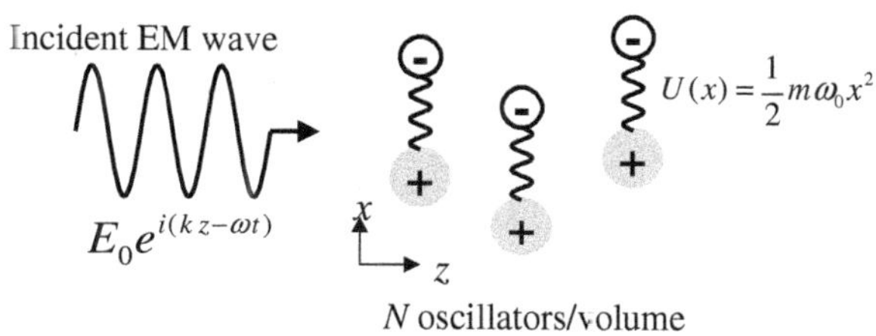

Fig. 3.26. The harmonic oscillator model accounts for the linear optical response of bound electrons in a dielectric medium.

Now we turn to the question, "What if the applied optical field is considerably strong?". A simple answer to this question is that the strong optical field gives rise to nonlinear optical phenomena, characterized by the field-induced changes of optical properties of the illuminated material. They arise from the nonlinear motions of electrons with relatively large amplitudes. The advances of modern laser technology—in particular, high power and ultrafast lasers—initiated the field of nonlinear optics, and optical THz generation and detection schemes have taken advantage of the evolving laser technology.

3.3.1 Nonlinear Optical Interactions with Noncentrosymmetric Media

The linear optical regime fails when the applied field is sufficiently strong to induce large electron displacements from equilibrium, yet the classical bound electron model is still useful to describe various nonlinear optical phenomena if the potential energy function is properly revised [32].

It is helpful to consider a real example for understanding the nonlinear optical processes which interests us. ZnTe is a widely used electro-optic (EO) crystal for THz generation. Figure 3.27 shows its crystal structure. ZnTe is noncentrosymmetric, meaning that it has no inversion symmetry. Since Te has a higher electronegativity than Zn, the electron charge distribution in a chemical bond inclines toward Te. The asymmetric charge distribution gives rise to an asymmetric potential energy along the chemical bond. A sensible

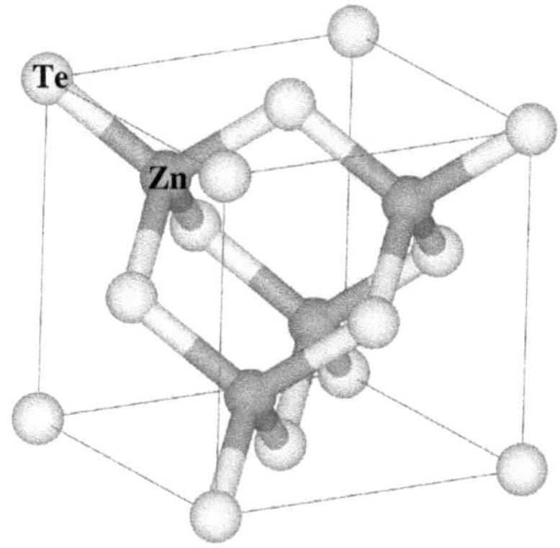

Fig. 3.27. ZnTe crystal structure

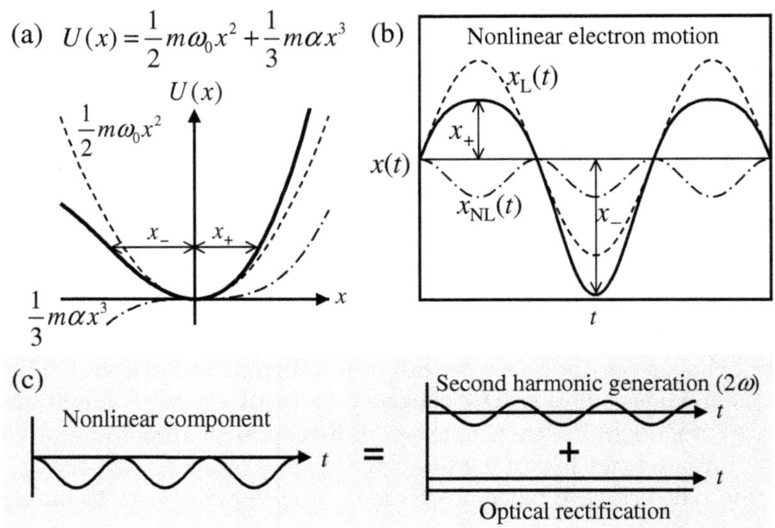

Fig. 3.28. Electric potential energy and nonlinear motion for an electron in a noncentrosymmetric medium.

approximation of the potential energy function is a Taylor series expansion about the equilibrium position. Figure 3.28(a) shows a model potential energy in which the cubic term of the Taylor series expansion is included to represent the asymmetry of the chemical bond. When the electron motion is sufficiently large, the discrepancy between the positive (x_+) and negative (x_-) displacements becomes substantial. Figure 3.28(b) illustrates the nonlinear electron motion, which is decomposed into linear ($x_\mathrm{L}(t)$, dashed line) and nonlinear ($x_\mathrm{NL}(t)$, dashed-dotted line) parts. The nonlinear part consists of two frequency components (Fig. 3.28(c)) representing the two prominent nonlinear optical processes: second harmonic generation (SHG) and optical rectification.

The equation of motion for the Lorentz model revised to incorporate the nonlinear responses has the form

$$\frac{d^2x}{dt^2} + \gamma\frac{dx}{dt} + \omega_0^2 x + \alpha x^2 = -\frac{e}{m}E(t). \tag{3.56}$$

Here we assume that the incident wave is monochromatic, i.e., $E(t) = E_0 e^{-i\omega t}$. In the perturbative regime where the nonlinear term αx^2 is much smaller than the linear term $\omega_0^2 x$, it is valid to expand $x(t)$ as

$$x(t) = \sum_{n=1}^{\infty} x^{(n)}(t), \;\; n = 1, 2, 3, \cdots \tag{3.57}$$

where $x^{(n)} \propto (E_0)^n$. We can apply a perturbation procedure to obtain the n-th order solution $x^{(n)}$, assuming the solution is convergent, i.e., $x^{(1)} \gg x^{(2)} \gg x^{(3)} \cdots$, in the limit of a relatively small nonlinearity. By substituting Eq. 3.57 into Eq. 3.56 and equating terms of equivalent frequency dependence, we obtain the equations for the first and second-order terms,

$$\frac{d^2x^{(1)}}{dt^2} + \gamma\frac{dx^{(1)}}{dt} + \omega_0^2 x^{(1)} = -\frac{e}{m}E(t), \tag{3.58}$$

$$\frac{d^2x^{(2)}}{dt^2} + \gamma\frac{dx^{(2)}}{dt} + \omega_0^2 x^{(2)} = -\alpha\left[x^{(1)}\right]^2. \tag{3.59}$$

It is obvious that the first-order (linear) response at ω is

$$x^{(1)}(t) = -\frac{e}{m}\frac{E_0 e^{-i\omega t}}{\omega_0^2 - \omega^2 - i\omega\gamma} + \text{c.c.} \tag{3.60}$$

By substituting Eq. 3.60 into Eq. 3.59, we obtain the second-order responses corresponding to SHG at 2ω and optical rectification:

$$\begin{aligned} x^{(2)}(t) &= x_{2\omega}^{(2)} + x_0^{(2)} \\ &= -\alpha\left[\frac{eE_0}{m}\right]^2 \frac{e^{-i2\omega t}}{(\omega_0^2 - \omega^2 - i\omega\gamma)^2\{\omega_0^2 - (2\omega)^2 - i2\omega\gamma\}} + \text{c.c.} \\ &\quad -2\alpha\left[\frac{e}{m\omega_0}\right]^2 \frac{|E_0|^2}{(\omega_0^2 - \omega^2)^2 + \omega^2\gamma^2}. \end{aligned} \tag{3.61}$$

The bulk polarization induced by optical rectification is

$$\begin{aligned} P_0^{(2)} &= -Nex_0^{(2)} = \frac{2\alpha e^2 N}{m^2\omega_0^2\left\{(\omega_0^2 - \omega^2)^2 + \omega^2\gamma^2\right\}}|E_0|^2 \\ &= 2\epsilon_0\chi^{(2)}(0, \omega, -\omega)|E_0|^2, \end{aligned} \tag{3.62}$$

where $\chi^{(2)}(0, \omega, -\omega)$ is the second-order nonlinear optical susceptibility corresponding to the optical rectification process. Note that the static nonlinear polarization is proportional to the applied light intensity.

Now we consider a rectified polarization induced by an optical pulse instead of a continuous wave. An optical pulse is expressed as $E(t) = E_0(t)e^{-i\omega t}$ with the time-dependent field amplitude $E_0(t)$. Assuming the pulse duration (τ_p) is much longer than the optical period ($\tau_p \gg \omega^{-1}$) and the dispersion of the nonlinear susceptibility is negligible near the optical frequency, the rectified nonlinear polarization replicates the optical pulse envelope. Figure 3.29 shows the electric field of a Gaussian optical pulse and corresponding nonlinear polarization induced by optical rectification. The time varying polarization is a source of electromagnetic radiation. Apparently the spectral bandwidth of the radiation is roughly the inverse of the optical pulse duration. The typical pulse duration of femtosecond laser pulses is in the range of 10-100 fs, thus ultrashort THz pulses can be produced by optical rectification of femtosecond pulses in a noncentrosymmetric medium.

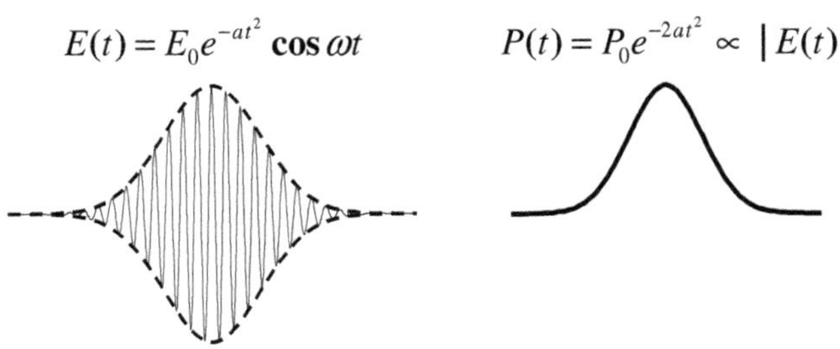

Fig. 3.29. Applied optical field of a Gaussian pulse and nonlinear polarization induced by optical rectification.

3.3.2 Second-Order Nonlinear Polarization and Susceptibility Tensor

Having periodic lattice structures, crystalline solids are not uniform media. It is necessary to introduce tensor formalism for properly describing the nonlinear susceptibility of a crystal system. We can express the nonlinear polarization induced by optical rectification as [32]

$$P_i^{(2)}(0) = \sum_{j,k} \epsilon_0 \chi_{ijk}^{(2)}(0, \omega, -\omega) E_j(\omega) E_k^*(\omega). \tag{3.63}$$

The indices i, j, and k indicate the cartesian components of the fields. $\chi_{ijk}^{(2)}$ is the second-order nonlinear susceptibility tensor element for the crystal system. When the indices are permutable, we can use the contracted notation [32]:

$$d_{il} = \frac{1}{2}\chi^{(2)}_{ijk}, \tag{3.64}$$

where

$$\begin{array}{lcccccc} l = & 1 & 2 & 3 & 4 & 5 & 6 \\ jk = & 11 & 22 & 33 & 23,32 & 31,13 & 12,21 \end{array} \tag{3.65}$$

Using the contracted matrix we can describe the nonlinear polarization as the matrix equation

$$\begin{pmatrix} P_x \\ P_y \\ P_z \end{pmatrix} = 2\epsilon_0 \begin{pmatrix} d_{11} & d_{12} & d_{13} & d_{14} & d_{15} & d_{16} \\ d_{21} & d_{22} & d_{23} & d_{24} & d_{25} & d_{26} \\ d_{31} & d_{32} & d_{33} & d_{34} & d_{35} & d_{36} \end{pmatrix} \begin{pmatrix} E_x^2 \\ E_y^2 \\ E_z^2 \\ 2E_yE_z \\ 2E_zE_x \\ 2E_xE_y \end{pmatrix} \tag{3.66}$$

If the crystal system is highly symmetric, many of the 18 tensor elements vanish and only a few nonvanishing elements are independent. Table 3.3 shows the d-matrices for several EO crystals, in which THz optical rectification has been demonstrated.

Table 3.3. d-matrices of EO crystals for THz generation

Material	Crystal class	d-matrix
ZnTe, GaAs, GaP, InP	$\bar{4}3m$	$\begin{pmatrix} 0 & 0 & 0 & d_{14} & 0 & 0 \\ 0 & 0 & 0 & 0 & d_{14} & 0 \\ 0 & 0 & 0 & 0 & 0 & d_{14} \end{pmatrix}$
GaSe	$\bar{6}2m$	$\begin{pmatrix} 0 & 0 & 0 & 0 & 0 & -d_{22} \\ -d_{22} & d_{22} & 0 & 0 & 0 & 0 \\ 0 & 0 & 0 & 0 & 0 & 0 \end{pmatrix}$
$LiNbO_3$, $LiTaO_3$	$3m$	$\begin{pmatrix} 0 & 0 & 0 & 0 & d_{15} & -d_{22} \\ -d_{22} & d_{22} & 0 & d_{15} & 0 & 0 \\ d_{15} & d_{15} & d_{33} & 0 & 0 & 0 \end{pmatrix}$

ZnTe, the most commonly used EO crystal for THz generation and detection, has the crystal class of $\bar{4}3m$. This crystal class has three nonvanishing contracted matrix elements and only one of them is independent: $d_{14} = d_{25} = d_{36}$. When an optical field interacts with ZnTe, its THz radiation power depends on the direction of the field in the crystal frame. An arbitrary field is expressed as

$$\mathbf{E}_0 = E_0 \begin{pmatrix} \sin\theta\cos\phi \\ \sin\theta\sin\phi \\ \cos\theta \end{pmatrix} \tag{3.67}$$

with the polar angle θ and the azimuthal angle ϕ. Using Eq. 3.66 we obtain the nonlinear polarization:

$$\begin{pmatrix} P_x \\ P_y \\ P_z \end{pmatrix} = 2\epsilon_0 d_{14} E_0^2 \begin{pmatrix} 0\,0\,0\,1\,0\,0 \\ 0\,0\,0\,0\,1\,0 \\ 0\,0\,0\,0\,0\,1 \end{pmatrix} \begin{pmatrix} \sin^2\theta\cos^2\phi \\ \sin^2\theta\sin^2\phi \\ \cos^2\theta \\ 2\sin\theta\cos\theta\sin\phi \\ 2\sin\theta\cos\theta\cos\phi \\ 2\sin^2\theta\sin\phi\cos\phi \end{pmatrix}$$
$$= 4\epsilon_0 d_{14} E_0^2 \sin\theta \begin{pmatrix} \cos\theta\sin\phi \\ \cos\theta\cos\phi \\ \sin\theta\sin\phi\cos\phi \end{pmatrix}. \qquad (3.68)$$

The consequent THz radiation field is parallel to the nonlinear polarization, therefore the intensity of the THz radiation has the angular dependence

$$I_{THz}(\theta,\phi) \propto |\mathbf{P}|^2 = 4\epsilon_0^2 d_{14}^2 E_0^4 \sin^2\theta\left(4\cos^2\theta + \sin^2\theta\sin^2 2\phi\right). \qquad (3.69)$$

The THz intensity is maximized when $\sin^2 2\phi = 1$ ($\phi = \frac{\pi}{4}$ or $\frac{3\pi}{4}$), i.e., the optical polarization lies in the $\{110\}$ plane.

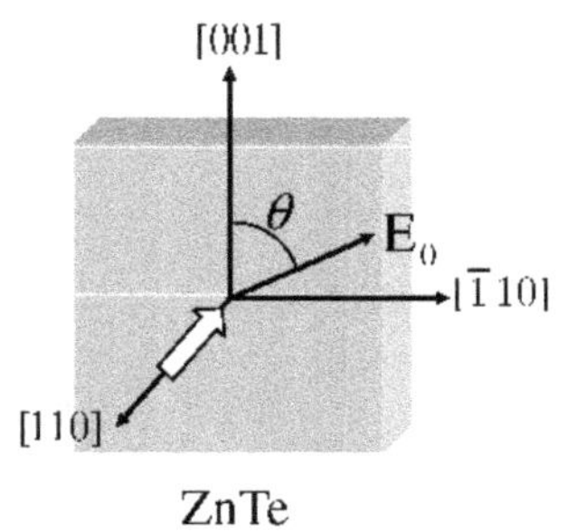

Fig. 3.30. A linearly polarized optical wave is incident on a (110) ZnTe crystal with normal angle. θ is the angle between the optical field and the [001] axis.

Figure 3.30 shows an optical field incident on a (110) ZnTe crystal. The linearly polarized optical beam is propagating along the [110] axis of the ZnTe crystal with an angle of θ between the optical field and the [001] axis. The radiated THz intensity as a function of θ is written as

$$I_{THz}(\theta) = \frac{3}{4} I_{THz}^{max} \sin^2\theta\left(4 - 3\sin^2\theta\right). \qquad (3.70)$$

The maximum intensity I_{THz}^{max} is obtained at $\theta = \sin^{-1}\sqrt{\frac{2}{3}}$. This angle corresponds to an optical field that is parallel to either the $[\bar{1}11]$ or $[1\bar{1}1]$ axis. In other words, we can optimize the THz intensity by aligning the optical field along the chemical bonds between the Zn and the Te (see the crystal structure of ZnTe in Fig. 3.27). Figure 3.31 shows the angle-dependent THz radiation intensity.

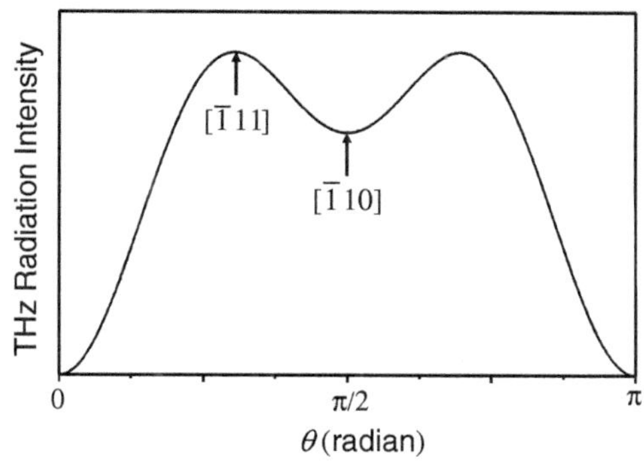

Fig. 3.31. THz radiation intensity vs θ in ZnTe.

We can deduce the expression for the angle-dependent THz field vector from Eq. 3.68 :

$$\mathbf{E}_{THz}(\theta) = \frac{\sqrt{3}}{2} E_{THz}^{max} \sin\theta \begin{pmatrix} \sqrt{2}\cos\theta \\ -\sqrt{2}\cos\theta \\ -\sin\theta \end{pmatrix} . \tag{3.71}$$

When the linearly polarized optical field is aligned along the [001], [$\bar{1}$11], or [$\bar{1}$10] axis, $\mathbf{E}_{THz}$ has the relations

$$\mathbf{E}_0 \;//\; [001] \quad \rightarrow \quad \mathbf{E}_{THz} = 0, \tag{3.72}$$

$$\mathbf{E}_0 \;//\; [\bar{1}11] \quad \rightarrow \quad \mathbf{E}_{THz} = \frac{1}{\sqrt{3}} E_{THz}^{max} \begin{pmatrix} 1 \\ -1 \\ -1 \end{pmatrix} \;//\; -\mathbf{E}_0, \tag{3.73}$$

$$\mathbf{E}_0 \;//\; [\bar{1}10] \quad \rightarrow \quad \mathbf{E}_{THz} = \frac{\sqrt{3}}{2} E_{THz}^{max} \begin{pmatrix} 0 \\ 0 \\ -1 \end{pmatrix} \;\perp\; \mathbf{E}_0. \tag{3.74}$$

The THz field is antiparallel to the optical field when the optical field is aligned along the $\langle 111 \rangle$ axes; they are perpendicular to each other when the optical field is parallel to the $\langle 110 \rangle$ axes. The angle α between the optical and the THz fields as a function of θ can be written as

$$\alpha(\theta) = \cos^{-1}\left\{ \frac{\mathbf{E}_0 \cdot \mathbf{E}_{THz}}{|\mathbf{E}_0||\mathbf{E}_{THz}|} \right\} = \cos^{-1}\left\{ -3\frac{\sin\theta|\cos\theta|}{\sqrt{4-3\sin^2\theta}} \right\}, \tag{3.75}$$

which, illustrated in Fig. 3.32, varies between 90° and 180°.

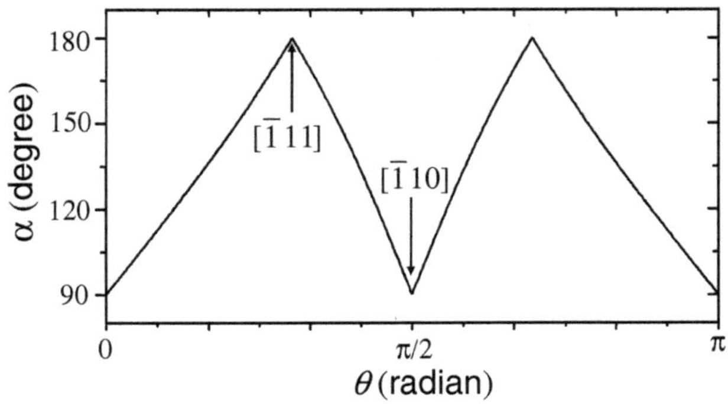

Fig. 3.32. Angle between the optical and the THz fields as a function of θ

3.3.3 Wave Equation for Optical Rectification

In the present section we shall discuss THz generation and propagation in a nonlinear medium. An optically induced nonlinear polarization depends not only on time but also on position. In order to describe THz generation and propagation in a bulk medium, we formulate the wave equation including the nonlinear polarization as a source term, assuming that a linearly polarized optical plane wave propagates in the z-axis:

$$\frac{\partial^2 E_T(z,t)}{\partial z^2} - \frac{n_T^2}{c^2}\frac{\partial^2 E_T(z,t)}{\partial t^2} = \frac{1}{\epsilon_0 c^2}\frac{\partial^2 P_T^{(2)}(z,t)}{\partial t^2} = \frac{\chi^{(2)}}{c^2}\frac{\partial^2 |E_0(z,t)|^2}{\partial t^2}, \quad (3.76)$$

where $E_T(z,t)$ and $P_T^{(2)}(z,t)$ are the THz field and the polarization, respectively. $E_0(z,t)$ is the optical field amplitude.

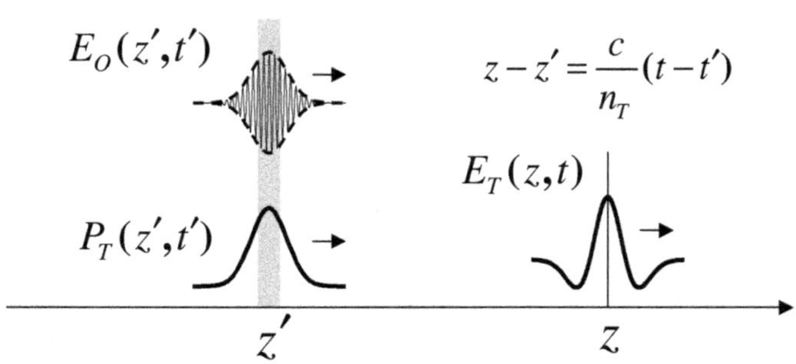

Fig. 3.33. THz dipole radiation from a thin layer of nonlinear polarization generated by a Gaussian optical pulse.

For a qualitative analysis of the wave equation we first consider a dispersionless medium at THz and optical frequencies, i.e., the refractive indices n_T

and n_O are independent of frequency. We present an intuitive way to find the solution of the wave equation. Imagine that a Gaussian optical pulse interacts with an infinitesimally thin layer of a nonlinear medium located at z' as illustrated in Fig. 3.33, which induces a Gaussian nonlinear polarization by optical rectification. In this one dimensional picture the dipole radiation field from the thin layer is proportional to the second-order time derivative of the Gaussian waveform. The total THz radiation field from a finite medium can be obtained by dividing the medium into thin layers and adding up the fields from them. Since the typical optical-to-THz conversion efficiency is less than 10^{-4}, we can assume that the optical pump is undepleted during the process.

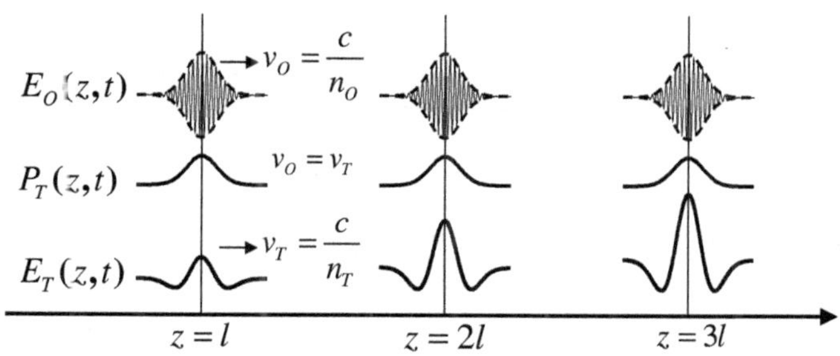

Fig. 3.34. Linear amplification of THz field in a medium satisfying the velocity matching condition.

An ideal case for THz generation is that the THz pulse propagates with the same velocity as the optical pulse, i.e., $n_T = n_O$. When the velocity matching condition is satisfied, the THz field is gradually amplified while propagating through the medium. The linear amplification process is illustrated in Fig. 3.34. Given the optical field amplitude

$$E_O(z,t) = E_0 \exp\left[-a\left(t - \frac{z}{v_O}\right)^2\right], \tag{3.77}$$

the total THz field generated from $z' = 0$ to l is the linear superposition,

$$E_T(z,t) = \int_0^l A\left[1 - 4a\left(t' - \frac{z'}{v_O}\right)^2\right] \exp\left[-2a\left(t' - \frac{z'}{v_O}\right)^2\right] dz', \tag{3.78}$$

where A is a constant, and

$$t' = t - \frac{z - z'}{v_T} = t - \frac{z - z'}{v_O}. \tag{3.79}$$

Due to the velocity matching condition $v_T = v_O$, the integrand of Eq. 3.78 is independent of z'. Thus,

$$E_T(z,t) = Al\left[1 - 4a\left(t - \frac{z}{v_O}\right)^2\right]\exp\left[-2a\left(t - \frac{z}{v_O}\right)^2\right]. \tag{3.80}$$

The THz field amplitude is proportional to the propagation length l.

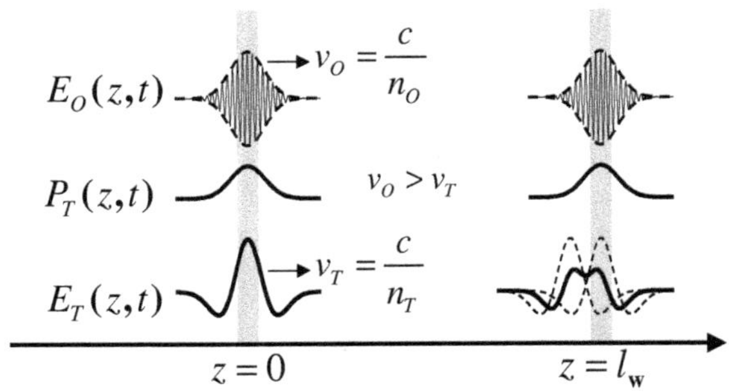

Fig. 3.35. Destructive interference between the THz radiation fields from two nonlinear layers separated by a walk-off length l_w. The dotted lines indicate the two THz fields.

In general, the velocity matching condition is difficult to satisfy. When the optical wave is faster than the THz wave ($n_T > n_O$), the optical pulse leads the THz pulse by the optical pulse duration, τ_p, after a walk-off length of

$$l_w = \frac{c\tau_p}{(n_T - n_O)}, \tag{3.81}$$

where c is the speed of light in vacuum. Imagine two thin layers of nonlinear media separated by the walk-off length, which is illustrated in Fig. 3.35. The superposition of the THz radiation fields from the two layers shows destructive interference near $z = l_w$. While propagating in a uniform nonlinear medium, the THz radiation field continuously undergoes destructive interference. When the thickness of the nonlinear medium is much longer than the walk-off length ($l >> l_w$), the total THz field is

$$\begin{aligned} E_{THz}(z,t) &\approx \int_{-\infty}^{\infty} A\left[1 - 2a\left(t' - \frac{z'}{v_O}\right)^2\right]\exp\left[-a\left(t' - \frac{z'}{v_O}\right)^2\right]dz' \\ &= \frac{cA}{n_T - n_O}\int_{-\infty}^{\infty}\left(1 - 2ax^2\right)\exp\left(-ax^2\right)dx \\ &= \frac{cA}{n_T - n_O}\left[x\exp\left(-ax^2\right)\right]_{-\infty}^{\infty} = 0, \end{aligned} \tag{3.82}$$

where

$$x = t - \frac{z'}{v_O} = \frac{1}{c}(n_T - n_O)z' - \frac{n_T}{c}z + t. \tag{3.83}$$

That is, the THz field is washed out in the infinitely long medium. For a medium with a finite thickness, radiation generated near the surfaces within a depth of l_w survives. Apparently, efficient THz generation requires that the nonlinear medium must have a long walk-off length and that its thickness must be shorter than the walk-off length.

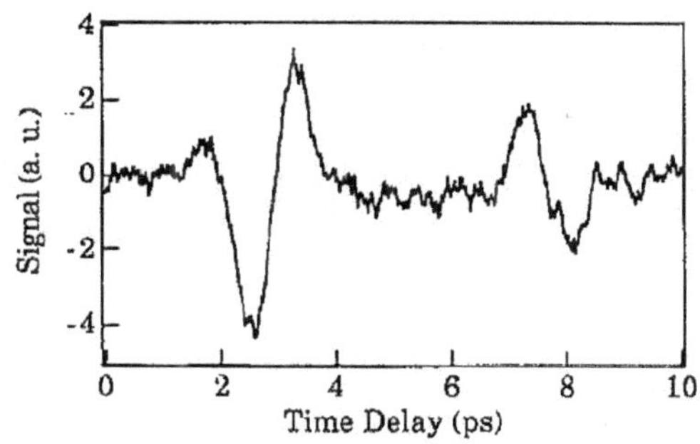

Fig. 3.36. Temporal waveform of the subpicosecond THz radiation from a 0.5-mm-thick $LiNbO_3$ crystal. (Reprinted with permission from [33]. ©1992, American Institute of Physics.)

Figure 3.36 shows a THz waveform from a lithium niobate ($LiNbO_3$) crystal generated by optical rectification of femtosecond pulses with a 620 nm wavelength [33]. The optical pulse is faster than the THz pulse in $LiNbO_3$: n_O=2.3 and $n_T = 5.2$ [34]. The two pulses in the waveform correspond to the radiation near the exit and entrance surfaces of the crystal. Due to the velocity mismatch between optical and THz pulses, the pulse from the exit surface arrived earlier at the detector. No contribution from the bulk region is noticeable. The time delay between the two THz pulses is consistent with the velocity mismatch between the optical and the THz pulses, $\Delta t = (n_T - n_O)l/c = 4.8$ ps. The pulse from the entrance surface is weaker due to linear absorption in the crystal.

3.3.4 Dispersion at Optical and Terahertz Frequencies

In practice, nonlinear crystals are dispersive both at optical and THz frequencies: the refractive index $n(\omega)$ varies with frequency. As a result, the group velocity v_{gr} differs from the phase velocity v_{ph} at most frequencies:

$$v_{gr}(\omega) = \frac{\partial \omega}{\partial k} \text{ and } v_{ph}(\omega) = \frac{\omega}{k}, \text{ where } k(\omega) = n(\omega)\frac{\omega}{c}. \tag{3.84}$$

Consequently, velocity matching in a dispersive medium can be achieved only for a certain THz frequency when the optical pulse envelope travels at the phase velocity of the monochromatic THz wave [35]. The optimal velocity-matching condition for a broadband THz pulse is that the optical group velocity is the same as the phase velocity of the central frequency of the THz spectrum.

ZnTe is the most widely used nonlinear crystal for THz generation because the group refractive index

$$n_{gr}(\lambda) = \frac{c}{v_{gr}} = n_O(\lambda) - \lambda\frac{\partial n_O}{\partial \lambda} \tag{3.85}$$

near the optical wavelength λ=0.8 μm (the operational wavelength of Ti:sapphire femtosecond lasers) matches well with the THz refractive index $n_T(\nu_{THz})$.

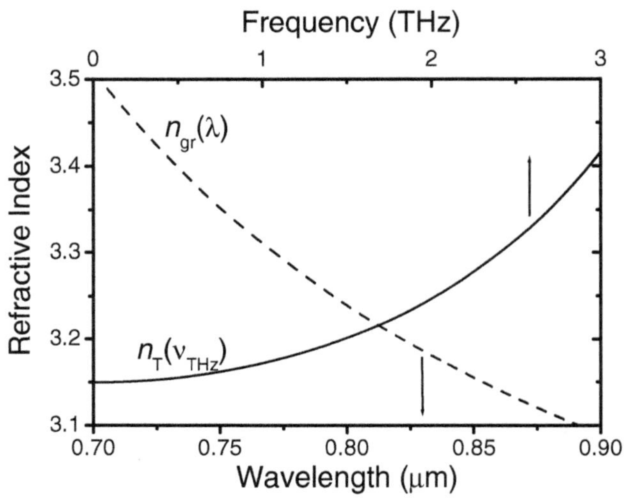

Fig. 3.37. Optical group refractive indexes $n_{gr}(\lambda)$ and THz refractive indexes $n_T(\nu_{THz})$ of ZnTe

Figure 3.37 shows the optical group refractive index and the THz refractive index of ZnTe. The refractive indices of ZnTe are calculated from the following equations [35]:

$$n_O^2(\lambda) = 4.27 + \frac{3.01\lambda^2}{\lambda^2 - 0.142}, \tag{3.86}$$

$$n_T^2(\nu_{THz}) = \frac{289.27 - 6\nu_{THz}^2}{29.16 - \nu_{THz}^2}, \tag{3.87}$$

where λ is in μm and ν_{THz} is in THz. The velocity-matching condition is satisfied when the optical wavelength is λ=812 nm and the Thz frequency is 1.69 THz: $n_{gr}(812\mu\text{m}) = n_T(1.69\ \text{THz}) = 3.22$.

Figure 3.38 shows the temporal waveform of a single-cycle THz pulse produced under a velocity-matching condition. It was generated by optical rectification in a 1.0-mm thick ⟨110⟩ ZnTe crystal using 100 fs optical pulses at 0.8 μm.

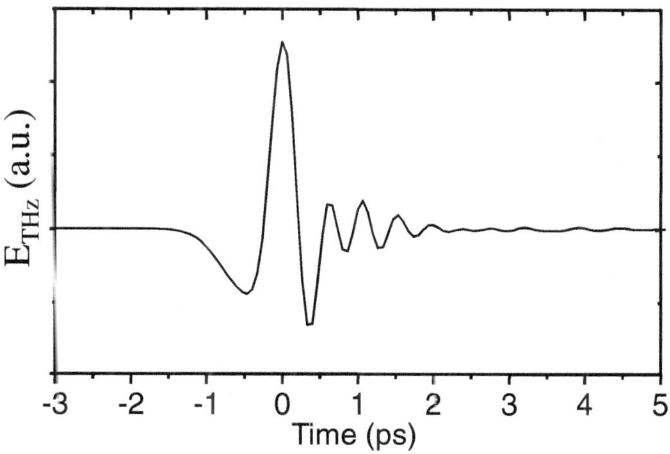

Fig. 3.38. Temporal waveform of a THz pulse generated by optical rectification in a 1.0-mm thick ⟨110⟩ ZnTe emitter. The THz electric fields are measured by EO sampling.

In addition to ZnTe, several other nonlinear crystals have been tested for THz generation by optical rectification. Some of these crystals meet the velocity-matching condition at certain optical frequencies, depending on material dispersion. The degree of velocity-matching is measured by the interaction length of optical rectification. For example, the interaction length is infinitely long for perfect velocity-matching. The effective interaction length is expressed as the coherence length:

$$l_c = \frac{c}{2\nu_{THz}|n_{gr} - n_T|}. \tag{3.88}$$

The coherence length is the distance over which the optical pulse propagates before leading or lagging the THz wave by a $\pi/2$ phase shift.

Figure 3.39 shows the coherence length of ZnTe, CdTe, GaP, InP, and GaAs at 2 THz as a function of optical wavelength. Table 3.4 lists the optical wavelengths at which the velocity-matching condition is satisfied in the zinc-blende crystals.

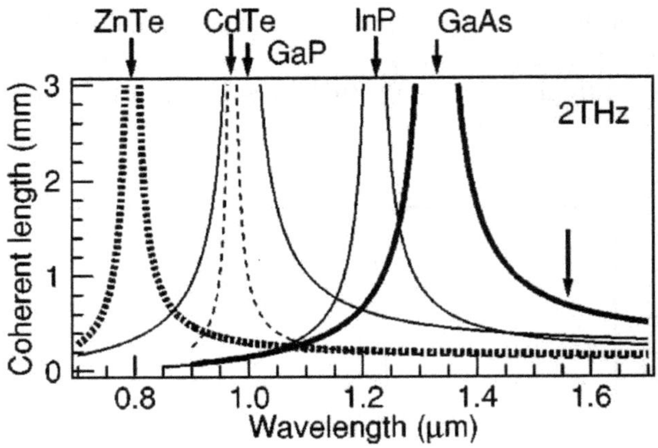

Fig. 3.39. Coherence length of ZnTe, CdTe, GaP, InP, and GaAs at 2 THz as a function of the optical wavelength. (Reprinted with permission from [36]. ©2004, American Institute of Physics.)

Table 3.4. Optical wavelengths for velocity-matching in zinc-blende crystals

	ZnTe	CdTe	GaP	InP	GaAs
Wavelength (μm)	0.8	0.97	1.0	1.22	1.35

3.3.5 Absorption of Electro-Optic Crystals at the Terahertz Frequencies

The spectral bandwidth of THz generation in a nonlinear crystal is limited by absorption in the THz frequency region. The dominant THz absorption processes in EO crystals are the transverse-optical (TO) phonon resonances, which usually lie in the the range from 5 to 10 THz. At lower frequencies, second-order phonon processes give rise to weak, yet complicated and broad absorption spectra.

Figure 3.40 shows the measured (solid line) absorption coefficient for ZnTe crystal [37] compared with the calculated (dashed line) absorption for the TO-phonon line. For the calculation we use the dielectric response of a harmonic oscillator for the TO-phonon mode [37]:

$$\epsilon(\nu) = \epsilon_{el} + \frac{\epsilon_{st}\nu_{TO}^2}{\nu_{TO}^2 - \nu^2 + 2i\gamma\nu} = (n + i\kappa)^2, \tag{3.89}$$

where ϵ_{st} and γ are the oscillator strength and the linewidth of the TO-phonon mode, respectively. ZnTe has a strong TO-phonon resonance at ν_{TO} =5.32 THz at room temperature. The other parameters are $\epsilon_{el} = 7.44$, $\epsilon_{st} = 2.58$, and $\gamma = 0.025$ THz. The absorption coefficient is expressed as

$$\alpha(\nu) = \frac{4\pi\nu\kappa(\nu)}{c}. \tag{3.90}$$

The wing of the TO-phonon line below 3.5 THz is insignificant, yet the absorption bands of two-phonon processes near 1.6 and 3.7 THz are prominent [8].

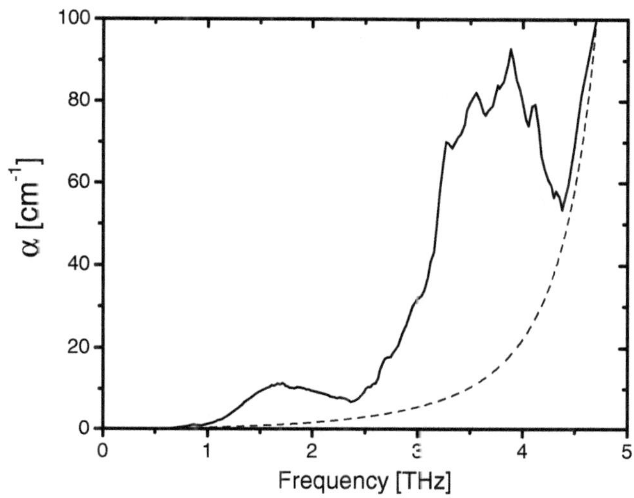

Fig. 3.40. Absorption coefficient $\alpha(\nu_{THz})$ of ZnTe from 0 to 5 THz at room temperature. The dashed line indicates the calculated absorption for the TO-phonon line centered at 5.32 THz. (Data from [37])

Table 3.5 lists the lowest resonant frequencies of TO-phonon modes for several EO crystals. Absorption in EO crystals between 5 and 10 THz is dominated by TO phonon lines. Figure 3.41 shows the absorption coefficients for some EO crystals in the low-frequency wing of the TO-phonon lines.

Table 3.5. Lowest TO-phonon frequencies of EO crystals

	ZnTe	CdTe	GaP	InP	GaAs	GaSe	$LiNbO_3$	$LiTaO_3$
ν_{TO} (THz)	5.3^a	4.3^a	11^b	9.2^c	8.1^d	6.4^d	7.7^e	4.2^f

[a]Reference [8]
[b]Reference [39]
[c]Reference [40]
[d]Reference [41]
[e]Reference [42]
[f]Reference [43]

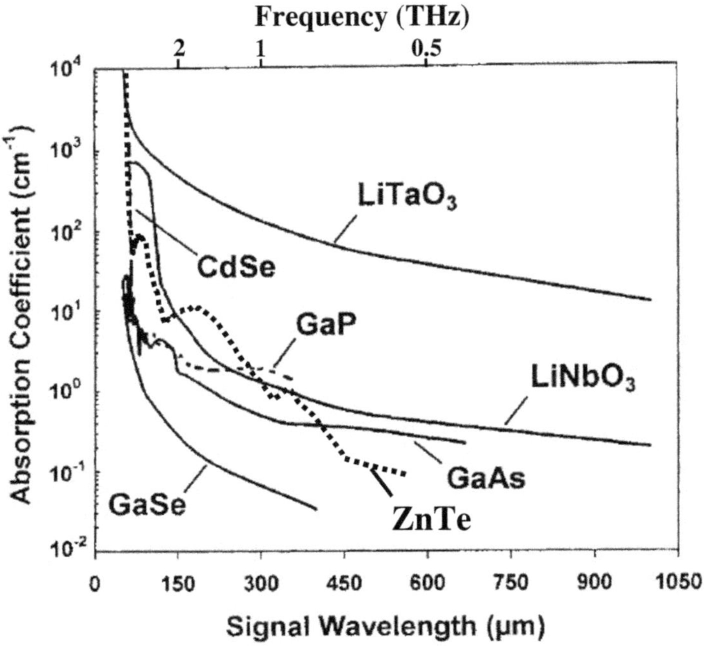

Fig. 3.41. Absorption spectra for an extraordinary wave in CdSe, $LiNbO_3$, GaSe, $LiTaO_3$, GaAs, GaP, and ZnTe. (Reprinted from [38], ZnTe data from [37])

3.4 Free-Space Electro-Optic Sampling

As with a PC detector, free-space EO sampling measures the actual electric field of THz pulses in the time domain, determining not only the amplitude, but also the phase with high precision ($< 10^{-2}$ rad). The underlying mechanism that EO sampling utilizes is the Pockels effect in EO crystals. The Pockels effect is closely related to optical rectification, which is apparent in the similarity between the expressions of their second-order nonlinear polarizations shown in Eqs. 3.63 and 3.91.

$$\begin{aligned}\text{Pockels Effect: } P_i^{(2)}(\omega) &= 2\sum_{j,k} \epsilon_0 \chi_{ijk}^{(2)}(\omega,\omega,0)E_j(\omega)E_k(0) \\ &= \sum_j \epsilon_0 \chi_{ij}^{(2)}(\omega)E_j(\omega),\end{aligned} \tag{3.91}$$

where $\chi_{ij}^{(2)}(\omega) = 2\sum_k \chi_{ijk}^{(2)}(\omega,\omega,0)E_k(0)$ is the field induced susceptibility tensor. In a lossless medium $\chi_{ijk}^{(2)}(0,\omega,-\omega) = \chi_{ijk}^{(2)}(\omega,\omega,0)$, thus the Pockels effect has the same nonlinear optical coefficients as optical rectification. Eq. 3.91 indicates that a static electric field induces birefringence in a nonlinear optical medium proportional to the applied field amplitude. Inversely, the applied field strength can be determined by measuring the field-induced birefringence.

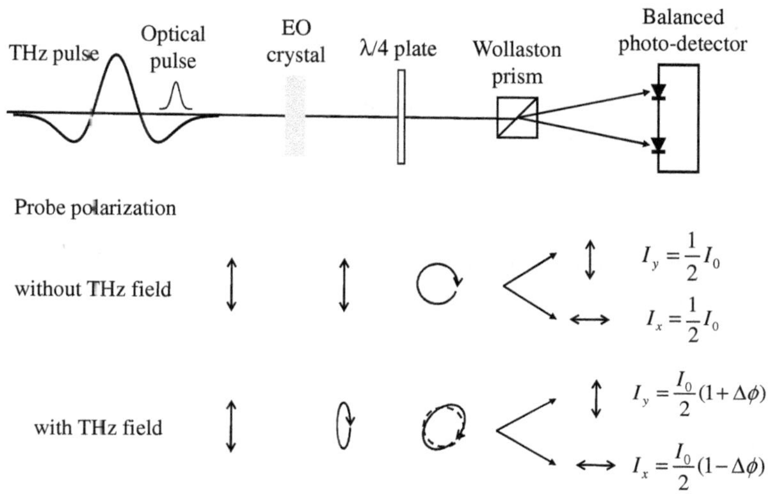

Fig. 3.42. Schematic diagram of a typical setup for free-space EO sampling. Probe polarizations with and without a THz field are depicted before and after the polarization optics.

Figure 3.42 illustrates a typical setup of free-space EO sampling to measure field-induced birefringence. Ideally, the optical group velocity matches well with the THz phase velocity in the EO crystal, then the optical pulse will feel a constant electric field of the THz pulse while propagating. In the lower part of Fig. 3.42, evolution of the probe polarization is shown in series for the steps of the polarization manipulation with or without a THz field. While the linearly polarized optical pulse and the THz pulse propagate through the EO crystal, the field-induced birefringence produces a slightly elliptical polarization of the probe pulse. The probe polarization evolves into an almost circular, but elliptical polarization after a $\lambda/4$-plate. A Wollaston prism splits the probe beam into two orthogonal components, which are sent to a balanced photodetector. The detector measures the intensity difference $I_s = I_y - I_x$ between the two orthogonal components of the probe pulse, which is proportional to the applied THz field amplitude.

The useful characteristics of ZnTe for THz generation—velocity-matching near 800 nm, high transparency at optical and THz frequencies, and large EO coefficient ($r_{41} = d_{14} = 4$ pm/V)—are also desirable for efficient EO sampling. A typical arrangement of the optical and THz polarizations for EO sampling is shown in Fig. 3.43. The field induced birefringence is maximized when both the THz electric field and the optical polarization are parallel to the $[1\bar{1}0]$ axis of an $\langle 110 \rangle$ oriented crystal. We can describe the nonlinear polarization

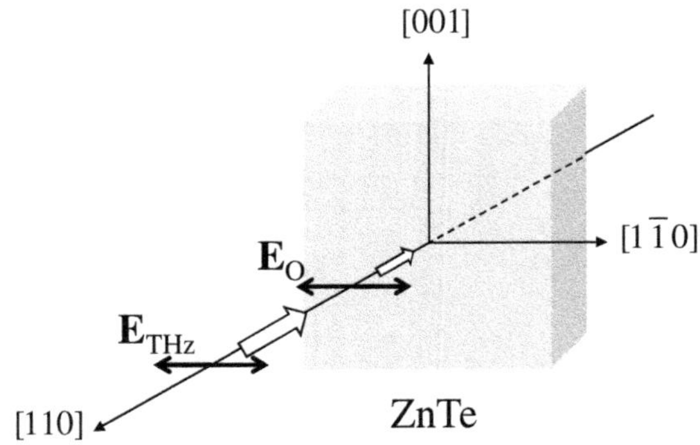

Fig. 3.43. Polarizations of the optical probe and the THz field are parallel to the [1$\bar{1}$0] direction of a ZnTe crystal in a typical EO sampling setup.

in Eq. 3.91 as the matrix equation

$$\begin{aligned}\begin{pmatrix} P_x \\ P_y \\ P_z \end{pmatrix} &= 4\epsilon_0 d_{14} \begin{pmatrix} 0\,0\,0\,1\,0\,0 \\ 0\,0\,0\,0\,1\,0 \\ 0\,0\,0\,0\,0\,1 \end{pmatrix} \begin{pmatrix} E_{O,x}E_{THz,x} \\ E_{O,y}E_{THz,y} \\ E_{O,z}E_{THz,z} \\ E_{O,y}E_{THz,z} + E_{O,z}E_{THz,y} \\ E_{O,z}E_{THz,x} + E_{O,x}E_{THz,z} \\ E_{O,x}E_{THz,y} + E_{O,y}E_{THz,x} \end{pmatrix} \\ &= -4\epsilon_0 d_{14} E_O E_{THz} \mathbf{e}_z \perp \mathbf{E}_O, \end{aligned} \tag{3.92}$$

where

$$\mathbf{E}_O = \frac{E_O}{\sqrt{2}} \begin{pmatrix} 1 \\ -1 \\ 0 \end{pmatrix} \quad \text{and} \quad \mathbf{E}_{THz} = \frac{E_{THz}}{\sqrt{2}} \begin{pmatrix} 1 \\ -1 \\ 0 \end{pmatrix}. \tag{3.93}$$

The nonlinear polarization at the optical frequency is orthogonal to the incident optical field, which implies that the linear polarization of the optical probe evolves into an elliptical polarization via propagation in ZnTe under the influence of the THz field.

The differential phase retardation $\Delta\phi$ experienced by the probe beam due to the Pockels effect over a propagation distance L is given as [44]

$$\Delta\phi = (n_y - n_x)\frac{\omega L}{c} = \frac{\omega L}{c} n_O^3 r_{41} E_{THz}, \tag{3.94}$$

where n_O is the refractive index at the optical frequency and r_{41} is the EO coefficient. The intensities of the two probe beams at the balanced photodetector are

$$I_x = \frac{I_0}{2}(1 - \sin\Delta\phi) \approx \frac{I_0}{2}(1 - \Delta\phi), \tag{3.95}$$

$$I_y = \frac{I_0}{2}(1 + \sin\Delta\phi) \approx \frac{I_0}{2}(1 + \Delta\phi), \tag{3.96}$$

where I_0 is the incident probe intensity. For the approximation we use $\Delta\phi \ll 1$ which is true for most cases of EO sampling. Thus, the signal of the balanced photo-detector measures the THz field amplitude:

$$I_s = I_y - I_x = I_0 \Delta\phi = \frac{I_0 \omega L}{c} n_O^3 r_{41} E_{THz} \propto E_{THz}. \tag{3.97}$$

In a realistic situation the temporal or spectral resolution of EO sampling is limited by three factors: (i) finite pulse duration of optical probe, (ii) dispersion of nonlinear susceptibility, and (iii) mismatch between optical group and THz phase velocity. Consequently, the EO signal is the convolution of the THz field with the detector response function $F(\omega, \omega_{THz})$ which combines these three effects [37]:

$$E_s(t) = \int_{-\infty}^{\infty} E_{THz}(\omega_{THz}) F(\omega, \omega_{THz}) e^{-i\omega_{THz} t} d\omega_{THz}, \tag{3.98}$$

where $E_s(t)$ is the time-resolved EO signal and $E_{THz}(\omega_{THz})$ is the complex spectral amplitude of the incoming THz pulse. The Fourier transform of both sides of this equation yields a simple relation between the EO signal and the THz field in the frequency domain:

$$E_s(\omega_{THz}) = F(\omega, \omega_{THz}) \cdot E_{THz}(\omega_{THz}). \tag{3.99}$$

The detector response function $F(\omega, \omega_{THz})$ is a product of the three frequency dependent factors:

$$F(\omega, \omega_{THz}) = A_{Opt}(\omega_{THz}) \cdot \chi^{(2)}(\omega; \omega_{THz}, \omega - \omega_{THz}) \cdot \Delta\Phi(\omega, \omega_{THz}), \tag{3.100}$$

where $A_{Opt}(\omega_{THz})$ is the autocorrelation of the optical electric field expressed as

$$A_{Opt}(\omega_{THz}) = \int_{-\infty}^{\infty} E^*_{Opt}(\omega' - \omega) E_{Opt}(\omega' - \omega - \omega_{THz}) d\omega' \tag{3.101}$$

and $\chi^{(2)}(\omega; \omega_{THz}, \omega - \omega_{THz})$ is the second order nonlinear susceptibility. The frequency-dependent velocity mismatching gives rise to the frequency filter function $\Delta\Phi(\omega, \omega_{THz})$.

Figure 3.44 shows the second-order nonlinear susceptibility in ZnTe. The sharp feature near 5 THz indicates that the dominant physical process which determines its value is the TO phonon mode at 5.3 THz.

$\Delta\Phi(\omega, \omega_{THz})$ represents the frequency filtering arising from the velocity mismatch, which is expressed as

$$\Delta\Phi(\omega, \omega_{THz}) = \frac{e^{i\Delta k(\omega, \omega_{THz})L} - 1}{i\Delta k(\omega, \omega_{THz})}. \tag{3.102}$$

The wave vector mismatch Δk is given by

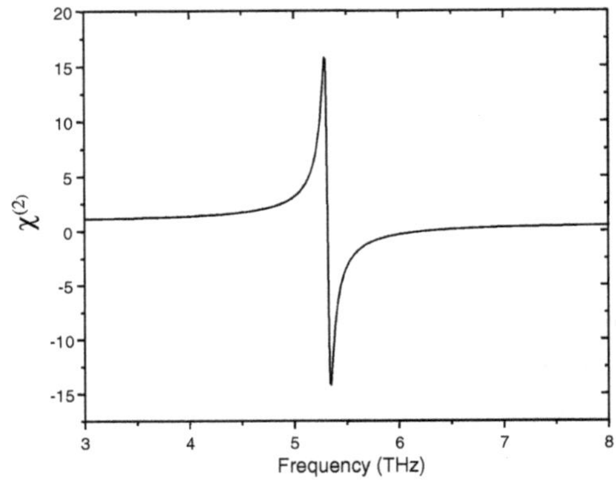

Fig. 3.44. Second-order nonlinear susceptibility in ZnTe.

$$\begin{aligned}\Delta k &= k(\omega) + k(\omega_{THz}) - k(\omega + \omega_{THz}) \\ &= \frac{\omega}{c} n_O(\omega) + \frac{\omega_{THz}}{c}\left\{n_T(\omega_{THz}) + i\kappa_T(\omega_{THz})\right\} \\ &\quad - \frac{\omega + \omega_{THz}}{c} n_O(\omega + \omega_{THz}),\end{aligned} \tag{3.103}$$

where n_O (Eq. 3.86) and $n_T + i\kappa_T$ (Eq. 3.89) are the optical and the THz refractive indices of ZnTe, respectively. Due to the nonlinear absorption involving second-order phonon processes, the imaginary part of the THz refractive index κ_T must be corrected by the measured absorption coefficient shown in Fig. 3.40.

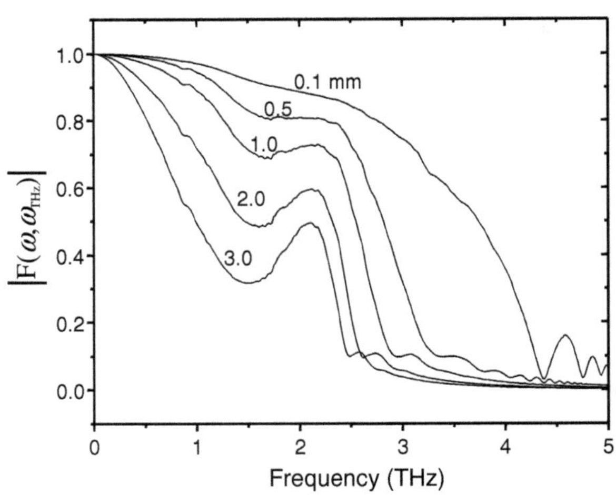

Fig. 3.45. Normalized amplitude of the detector response function $F(\omega, \omega_{THz})$ for ZnTe. The lines are for the crystal thickness of 0.1, 0.5, 1.0, 2.0, and 3.0 mm when the optical pulse duration is 100 fs.

Figure 3.45 shows the normalized amplitude of the detector response function between 0 and 5 THz for ZnTe with various crystal thicknesses when the probe pulse duration is 100 fs. Because of the TO phonon resonance at 5.3 THz, the detector response is negligible above 4 THz. While the detection sensitivity increases linearly as the crystal thickness increases, the detection bandwidth reduces because of the velocity mismatch. The absorption induced spectral modulation also becomes more severe for a thicker crystal.

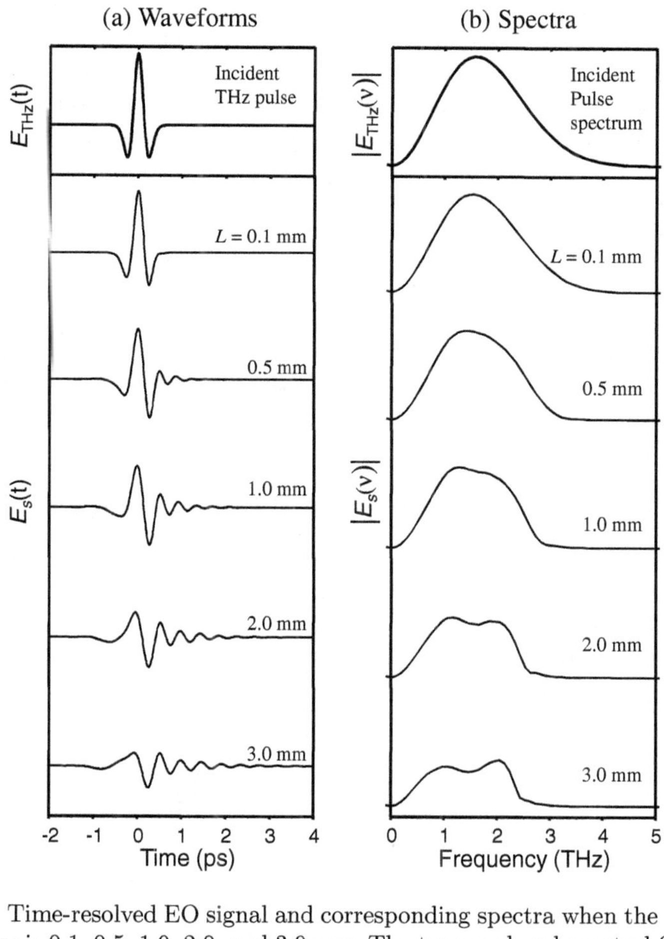

Fig. 3.46. Time-resolved EO signal and corresponding spectra when the ZnTe crystal thickness is 0.1, 0.5, 1.0, 2.0, and 3.0 mm. The temporal and spectral field amplitudes are normalized by crystal thickness. Incident THz waveform and its spectrum is shown at the top panels.

EO signals in ZnTe for a typical single-cycle THz pulse and corresponding spectra for the crystal thickness of 0.1, 0.5, 1.0, 2.0, and 3.0 mm are calculated using the above equations, and the results are shown in Fig. 3.46. It is notable that the ringing tail of the EO signal is extended and intensified as the crystal thickness increases. Consequently, the spectrum becomes narrower and weaker, and the spectral undulation between 1 and 3 THz is more prominent for a longer crystal.

3.5 Ultrabroadband Terahertz Pulses

The spectral range of the THz emission and detection in the experimental observations discussed in the previous sections is limited, extending only from 0 to 5 THz, while the bandwidth of typical femtosecond pulses well exceeds 10 THz. The deprivation of the high-frequency band stems from the absorption of THz waves by optical phonon resonances in dielectrics and semiconductors. This prevents PC switching, optical rectification, and EO sampling from fully exploiting the optical bandwidth.

Substantial extension of the THz bandwidth has been accomplished using even shorter femtosecond pulses of $\sim$10-fs pulse duration. For PC switching, the shorter pulse duration does not change the carrier lifetime of PC materials, yet the shorter rise time of carrier population brings in broader emission and detection spectra. In order to curtail the parasitic phonon effects, thin layers of nonlinear crystals, together with ultrashort pulses, are being used for optical rectification and EO sampling.

3.5.1 Optical Rectification and Electro-Optic Sampling

In a dispersionless and lossless nonlinear medium, 10-fs optical pulses can produce and detect radiation of $\sim$100 THz bandwidth, yet no nonlinear medium has a uniform optical response over such a wide spectral range. While ZnTe has relatively simple lattice vibration bands in the THz region, its nonlinear optical properties are strongly affected by the TO-phonon resonance at

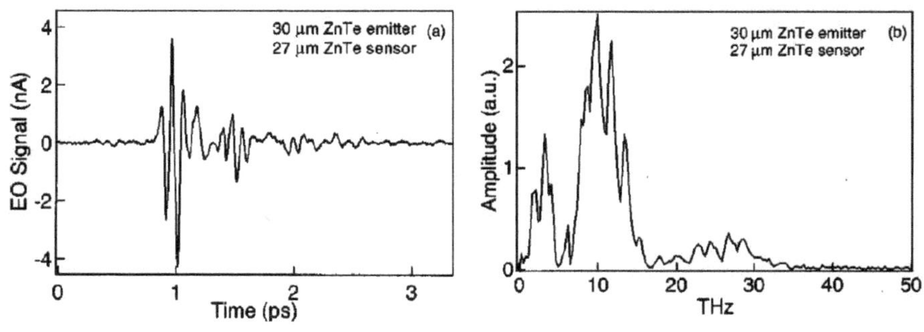

Fig. 3.47. (a) Temporal waveform of the THz radiation from a 30 μm ZnTe emitter measured by a 27 μm ZnTe sensor and (b) amplitude spectrum. (Reprinted with permission from [45]. ©1998, American Institute of Physics.)

5.3 THz. The effects of absorption near the resonance and coinciding dispersion are clearly seen in the experimental data shown in Fig. 3.47 [45]. The data were taken with 30-μm emitter and 27-μm sensor crystals. The spectrum covers the range from 0 to 35 THz, in which the phonon resonance at 5.3 is discernible. The dip near 17 THz is accounted for by the difference between the optical group and the mid-infrared refractive indices, 0.6. The velocity mismatch between the ultrashort THz and optical pulses are much greater than that of longer pulses.

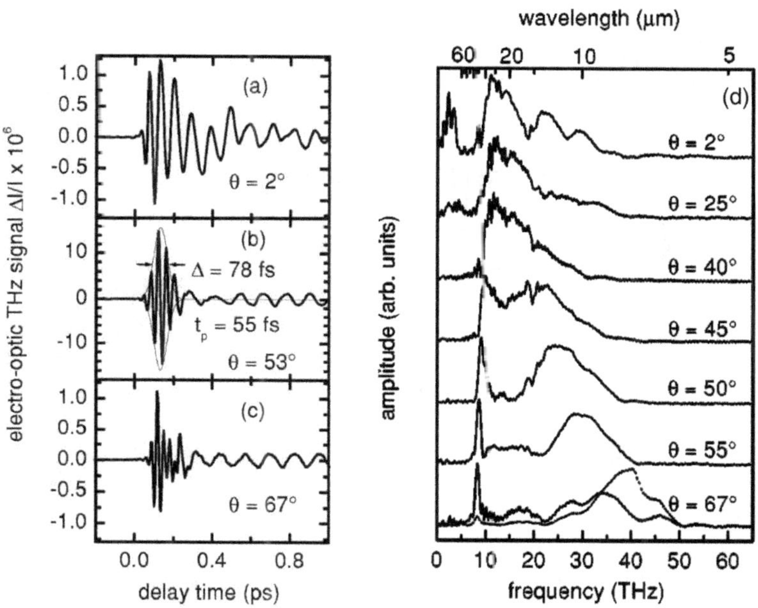

Fig. 3.48. THz pulses generated in a 90-μm-thick GaSe crystal by optical rectification of 10-fs laser pulses when the phase matching angles are (a) 2°, (b) 53°, and (c) 67°. The electro-optic detection is carried out in a 10.3-μm-thick ZnTe crystal. (d) Normalized amplitude spectra of the measured electro-optic signal at various phase matching angles. (Reprinted with permission from [46]. ©2000, American Institute of Physics.)

An alternative, yet even better EO crystal for ultrabroadband applications is GaSe. GaSe belongs to the point group $6m2$: its nonlinear susceptibility has three nonvanishing elemements, and only one of them, d_{22}, is independent (see Table 3.3). Some outstanding properties of GaSe for nonlinear optical applications in the mid-IR are notable: the nonlinear coefficient (d_{22}=54 pm/V at 10.6 μm) is large, the damage threshold is high, and the transparency range extends from 0.62 μm to 20 μm. Another important property is that velocity matching is attainable in the spectral range between 15 and 50 THz due to its very large birefringence (the theoretical description for the angle phase-matching in a birefringent nonlinear crystal will be presented in section 4.2.2). Figure 3.48 shows the ultrabroadband THz pulses generated by optical rectification of 10-fs pulses in a 90-μm-thick GaSe crystal at various phase-matching angles. The spectra at the small phase-matching angles (θ=2 and 25°) cover the broad spectral range of 0.1-40 THz with a ~2-THz gap near the optical phonon resonance at 6.4 THz. The central frequency of the spectrum shifts to the higher frequency side as the phase-matching angle is increased, indicating the angular dependence of the phase-matching frequency.

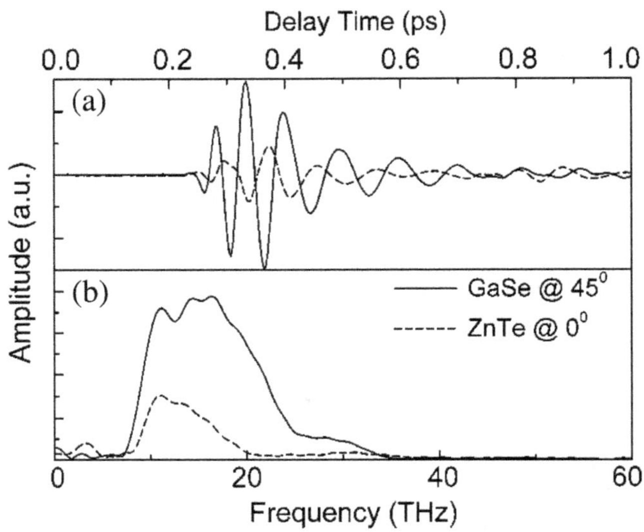

Fig. 3.49. (a) Temporal waveform and (b) amplitude spectrum of the THz pulses generated by optical rectification of 10-fs pulses in a 43-μm z-cut GaSe. The THz fields are measured by EO sampling in a 37-μm z-cut GaSe crystal (solid line) and a 21-μm ⟨110⟩-oriented ZnTe crystal (dotted line). (Reprinted with permission from [47]. ©2004, American Institute of Physics.)

As for sensing, the EO sampling characteristics of GaSe are superior to those of ZnTe in the high-frequency range from 7 to 30 THz. Figure 3.49 shows a comparison between EO signals of GaSe and ZnTe crystals [47]. The THz pulses are generated by optical rectification of 10-fs laser pulses in a 43-μm z-cut GaSe. The EO sensor crystals are a 37-μm z-cut GaSe crystal (solid line) and a 21-μm ⟨110⟩-oriented ZnTe crystal. The GaSe crystal is tilted 45° with respect to the incident probe beam to optimize velocity matching in the broad spectrum. The EO signal measured by the GaSe crystal is greater than that of the ZnTe crystal. The spectrum of the GaSe EO signal is also broader than that of the ZnTe signal.

3.5.2 Photoconductive Antennas

The theoretical discussions in sections 3.2.2 and 3.2.5 indicate that the temporal waveform from a PC emitter and the spectral response function of a PC detector is governed by the carrier lifetime as well as the carrier population rise time. The carrier lifetime is an intrinsic material property, thus we can do little about it. At present, the best PC material available for a device is LT-GaAs. On the other hand, the rise time is flexible to an extent, because it is mainly determined by the incident optical pulse duration, i.e., shorter optical pulses may expedite generation of shorter THz pulses and accommodate a broader spectral range of detection.

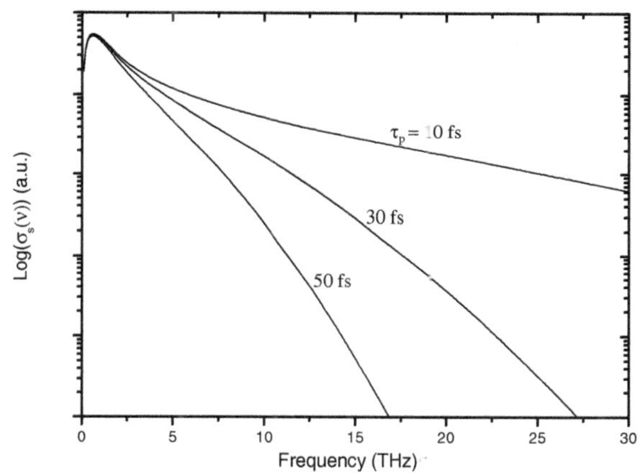

Fig. 3.50. Detector response function in log scale versus frequency when the optical pulse duration τ_p is 10, 30, and 50 fs. The carrier lifetime and the momentum relaxation time are 0.5 ps and 0.03 ps, respectively.

Figure 3.50 shows the frequency-dependent response function of a PC detector when the probe pulse duration is 10, 30, and 50 fs. The curves are obtained from Eq. 3.54, including the effect of the diffraction limit. The optical pulse duration has little impact on the overall detection bandwidth, which is roughly 1 THz, yet the high-frequency tail behaves quite differently. Indeed, shorter pulses provide substantially broader detectable ranges. If the dynamic range of the detector is three orders of magnitude, the highest detectable frequencies are 12, 20, and 50 THz for 50, 30, and 10 fs pulses, respectively.

Figure 3.51 shows an example of broadband sensing by a LT-GaAs PC antenna [48]. The THz radiation is generated from the surface of an InP wafer and measured by a LT-GaAs PC receiver gated with 15-fs optical pulses. The temporal waveform includes rapidly oscillating components on top of the slowly varying signal. The slow part arises from the transient currents in the InP substrate, while the origin of the fast oscillations is not clearly identified. Regardless of the ambiguity, the important aspect of the data is the detector's

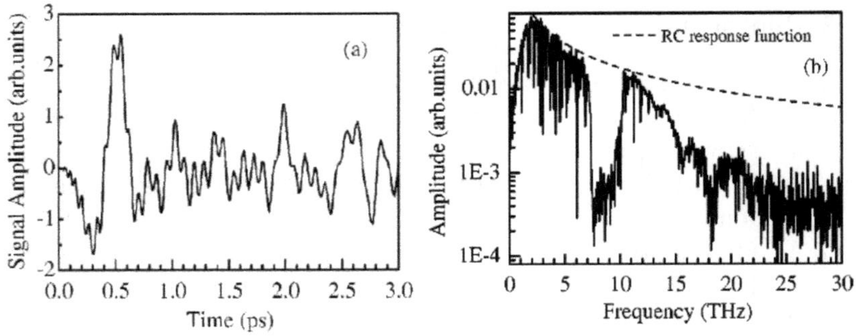

Fig. 3.51. (a) Temporal waveform of THz radiation from a semi-insulating InP wafer measured by a LT-GaAs PC receiver and (b) its Fourier amplitude spectrum. (Reprinted with permission from [48]. ©2000, American Institute of Physics.)

capability to resolve the high frequency signal. The Fourier transformed spectrum shown in Fig. 3.51(b) indicates that the detection bandwidth extends up to 20 THz.

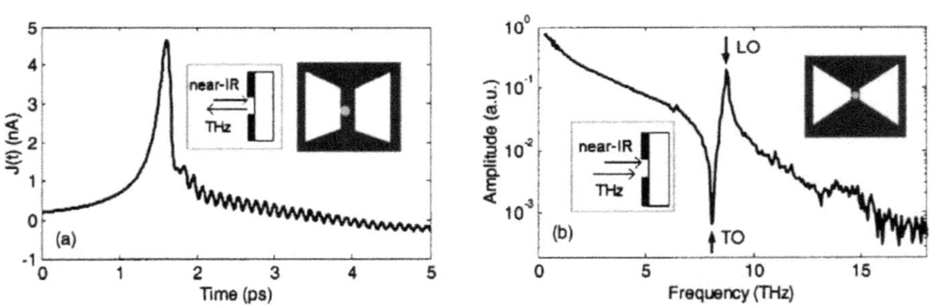

Fig. 3.52. (a) Temporal waveform of THz radiation from a LT-GaAs PC emitter measured by a LT-GaAs PC receiver and (b) its Fourier amplitude spectrum. The antenna structures are shown in the insets. (Reprinted with permission from [49]. ©2004, American Institute of Physics.)

PC antennas can be used for generating ultrabroadband THz pulses. The available spectral range, however, is significantly smaller than that of the pulses generated by optical rectification. Generation of ultrabroadband THz radiation using a LT-GaAs PC antenna is shown in Fig. 3.52. The TO and LO phonon modes of GaAs at 8.1 and 8.8 THz gives rise to the fast oscillations of the temporal waveform. The spectrum of the radiation extends up to 15 THz as shown in Fig. 3.52(b).

3.6 Terahertz Radiation from Electron Accelerators

Electron accelerators are outstanding light sources characterized by high brightness and broad tunability. The THz radiation power of the accelerator-based sources is several orders of magnitude higher than those of the table-top sources we have discussed so far.

A relativistic electron subject to accelerations emits radiation beamed in a narrow cone in the direction of its velocity. The power radiated per unit solid angle is given as

$$\frac{dP}{d\Omega} = \frac{e^2}{16\pi^2 c\epsilon_0} \frac{|\mathbf{n} \times \{(\mathbf{n} - \mathbf{v}/c) \times \dot{\mathbf{v}}/c\}|^2}{(1 - \mathbf{n} \cdot \mathbf{v}/c)^5}, \tag{3.104}$$

where $\mathbf{v}$ is the electron velocity and $\mathbf{n}$ is a unit vector in an arbitrary direction. Suppose $\mathbf{v}$ and $\dot{\mathbf{v}}$ are parallel in linear motion. If θ is the angle measured from the direction of $\mathbf{v}$ and $\dot{\mathbf{v}}$, Eq. 3.104 reduces to

$$\frac{dP}{d\Omega} = \frac{e^2 \dot{\beta}^2}{16\pi^2 c\epsilon_0} \frac{\sin^2\theta}{(1 - \beta\cos\theta)^5}, \tag{3.105}$$

where $\beta = v/c$. For a relativistic electron ($\beta \approx 1$), most of the radiation is concentrated in the forward direction within the angle, $\Delta\theta \approx 1/\gamma = \sqrt{1-\beta^2} \ll 1$. The total radiation power is obtained by integrating Eq. 3.105 over all angles:

$$P = \frac{e^2 \dot{\beta}^2 \gamma^6}{6\pi c\epsilon_0}. \tag{3.106}$$

The factor γ^6 indicates that the radiation power increases exceedingly fast as the velocity approaches the speed of light. One method to generate radiation using high speed electrons is to shoot the electron beam at a metal target, which rapidly decelerates the electrons. The radiation produced by the deceleration of the electrons is called *bremsstrahlung*, meaning "breaking radiation".

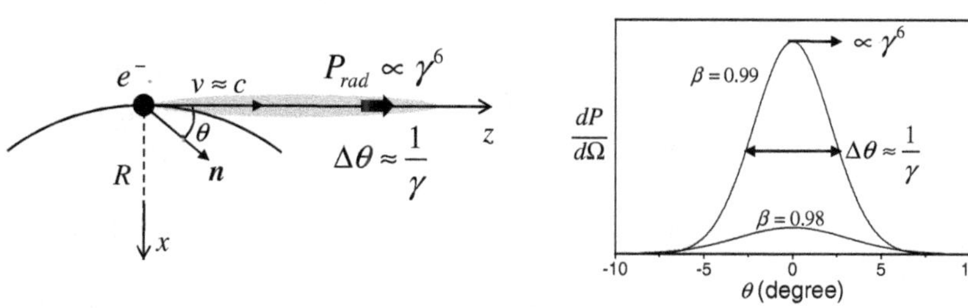

Fig. 3.53. Radiation emitted by a relativistic electron in circular motion

The radiation from relativistic electrons undergoing circular motion is called **synchrotron radiation** because it was first observed in electron synchrotrons. Figure 3.53 sketches the radiation mechanism for circular motion. The radiation power in the plane of motion (i.e., at ϕ=0 or π) is expressed as

$$\frac{dP}{d\Omega} = \frac{ce^2\beta^4}{16\pi^2\epsilon_0 R^2}\frac{(\cos\theta-\beta)^2}{(1-\beta\cos\theta)^5}, \tag{3.107}$$

where R is the radius of curvature. The radiation is predominantly in the forward direction: the angle of the radiation cone is $\Delta\theta \approx 1/\gamma$, which is similar to linear motion. The total radiation power is given as

$$P = \frac{ce^2\beta^4\gamma^4}{6\pi\epsilon_0 R^2}. \tag{3.108}$$

The radiation power versus θ shown in Fig. 3.53 demonstrates that a mere increase of β from 0.98 to 0.99 gives rise to an eight-fold enhancement of the peak power. The cutoff frequency ν_c, beyond which the radiation power is negligible has the relation,

$$\nu_c \approx \left(\frac{E}{m_e c^2}\right)^3 \frac{c}{R}. \tag{3.109}$$

Given that $R \approx$100 m, the electron energy E should be 10-100 MeV for ν_c to be in the THz region.

For the generation of pulsed radiation, many electrons are bunched together and move in packets. When the size of the electron bunch is comparable to the radiation wavelength, emissions from individual electrons superimpose in phase, and the total radiation power is proportional to the square of the number of electrons. Consequently, the coherent THz radiation from an electron bunch comprised of many electrons is potentially very powerful. Figure 3.54 sketches the schematic of the THz radiation from a linear particle accelerator (linac). Femtosecond laser pulses trigger an electron source—photocathode electron guns or semiconductor surfaces have been used—to produce ultrashort electron bunches. The electrons gain a relativistic energy, 10-100 MeV, in the accelerator, and coherent THz radiation is generated either by smashing the electron bunch into a metal target or by a magnetic field bending the electron path.

The strongest THz radiation has been generated in the Source Development Lab (SDL) of the Brookhaven National Laboratory, a branch of the National Synchrotron Light Source(NSLS) [50]. The NSLS SDL operates a photo-injected linac producing sub-picosecond electron bunches and accelerating the electrons to ~100 MeV. The high energy electrons smash a metal target that generates the bremsstrahlung, including 0.3-ps, single-cycle THz radiation. The pulse energy of the coherent THz radiation reaches up to 100 μJ, and the electric field amplitude approaches ~1 MV/cm. The operational repetition rate of this facility is ~10 Hz.

The energy-recovered linac (ERL) in the Jefferson Laboratory has been utilized to produce very high average power, ~20 W, of broadband THz radiation [51]. The most notable characteristic of the ERL is that the operational

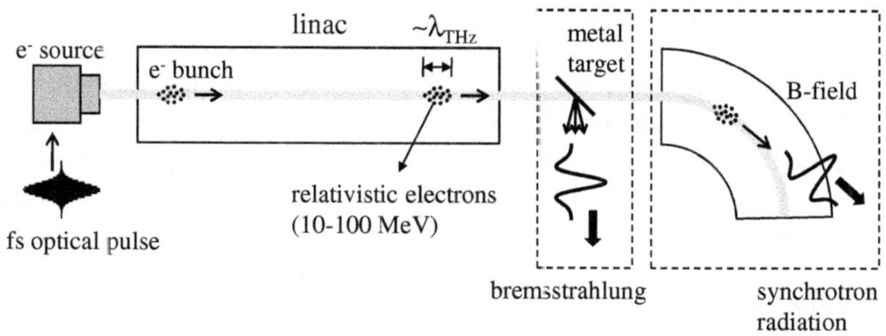

Fig. 3.54. Coherent THz radiation from relativistic electrons in a linac

repetition rate is very high, up to 75 MHz, and hence the average current is much higher than that in conventional linacs. Ultrashort electron bunches, produced by femtosecond pulses incident on a GaAs surface, are accelerated in the linac and bent by a magnetic field, emitting THz pulses with ~1 μJ pulse energy.

Synchrotron storage rings are also a promising source of broadband THz radiation. The main technical challenge for generating single-cycle THz pulses in a storage ring is to stabilize the electron bunch. Stable, coherent synchrotron radiation has been produced at Berliner Elektronenspeicherring - Gesellschaft für Synchrotronstrahlung m.b.H. (BESSY) [52] and at MIT-Bates South Hall Ring [53].

Table 3.6. Characteristics of the THz Radiation from Electron Accelerators

Accelerator	Pulse Duration	Pulse Energy	Rep Rate	Average Power
NSLS SDL	~0.3 ps	~100 μJ	~10 Hz	~1 mW
JLab ERL	~0.3 ps	~1 μJ	75 MHz	~20 W
BESSY	~1 ps	~1 nJ	500 MHz	~1 W

The performance characteristics of the accelerator-based sources are summarized in Table 3.6. Several optimization projects are in progress. When they are completed, the bandwidth will be extended up to ~10 THz and the output power will be augmented many fold. Free-electron lasers, which are also accelerator-based light sources, have relativistic electrons passing through a periodically alternating magnetic field. The subsequent periodic acceleration of the electrons produces quasi-continuous-wave THz radiation. The narrowband THz generation by free-electron lasers will be discussed in section 4.8.

3.7 Novel Techniques for Generating Terahertz Pulses

The majority of applications using broadband THz radiation utilize the two predominant techniques of THz generation, photoconductive switching and optical rectification in electro-optic crystals. The prevalence of these techniques is justifiable as they are relatively compact, easy to use, and sufficiently bright to get an excellent signal-to-noise ratio. Nevertheless, it is also true that the rapid development of new applications demands a variety of sources with higher efficiency, higher intensity, more complicated waveforms, etc. In fact, many different types of broadband THz sources have been developed so far. For example, generation of THz radiation by illuminating semiconductor surfaces with femtosecond pulses has been intensely studied not only for developing sources, but also for understanding ultrafast and microscopic mechanisms of electron and phonon dynamics [54]. It is notable that the power of THz radiation from semiconductor surfaces is intensified in the presence of strong magnetic fields [55]. In the present section, we shall survey a few unique techniques which have the potential to provide new opportunities in THz technology.

3.7.1 Phase-Matching with Tilted Optical Pulses in Lithium Niobate

Lithium niobate ($LiNbO_3$) is one of the most widely used nonlinear optical crystals because of its unique properties such as its high optical transparency over a broad spectral range (350-5200 nm), its strong optical nonlinearity, ferroelectricity, and piezoelectricity. In particular, the large electro-optic coefficient, $d_{33} = 27$ pm/V, is an attractive trait for THz generation. The conventional method of optical rectification, however, is ineffective for generating THz radiation because of the large mismatch between the optical group and the THz phase velocities (Extraordinary optical group and THz refractive indices are n_O=2.3 and n_T=5.2. See Fig. 3.36).

An ingenious way to overcome the velocity mismatching problem is to steer the THz radiation to the direction normal to the Cherenkov cone by tilting the optical pump pulses. Figure 3.55(a) illustrates the scheme of velocity matching with tilted optical pulses. A well-known analogy of the Cherenkov radiation is the sonic boom of a supersonic object: as the object, or the source of the sound waves, moves faster than the waves, the collapsed waves form a shock front in the shape of a cone. The shock wave is emitted under a constant angle θ_c with the object trajectory, given by $\cos\theta_c = v_{wave}/v_{object}$. Similarly, in a $LiNbO_3$ crystal, a femtosecond laser pulse with a small beam size—more precisely, the beam diameter is considerably smaller than the wavelength of the THz radiation—acts like a point source moving faster than the THz waves. Under this condition, the Cherenkov angle is given by

$$\theta_c = \cos^{-1}\left(\frac{v_T}{v_O}\right) = \cos^{-1}\left(\frac{n_O}{n_T}\right) \cong 64^\circ. \tag{3.110}$$

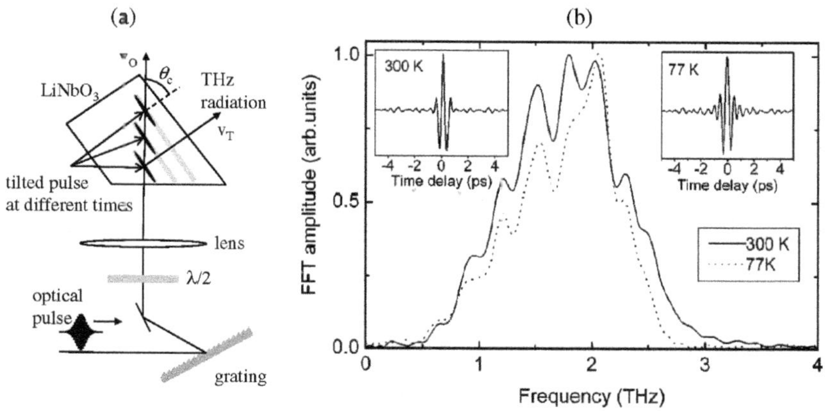

Fig. 3.55. THz generation by optical rectification of tilted femtosecond pulses in a bulk $LiNbO_3$ crystal. (a) Schematic diagram of the experimental setup. (b) Power spectra of the THz radiation obtained by Fourier transform of the inteferograms (shown in the insets) at 77 and 300 K. (Reprinted with permission from [56]. ©2003, American Institute of Physics.)

If the optical beam size is larger than the wavelength of the THz radiation and the optical pulse front is aligned with the Cherenkov cone, the optical pulse front, copropagating with the THz waves at the same speed ($v_T = v_O \cos\theta_c$), constantly supplies THz radiation in phase. In other words, the THz Cherenkov radiation is amplified coherently while the optical and THz pulses propagate through the $LiNbO_3$ crystal. The tilt of the optical pulse front is obtained by a diffraction grating. After the diffraction, the laser beam is collimated horizontally and focused vertically by a lens or an imaging system, then brought to the $LiNbO_3$ crystal.

Figure 3.55(b) shows the power spectra of the THz pulses generated by optical rectification of tilted femtosecond pulses in a $LiNbO_3$ crystal at 77 and 300 K. The spectra are obtained by Fourier transforms of the interferograms (see insets) which are measured by a detection system which consists of a Michelson interferometer and a bolometer. The THz pulse energy is measured as 98 pJ at 77 K and 30 pJ at 300 K with an input optical pulse energy of 2.3 μJ, which corresponds to an energy conversion efficiency of 4.3×10^{-5} at 77 K and 1.3×10^{-5} at 300 K. The conversion efficiency is significantly lower at room temperature, because the majority of the THz radiation is absorbed in the crystal (The absorption coefficient increases from $\sim$1 cm^{-1} at 1 THz to $\sim$10 cm^{-1} at 2 THz. See Fig. 3.41.). The dominant absorption mechanism, anharmonic decay of the optical phonon into two acoustic phonons, is strongly suppressed at the lower temperature. A more recent study reports that 10-μJ THz pulses centered at 0.5 THz have been generated in MgO-doped $LiNbO_3$ crystals using a high-power Ti:sapphire amplifier system producing 20-mJ pulses at 10 THz [57].

3.7.2 Terahertz Generation in Air

It is amazing that THz pulses can literally be generated out of "thin air" [58]. In fact, it is an important subject in the field of plasma science to produce various kinds of electromagnetic radiation from THz waves to x-rays by using high-power laser pulses interacting with a photogenerated plasma. The underlying mechanisms behind THz radiation are the ponderomotive forces which create a density difference between electrons and ions. In the presence of a strong bias field, the efficiency of THz generation reaches an unexpectedly high level, comparable to that of THz radiation from semiconductor surfaces [59].

An appealing development is that mixing a fundamental optical wave of frequency ω with a second-harmonic (SH) wave of frequency 2ω creates a huge enhancement of the efficiency of THz generation [60, 61, 62]. The experimental arrangement is illustrated in Fig. 3.56. The second-harmonic generation is usually carried out in a type-I β-barium borate (BBO) crystal. The THz radiation intensity is maximized when the fundamental and SH polarizations are parallel, while it is negligible when they are perpendicular [61, 63].

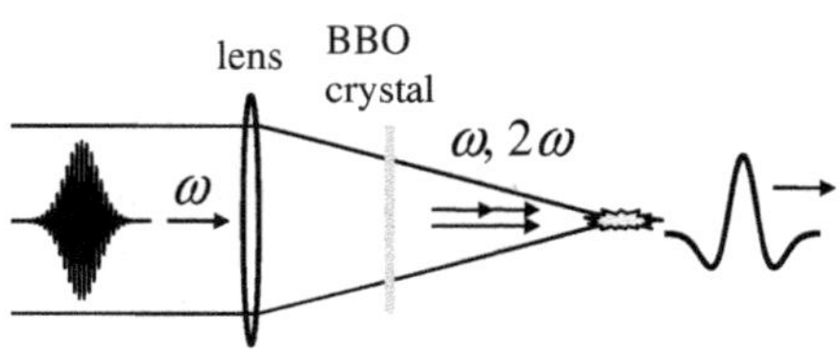

Fig. 3.56. Schematic diagram of the experimental setup for generation of THz radiation by mixing the fundamental and SH optical pulses. The SH generation is executed in a BBO crystal.

The THz radiation originates from the transient photocurrent of ionizing electrons driven by the asymmetric electric fields of the superposition of the fundamental and the SH waves [64]. A phenomenological model describes this process as four wave mixing, which is a third-order nonlinear optical process. Note that the optical fields induce a nonlinear current, not a nonlinear polarization. This is an important point because the THz field is subject to the phase difference between the fundamental and the SH waves. We will see how this plays out shortly. The fundamental and the SH waves are expressed as $E_1(t) = E_\omega e^{-i\omega t}$ and $E_2(t) = E_{2\omega} e^{-i(2\omega t+\phi)}$, respectively, where ϕ is the phase difference. The THz field amplitude, $E_{THz}(t)$, is proportional to the time derivative of the rectified nonlinear current, $J^{(3)}(t)$:

$$E_{THz}(t) \propto \frac{\partial}{\partial t} J^{(3)}(t), \tag{3.111}$$

and, in the frequency domain, we can write

$$E_{THz}(\omega_T) \propto -i\omega_T J^{(3)}(\omega_T) + c.c., \tag{3.112}$$

where $J^{(3)} = \sigma^{(3)} E_{2\omega} E_\omega E_\omega e^{-i\phi}$ with the third-order nonlinear conductivity, $\sigma^{(3)}$. Consequently,

$$E_{THz} \propto \sigma^{(3)} E_{2\omega} E_\omega^2 \sin\phi. \tag{3.113}$$

The $\sin\phi$ dependence has been confirmed by experimental studies [61, 64]. It is also shown that $E_{THz} \propto E_{2\omega}$ and $E_{THz} \propto E_\omega^2$ when the optical pulse energy is sufficiently low (<100 μJ) [63]. For a higher optical power, a nonperturbative analysis is more appropriate to describe the highly nonlinear process.

Figure 3.57 shows the THz field amplitudes of plasma-based emitters versus input optical pulse energy, compared with those of large-aperture PC antennas. Note that large-aperture PC antennas have been used for generating high-power THz pulses (see section 3.2.4). With ~1-mJ optical pulses, the THz field amplitude of the SH-based plasma emitter reaches up to 10 kV/cm, close to the output of the DC-biased PC emitter. A field amplitude greater than 100 kV/cm is achievable with 20-mJ optical pulses [64].

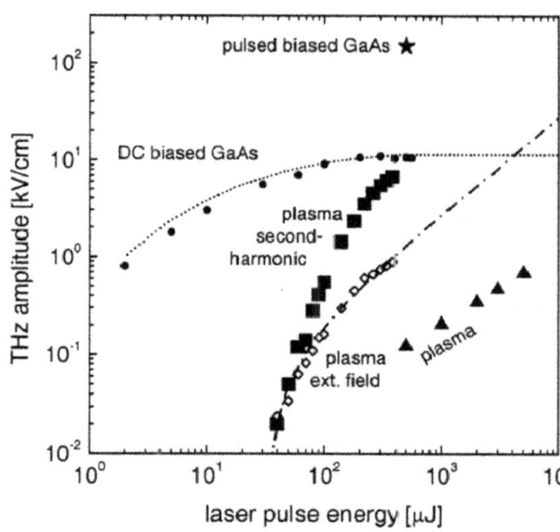

Fig. 3.57. THz field amplitude versus optical pulse energy. The output of the plasma-based THz emitters—photogenerated plasma (filled triangles), plasma with external bias (open diamonds), and plasma with SH waves (filled squares)—are compared to that of the large-aperture PC emitters with 1-kV/cm DC bias (filled circles) and 10-kV/cm pulsed bias (filled star). (Reprinted from [61])

3.7.3 Narrowband Terahertz Generation in Quasi-Phase-Matching Crystals

Up to this point, the underlying premise of this chapter has been anchored to the techniques of generating and detecting single-cycle, broadband THz pulses. A wide range of applications in spectroscopy, sensing, communication, and imaging, however, need spectrally narrow and bright THz sources. One promising technique for generating narrowband THz pulses is to use optical rectification of femtosecond pulses in a quasi-phase-matching (QPM) nonlinear crystal [34]. QPM structures consist of a system of periodic domains of alternating crystal orientation. Periodically-poled lithium niobate (PPLN) crystals are the most common QPM system for generating THz pulses.

The scheme of narrowband THz generation is shown in Fig. 3.58, which illustrates optical rectification in the pre-engineered periodic domain structure of a QPM crystal. The second order nonlinear susceptibility $\chi^{(2)}$ of the crystal reverses sign between neighboring domains. The thick vertical arrows in Fig. 3.58 indicate the direction of the optic axis. When a femtosecond optical pulse propagates through a QPM crystal, a THz nonlinear polarization is generated via optical rectification. Since the optical group velocity exceeds the THz phase velocity (e.g., the optical group and THz refractive indices of $LiNbO_3$ are $n_O = 2.3$ and $n_T = 5.2$, respectively), the optical pulse will lead the THz pulse by the optical pulse duration τ_p after the walk-off length l_w (see Eq. 3.81). If the domain length d of the QPM crystal is comparable to the walk-off length, each domain in the crystal contributes a half cycle to the radiated THz field. Hence, the THz wave consists of N cycles, where N is the total number of periods over the length of the QPM structure. If the domains are perfectly periodic, a narrowband THz pulse is generated, propagating in the forward direction with a frequency

$$\nu_T = \frac{c}{\Lambda(n_T - n_O)}, \tag{3.114}$$

where $\Lambda = 2d$ is the QPM period. Frequency tuning can be accomplished by adjusting Λ. In the absence of absorption and domain-width fluctuations, the relative bandwidth $\Delta\nu_T/\nu_T$ of the THz field is simply $1/N$. The nonlinear polarization radiates THz waves not only in the forward but also in the backward direction. The phase-matching condition for the backward wave leads to the radiation frequency $\nu_T = c/\Lambda(n_T + n_O)$.

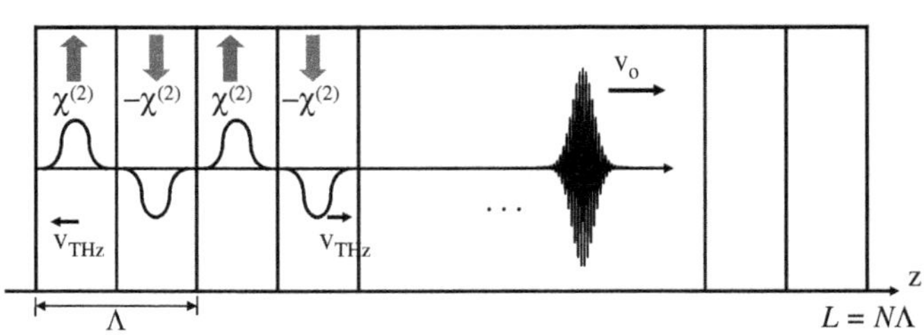

Fig. 3.58. Narrowband THz generation in a quasi-phase matching crystal

Figure 3.59(a) and (b) shows the THz waveforms and the respective spectra from a PPLN crystal for Λ = 60, 80, 100, and 120 μm at the crystal temperature of 115 K [65]. The pulse duration and bandwidth are ~100 ps and ~0.01 THz, respectively. The sample is a z-cut PPLN crystal, which is laterally chirped, i.e., multiple domain structures of slightly different domain width were fabricated side by side at a regular distance from one to another. As shown in Fig. 3.59(a), tuning of the THz frequency is accomplished by simply scanning the sample laterally to the beam propagation direction. Continuous THz frequency tuning also has been demonstrated using fanned-out PPLN crystals in which the domain width varies continuously across the lateral direction [66].

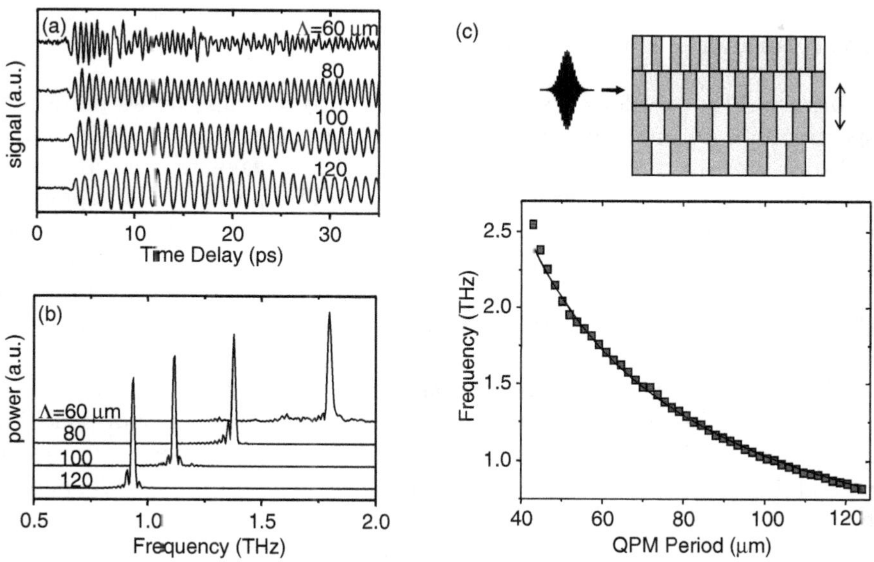

Fig. 3.59. (a) THz waveforms and (b) power spectra at $T = 115$ K when the PPLN QPM period is 60, 80, 100, and 120 μm. The sample is a laterally chirped z-cut PPLN crystal: multiple domain structures of slightly different QPM periods are fabricated side by side at a regular distance from one to another. (c) Measured (solid square) and calculated (solid line) frequency of the THz waves as a function of QPM period using a 10-mm-thick z-cut PPLN at room temperature [65].

The THz generation methods of phase-matching in ZnTe and quasi-phase-matching in PPLN have similar optical-to-THz conversion efficiencies: the power and photon conversion efficiencies are about 10^{-5} and 10^{-3}, respectively. A significant enhancement of the conversion efficiency is achieved by using a new material system, periodically structured QPM GaAs: the optical-to-THz photon conversion efficiency reaches up to 0.03 [67, 68]. In fact, GaAs contains several outstanding properties for THz applications. First, GaAs is highly transparent at THz frequencies (below 1.5 THz the absorption coefficient is less than 1 cm^{-1}). Second, the nonlinear optical coefficient is large:

d=47 pm/V [69]. Third, the mismatch between the optical group velocity and THz phase velocity is small (the corresponding group and refractive indices are n_g=3.431 at 2 μm and n=3.61 at 1 THz, respectively), thus the walk-off length, i.e., the length of collinear interaction of THz and optical pulses, is relatively long. There is, however, a limitation on optical sources for optical rectification in GaAs. Ti:sapphire lasers are of no use because GaAs is opaque near 800 nm. Besides, the second-order nonlinear absorption for $\lambda < 1.8$ μm is too strong for efficient THz generation. In order to suppress the parasitic absorption processes in GaAs, it is necessary to use mid-IR femtosecond pulses with wavelengths longer than 2 μm.

3.7.4 Terahertz Pulse Shaping

Arbitrary THz waveform generators can greatly extend the scope of THz spectroscopy, in analogy with the user specified waveforms used in optical pulse shaping, or the RF waveforms used in nuclear magnetic resonance spectroscopy. For example, THz-TDS with shaped THz pulses can be applied to investigating quantum coherence and nonlinear response of material systems by manipulating and monitoring temporal evolution of the quantum transitions. The precise control and manipulation of quantum systems by THz pulse shaping techniques has great potential for quantum information processing and communication.

One method of generating arbitrary THz waveforms is to employ non-periodic poled $LiNbO_3$ crystals, a twist from the technique of narrowband THz generation in QPM structures. The eventual THz pulse shape replicates the crystal domain structure. Figure 3.60 shows the experimental results and numerical simulations for two different types of poled $LiNbO_3$ structures, which demonstrates the feasibility and versatility of the THz pulse shaping scheme. First, a zero-area double pulse (Fig. 3.60(a)) consisting of two pulses with a phase shift is generated from a domain structure in which a single domain (100 μm) is placed between two sets of multiple domains (50 μm). The corresponding spectrum clearly shows the signature of the interference fringes of two coherent pulses. Second, a negatively chirped THz waveform is shown in Fig. 3.60(b). The domain structure for the chirped pulse includes multiple domains ranging from 20 to 90 μm with 1 μm gradual increment. The corresponding spectrum shows a broad feature ranging from 0.5 to 2 THz. The experimental results demonstrate that shaped THz pulses can be generated in poled $LiNbO_3$ structures. Most of the quantitative discrepancy between experiment and simulation comes from the temporal decay of the THz pulses in the experiments because of THz absorption in $LiNbO_3$ by optical phonons. The THz absorption can be suppressed by cooling down the samples below 100 K [71].

A drawback of using poled $LiNbO_3$ crystals is that the THz pulse shape is predetermined by the crystal domain structure; thus pulse shaping is not

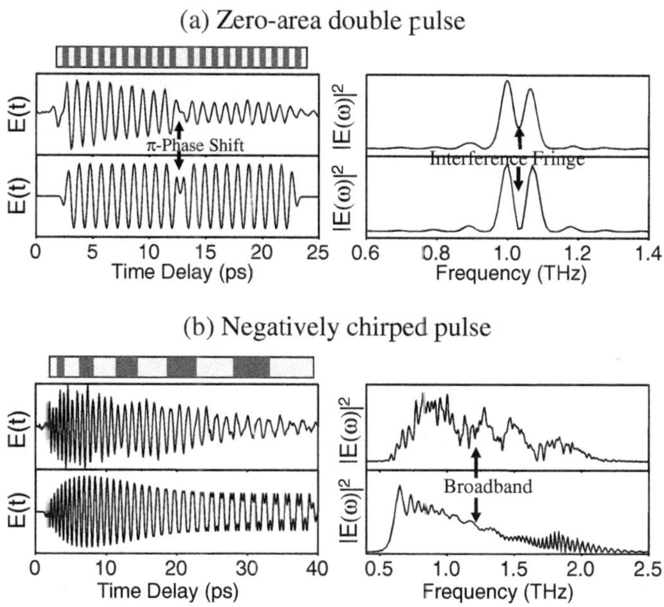

Fig. 3.60. Experimental data and numerical solutions of (a) zero area double and (b) chirped THz waveforms and corresponding spectra from poled $LiNbO_3$ structures [70].

adaptive in this scheme. Flexible THz pulse shaping is achievable by manipulating spatially dispersed multifrequency components generated by optical rectification in a fanned-out (FO) PPLN crystal [72]. The domain width of the FO-PPLN crystal varies continuously across the lateral direction; thus different regions emit THz radiation of different wavealengths when illuminated. The schematic of this technique is illustrated in Fig. 3.61(a). The laser beam is line focused onto the FO-PPLN crystal to generate spatially chirped THz pulses. A spatial mask is placed in front of the FO-PPLN crystal in order to manipulate the spatial intensity pattern of the incident optical beam, thus controlling the amplitudes of spatially dispersed THz frequency components. After the spherical mirror assembles the various frequencies into a single collimated beam, a shaped THz pulse can be obtained with the pulse shape determined by the Fourier transform of the pattern transferred by the mask. The beam collimation is accomplished by line-to-point imaging with a spherical mirror.

Fig. 3.61(b)-(d) shows the THz waveforms determined by the spatial patterns of the masks and corresponding spectra. The spatial patterns of the masks are shown in the insets: blocked regions are in black. The dotted lines in the spectra indicate the unmasked spectrum. The low-pass filter blocks high frequency parts (>0.65 THz) of the spectrum, as shown in Fig. 3.61(b). Consequently, the narrowband THz radiation has a long pulse duration. Fig. 3.61(c) shows the spectrum of the high-pass filter, which blocks frequency components lower than 0.65 THz. The asymmetric waveform reflects the lopsided spectrum. In Fig. 3.61(d), the double-slit mask suppresses the middle part of the spectrum, especially the main peak at 0.7 THz. A 15-ps beating is clearly visible in the waveform, demonstrating interference between the lower and higher parts of the spectrum.

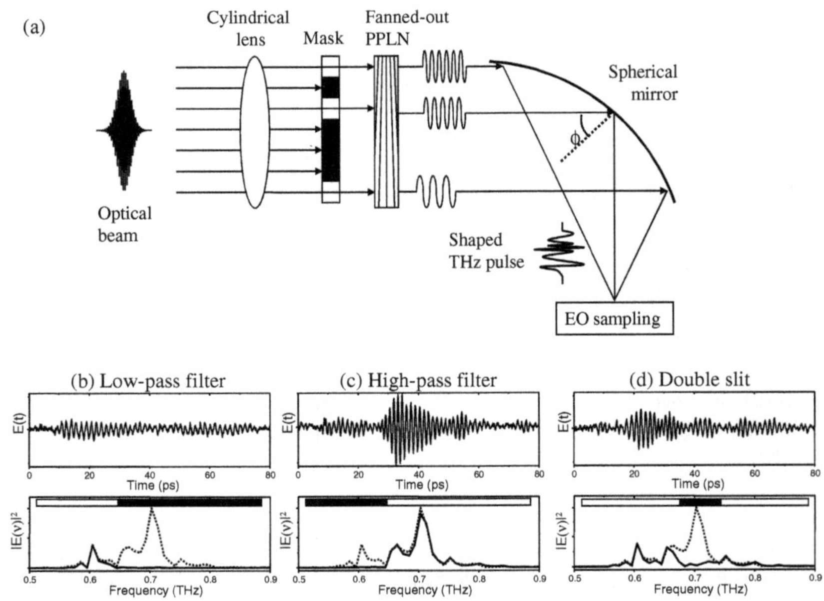

Fig. 3.61. (a) Experimental layout for arbitrary pulse shaping in a fanned-out PPLN crystal. ϕ is the incident angle of the terahertz beam on the spherical mirror. THz waveforms and corresponding spectra with metal masks: (b) high-pass filter, (c) low-pass filter, and (d) double-slit filter. The insets show the spatial patterns of the masks scaled to the corresponding spectra. The dotted lines in the spectra indicate the unmasked spectrum [72].

Another approach used to obtain complicated THz waveforms is to employ an optical pulse shaping technique. The underlying concept of this method is straightforward: the THz radiation replicates the exciting optical intensity profiles impinging on the THz emitter, either a photoconductive switch or an electro-optic crystal. Figure 3.62(a) illustrates the schematic of the conventional optical pulse shaper, and its application to shaping THz pulses. The optical pulse shaper consists of a pair of gratings placed at the focal planes of a lens pair. A spatial mask (e.g., a programmable liquid crystal modulator) controls the output pulse shape, inflicting amplitude and/or phase modulations on the spectrum. A few examples of complicated THz waveforms are demonstrated in Fig. 3.62. Figure 3.62(b) shows that two different femtosecond pulse sequences produce THz waveforms encoded as two different 4-bit words [73]. A photoconductive switch is used as a THz emitter. The optical pulse trains consist of up to four pulses, among which the third one at 2 ps is either missing or present, representing 1111 and 1101 4-bit sequences (Fig. 3.62(b-2)). These optical signals are directly transferred to THz waveforms as shown in Fig. 3.62(b-2). Figure 3.62(c) shows that the THz pulse shaper can also be used as a tunable narrowband THz source [74]. The frequency is continuously tunable from 0.5 to 3 THz. A photoconductive antenna is used for THz generation. Figure 3.62(d) demonstrates two types of shaped THz waveforms, chirped and zero-area double pulses, generated by optical rectification in a ZnTe crystal.

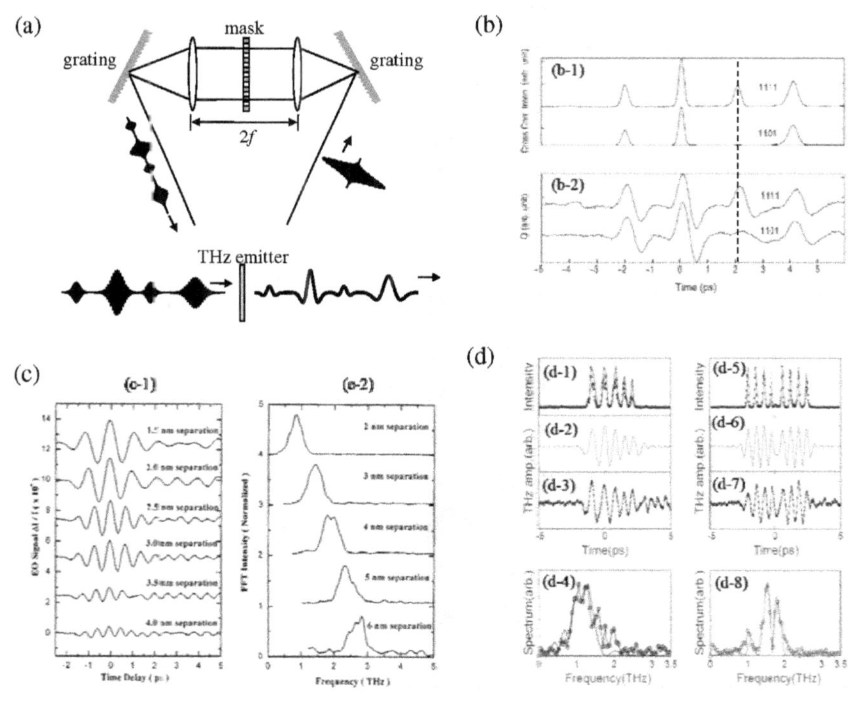

Fig. 3.62. (a) Schematic diagram of THz pulse shaping by use of shaped optical pulses. (b) THz time-domain multiplexing of two 4-bit words: (b-1) input optical pulses and (b-2) THz waveforms as a result of the two encoded sequences. (Reprinted from [73]) (c) Continuous frequency tuning: (c-1) waveforms and (c-2) spectra. (Reprinted with permission from [74]. ©2002, American Institute of Physics.) (d) Chirped and zero-area THz pulses. The optical pulse trains programmed via Gerchberg-Saxton algorithm. (Reprinted from [75])

The THz pulse shaping technique shown in Fig. 3.63 is a variation of the scheme of THz generation in a $LiNbO_3$ crystal with tilted optical pulses [76].

Two femtosecond laser pulses propagating in a $LiNbO_3$ crystal at a small angle create interference fringes and thus imprint a transient polarization grating. The Cherenkov radiation from the periodic polarization forms a narrowband pulse as shown on the top right corner of Fig. 3.63. Since the THz waveform maps out the spatial pattern of the interference fringes, pulse shaping is attainable by controlling the laser intensity distribution. Figure 3.63 shows a series of pictures demonstrating the pulse shaping control that is attainable with spatial masking.

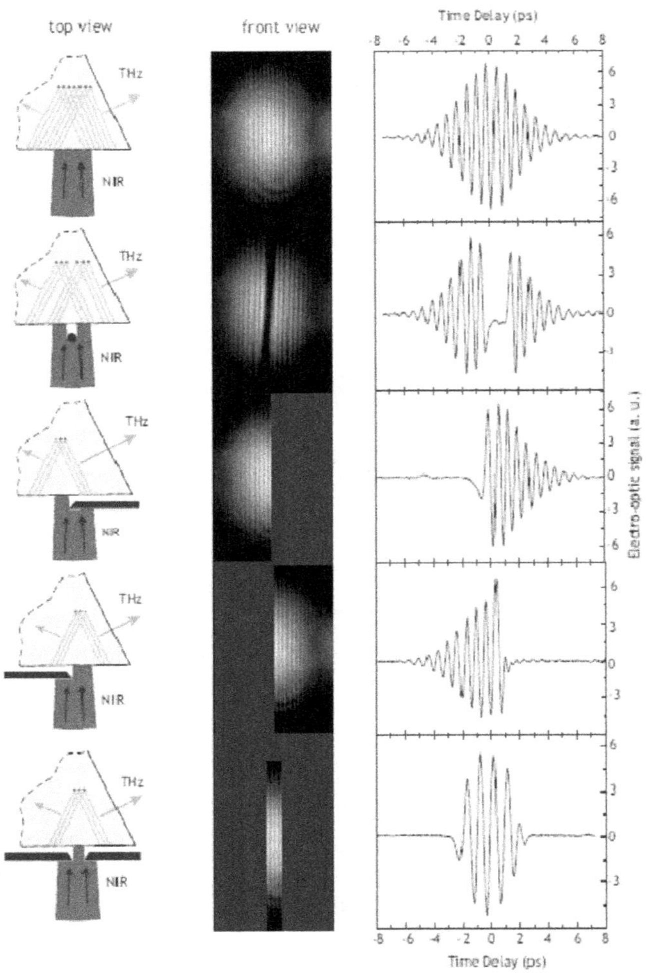

Fig. 3.63. THz pulse shaping by two-beam excitation in a $LiNbO_3$ crystal. (Reprinted from [76])

4

Continuous-Wave Terahertz Sources and Detectors

The technology behind continuous-wave (CW) THz emitters and sensors has a long history involving many different types of technical schemes unlike the methods of broadband THz radiation, mainly relying on ultrafast optical technology. The efforts to integrate optical technology with electronics result in THz optoelectronic devices such as photomixers and difference frequency generators. Optically pumped THz gas lasers produce high power THz radiation using the rotational transitions of heteropolar molecules in the gas phase. The implementation of actual devices requires compact and portable all-solid-state THz sources. Diode-based frequency multipliers, p-type germanium lasers, and quantum cascade lasers belong to this category. The practical demand has also been encouraging the development of solid-state sensors such as intersubband detectors and Schottky diodes. Free-electron based sources include small scale devices such as backward wave oscillators and large facilities such as free-electron lasers. Commonly used THz sensors are thermal detectors such as bolometers, Golay cells, and pyroelectric devices. In this chapter, we shall briefly review various methods of generating and detecting CW THz radiation and look into the underlying mechanisms under which the devices are operating.

4.1 Photomixing

Photomixing, also known as optical heterodyne downconversion, is a technique to generate CW THz radiation with a PC switch. LT-GaAs is the prevailing PC material for this technique because of its high mobility and short lifetime. A photomixer is a compact, solid-state device. The tuning range can be exceptionally broad provided a high-quality, tunable, dual-frequency laser system is available. The primary disadvantage of this method is that the output power is relatively low compared with other techniques of CW THz generation. Its optical-to-THz conversion efficiency is 10^{-6} – 10^{-5}, and the typical output power is in the microwatt range. Carrier transport in LT-GaAs, described in section 3.2.1, is an important factor governing photomixing processes. Because photomixing requires continuous optical excitation, the maximum THz output power is limited by the low thermal conductivity of LT-GaAs (~15 W/mK). The damage threshold of a 1-μm LT-GaAs layer with biased electrodes is less than 10^5 W/cm^2 of optical excitation [77].

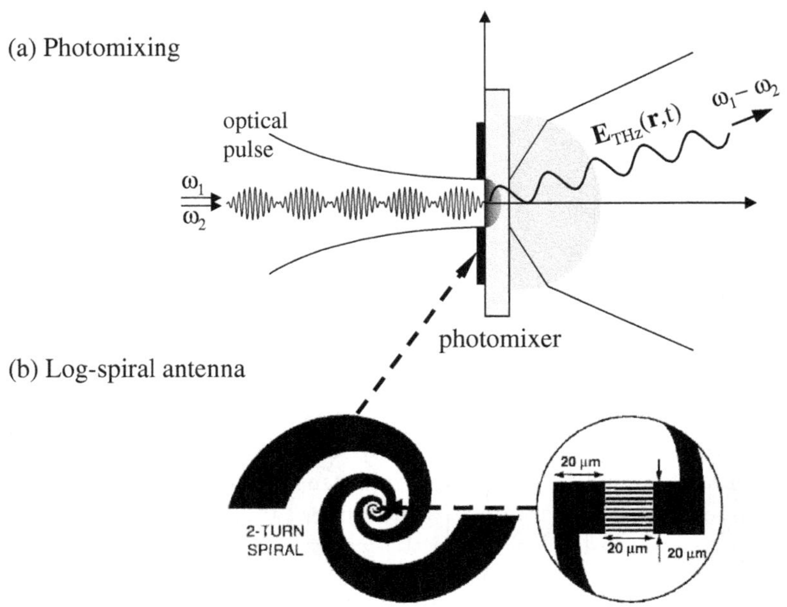

Fig. 4.1. Schematic diagram of photomixing. (a) THz radiation from a photomixer (b) Log-spiral antenna with interdigitated electrode fingers. (Reprinted from [78]. ©1997 IEEE)

At a fundamental level, photomixing shares several essential features with the THz radiation from a PC emitter as discussed in section 3.2.2. Figure 4.1 illustrates the principle of CW THz generation from a photomixer. A typical photomixer includes an antenna structure of metal on a LT-GaAs layer grown on a SI-GaAs substrate. A silicon hyper-hemispherical lens is attached to the back side of the substrate. A commonly used antenna structure for photomixing is the logarithmic uniplanar spiral antenna (log-spiral antenna) with interdigitated electrode fingers shown in Fig. 4.1(b) [78]. It has the advantage of a broad tuning range: its radiation pattern, impedance, and polarization remain virtually unchanged below 1 THz.

The optical excitation of photomixing utilizes a beat between two CW laser beams with slightly different frequencies. The most common light sources are diode lasers in the spectral range between 800 and 850 nm. The optical field at the antenna is expressed as:

$$E_{opt}(t) = E_1 e^{-i\omega_1 t} + E_2 e^{-i\omega_2 t}, \tag{4.1}$$

where E_1, E_2, ω_1, and ω_2 are the electric field amplitudes and the angular frequencies of the two CW laser beams, respectively. Thus, the optical intensity at the photomixer is given by

$$I_{opt}(t) = \frac{1}{2} c\epsilon_0 |E_{opt}(t)|^2 = I_0 + I_B \cos\omega t, \tag{4.2}$$

where $I_0 = I_1 + I_2$ is the average intensity, $I_B = 2\sqrt{I_1 I_2}$ is the beat intensity, and $\omega = \omega_1 - \omega_2$ is the difference frequency. Due to the periodic modulation of the optical intensity, the induced photocurrent oscillates with the beat frequency ω. Therefore, the antenna, an oscillating dipole in this picture, radiates CW electromagnetic waves of the frequency ω. THz radiation is obtained when the difference frequency is tuned to the THz frequency range.

The Drude-Lorentz model (see section 3.2.2) provides a simple, yet accurate picture of the essential properties of THz radiation from a photomixer. According to Eqs. 3.24, 3.25, and 3.27, the induced photocurrent is the convolution of the optical intensity and the impulsive current density governed by the carrier lifetime τ_c and the momentum relaxation time τ_s:

$$\begin{aligned} I_{PC}(t) &= \int I_{opt}(t-t')\,[e\,n(t')v(t')]\,dt' \\ &= u_e E_{DC} \int_0^\infty \{I_0 + I_B \cos[\omega(t-t')]\}\, e^{-t'/\tau_c} \left[1 - e^{-t'/\tau_s}\right] dt'. \end{aligned} \tag{4.3}$$

The momentum relaxation time in LT-GaAs ($20-30$ fs) is much shorter than the carrier lifetime ($200-500$ fs), thus Eq. 4.3 reduces to

$$I_{PC}(t) = I_D + \frac{I_A}{\sqrt{1+\omega^2\tau_c^2}} \cos(\omega t - \phi), \tag{4.4}$$

where $I_D = \tau_c \mu_e E_{DC} I_0$ is the DC photocurrent, $I_A = \tau_c \mu_e E_{DC} I_B$ is the amplitude of the AC photocurrent at the frequency ω, and $\phi = \tan^{-1}(\omega\tau_c)$ is the phase delay induced by the finite carrier lifetime.

The THz field radiated from the oscillating current is

$$E_{THz}(t) \propto \frac{dI_{PC}(t)}{dt} = -\frac{\omega I_A}{\sqrt{1+\omega^2\tau_c^2}} \sin(\omega t - \phi), \tag{4.5}$$

which leads to THz radiation power of frequency ω

$$P_{THz}(\omega) = \frac{1}{2}\frac{I_A^2 R_A}{1+\omega^2\tau_c^2} = \frac{1}{2} R_A \cdot (E_{DC}^2 I_B^2) \cdot \frac{\tau_c^2 \mu_e^2}{1+\omega^2\tau_c^2}, \tag{4.6}$$

where R_A is the radiation resistance. The radiation power increases quadratically with the optical intensity I_B and the bias field E_{DC}. The last term in

Eq. 4.6 represents the effects of the intrinsic material properties, mobility and carrier lifetime, of the PC substrate.

The radiation resistance of a Hertzian dipole is given by $R_A = 197\left(\frac{d}{\lambda}\right)^2 \propto \omega^2$, where d is the size of the dipole and $\lambda = 2\pi c/\omega$ is the free-space wavelength. For a practical PC switch, the radiation resistance corresponds to the real part of the antenna load impedance $Z_R = R_A/[1+(\omega R_A C_A)^2]$, where C_A is the electrode capacitance. Consequently, the THz radiation power depends on the antenna circuit design [77]:

$$P_{THz}(\omega) = \frac{1}{2}\frac{I_A^2}{[1+(\omega\tau_c)^2]}\frac{R_A}{[1+(\omega R_A C_A)^2]}. \tag{4.7}$$

The THz radiation power declines at high frequencies because the carrier response time decreases with an increase of the modulation frequency.

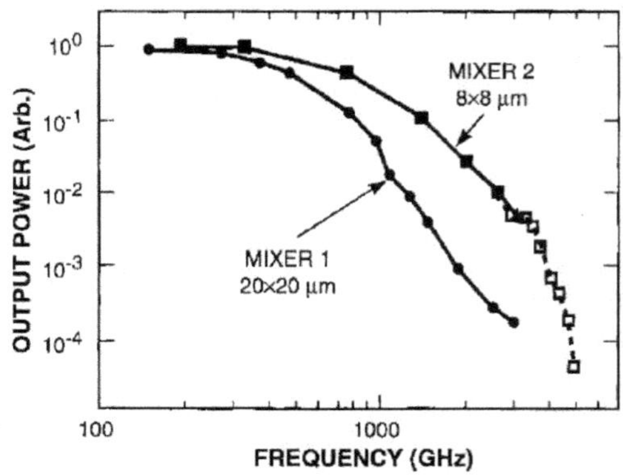

Fig. 4.2. THz radiation power of the photomixers with log-spiral antennas versus frequency. The antenna radiation resistance R_A is 72 Ω. Mixer 1 comprises a 20×20 μm^2 active area with 1.8-μm gaps between 0.2-μm-wide metallic electrodes (C_A=2.9 fF). Mixer 2 uses the same electrode geometry, but the area is only 8×8 μm^2 ($C_A = 0.5$ fF). The normalized bandwidth curves were measured with a 4.2-K bolometer. (Reprinted from [78]. ©1997 IEEE)

Figure 4.2 shows the frequency-dependent output power of the two photomixers with log-spiral antennas [78]. The radiation resistance is $R_A = 60\pi\sqrt{\epsilon_{eff}} = 72\ \Omega$ for a log-spiral antenna on a semi-infinite GaAs substrate. The carrier lifetime of LT-GaAs and the RC time constant of a typical log-spiral photomixer are a few hundred femtoseconds, thus the frequency roll-off starts around 1 THz. The radiation power is proportional to ω^{-4} at higher frequencies. Mixer 2 has broader bandwidth than Mixer 1 because the RC time constant of Mixer 2 is shorter.

At the expense of a broad tuning range, the log-spiral antenna has low output power due to the relatively low radiation resistance. The radiation power can be enhanced by using resonant antenna structures. Figure 4.3 shows the structure of a dipole PC antenna and the spectrum of its radiation power in the THz spectral range, where L is the distance between strip lines [79]. The radiation resistance of a dipole antenna on a GaAs substrate peaks around $L/\lambda_R = 0.3$ [80], which corresponds to the resonant frequency $\nu_R \approx 1.8$ THz for the 50-μm dipole. The measured output power is maximized near 1.2 THz and extends throughout the broad spectral range from 0.5 to 2 THz. The peak radiation resistance is estimated as 360 Ω, which is several times larger than that of the log-spiral antenna, 72 Ω.

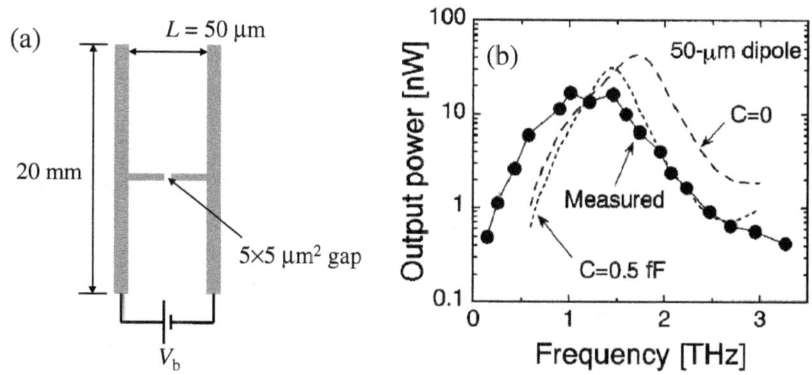

Fig. 4.3. (a) Schematic diagram of a dipole PC antenna. (b) Output radiation spectrum for the 50-μm dipole photomixer. The dashed curves show calculated radiation spectra for C_A=0 and 0.5 fF. (Reprinted with permission from [79]. ©1997, American Institute of Physics)

Higher output powers are achievable using more sophisticated antenna designs. The dual dipole antennas illustrated in Fig. 4.4(a) have several advan-

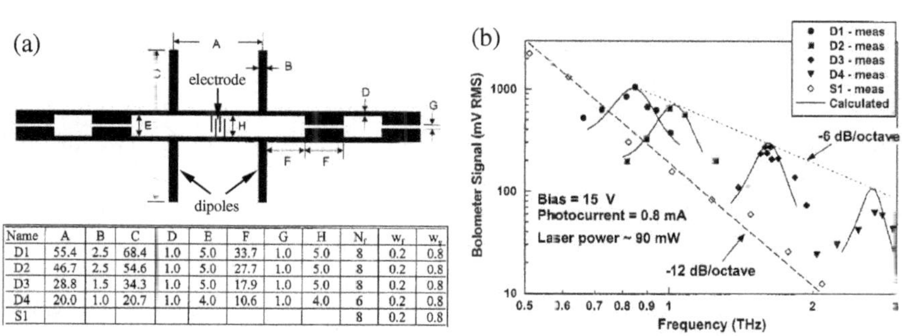

Name	A	B	C	D	E	F	G	H	N_f	w_f	w_g
D1	55.4	2.5	68.4	1.0	5.0	33.7	1.0	5.0	8	0.2	0.8
D2	46.7	2.5	54.6	1.0	5.0	27.7	1.0	5.0	8	0.2	0.8
D3	28.8	1.5	34.3	1.0	5.0	17.9	1.0	5.0	8	0.2	0.8
D4	20.0	1.0	20.7	1.0	4.0	10.6	1.0	4.0	6	0.2	0.8
S1									8	0.2	0.8

Fig. 4.4. (a) Geometry of dual dipoles, with parameters given in the table below. (b) The output power spectra for dual dipoles D1, D2, D3, and D4 and a spiral S1. Calculated results are shown by solid lines. (Reprinted from [81]. ©2001 IEEE)

tages over simpler antenna designs: the radiation pattern is more symmetric and the radiation resistance is higher [81]. The quintessential feature of the dual antenna design is that the electrode capacitance is cancelled out by the inductive tuning when the transmission line's length is adjusted to the resonant frequency. Thus, the radiation resistance is determined mainly by the carrier lifetime. The output spectra of the dual-dipole antennas in Fig. 4.4(b) demonstrate significantly higher powers near their resonant frequencies when compared to the output from the log-spiral antenna [81]. The maximum output powers of the dual-dipole antennas are 3, 2, 0.8, and 0.3 μW at 0.9, 1.0, 1.6, and 2.7 THz, respectively. The peak power declines with increase in frequency as ω^{-2}, while the log-spiral output is proportional to ω^{-4}. The resonant frequencies are determined by the length of the dipole C and of the transmission line $A + F$.

A few interesting ideas to improve radiation power are noteworthy. One way to avoid the limitation due to the low threshold for thermal damage of LT-GaAs, as well as to increase photocurrent, is to make the optical-excitation area large and to illuminate it with an extended beam from a high-power laser. If the dimensions of the illumination area are comparable to or larger than the wavelength of radiation, the optical excitation should be phase-matched with the THz radiation. A travelling-wave photomixer contains a long, thin active area between two electrodes. The active area is structured to maintain the coherent superposition between photocurrent and THz radiation [82]. Another approach to enhance photocurrent is to replace LT-GaAs substrates with semiconductor heterostructures such as p-i-n photodiodes. A uni-travelling-carrier photodiode (UTC-PD) contains a collection layer of InP which takes advantage of the exceptionally high electron mobility in the material and of the optical excitation at the optical communication wavelength, 1.55 μm [83].

4.2 Difference Frequency Generation and Parametric Amplification

Difference frequency generation (DFG) is a second-order nonlinear optical process which produces an electromagnetic wave of frequency ω_T when two optical beams at frequencies ω_1 and ω_2 are incident upon a nonlinear crystal, such that the output frequency is the difference between the two input frequencies: $\omega_T = \omega_1 - \omega_2$. We have discussed the basic concepts and theoretical formulations of second-order nonlinear optical interactions in sections 3.3.1 and 3.3.2 to describe optical rectification. Optical rectification can be interpreted as DFG of the different frequency components within the broad bandwidth of ultrashort laser pulses. In this section, we shall explore two schemes of narrowband THz generation: DFG with two input beams and parametric generation with a single optical pump.

4.2.1 Principles of Difference Frequency Generation

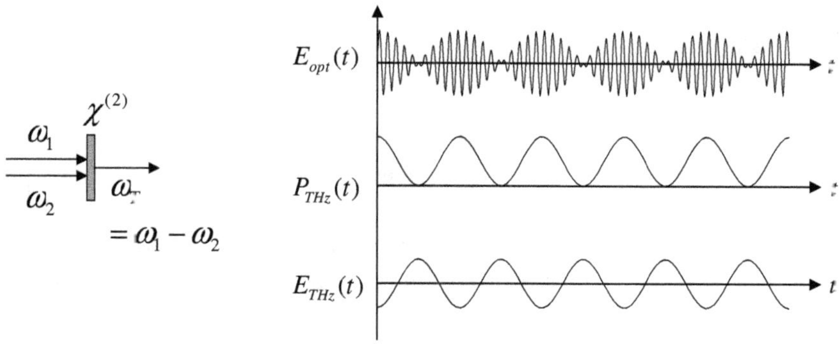

Fig. 4.5. Schematic diagram of difference frequency generation in a thin nonlinear crystal

A simple physical picture of DFG is illustrated in Fig. 4.5. DFG is a second-order nonlinear optical process; thus it requires a noncentrosymmetric crystal. For simplicity, we assume a very thin crystal and neglect any propagation effects. Like photomixing, the optical source for DFG is two narrowband laser beams with slightly different frequencies (ω_1 and ω_2). When the optical beams co-propagate and are linearly polarized in the same direction, their interference manifests a beat, which oscillates with the difference frequency or beat frequency ($\omega_T = \omega_1 - \omega_2$, $\omega_1 > \omega_2$):

$$\begin{aligned} E_O(t) &= E_1(t) + E_2(t) = E_0\left(\sin\omega_1 t + \sin\omega_2 t\right) \\ &= 2E_0 \cos\left(\frac{\omega_T}{2}t\right)\sin\omega_O t, \end{aligned} \tag{4.8}$$

where $\omega_O = (\omega_1 + \omega_2)/2$ is the average optical frequency. The second-order nonlinear polarization of DFG is proportional to the beat intensity:

$$P_T(t) = \chi^{(2)} E_0^2 \left[\cos\left(\frac{\omega_T}{2}t\right)\right]^2 = \frac{1}{2}\chi^{(2)} E_0^2 \left[1 + \cos\left(\omega_T t\right)\right]. \tag{4.9}$$

Consequently, the THz radiation field induced by the nonlinear polarization is given by

$$E_T(t) \propto \frac{\partial^2 P_T(t)}{\partial t^2} = -\frac{1}{2}\chi^{(2)}\omega_T^2 E_0^2 \cos\left(\omega_T t\right). \tag{4.10}$$

The THz field oscillates at the difference frequency, ω_T.

We next discuss the generation and propagation of THz waves in a bulk medium. The wave equation formulated in section 3.3.3 is still valid. We assume that a linearly polarized optical plane wave propagates in the z direction.

$$\frac{\partial^2 E_T(z,t)}{\partial z^2} - \frac{n_T^2}{c^2}\frac{\partial^2 E_T(z,t)}{\partial t^2} = \frac{1}{\epsilon_0 c^2}\frac{\partial^2 P_T^{(2)}(z,t)}{\partial t^2}, \tag{4.11}$$

where $E_T(z,t)$ and $P_T^{(2)}(z,t)$ are the THz field and the nonlinear polarization. The THz field is expressed as

$$E_T(z,t) = A_T(z)e^{i(k_T z - \omega_T t)} + c.c., \tag{4.12}$$

where

$$k_T = \frac{n_T \omega_T}{c}. \tag{4.13}$$

n_T is the refractive index at the THz frequency. We assume that the field amplitude $A_T(z)$ is a slowly varying function of z. The nonlinear polarization can be written as

$$P_T^{(2)}(z,t) = P_T(z)e^{-i\omega_T t} + c.c. \tag{4.14}$$

The amplitude of the nonlinear polarization induced by two monochromatic optical waves is given by

$$P_i^{(2)}(z,\omega_T) = \sum_{j,k} \epsilon_0 \chi_{ijk}^{(2)}(\omega_T, \omega_1, -\omega_2) E_j(z,\omega_1) E_k^*(z,\omega_2). \tag{4.15}$$

For fixed propagation and polarization directions, this equation can be simplified as a scalar relation [84]:

$$P_T(z) = 4\epsilon_0 d_{\text{eff}} E_1(z) E_2^*(z) = 4\epsilon_0 d_{\text{eff}} A_1 A_2^* e^{i(k_1-k_2)z}, \tag{4.16}$$

where $d_{\text{eff}} = \frac{1}{2}\chi_{\text{eff}}^{(2)}$ is the effective nonlinear coefficient, and A_1 and A_2 are the amplitudes of the optical fields.

Substituting Eqs. 4.12, 4.14, and 4.16 into the wave equation, Eq. 4.11, we obtain

$$\frac{d^2 A_T}{dz^2} + 2ik_T\frac{dA_T}{dz} = -\frac{4d_{\text{eff}}\omega_T^2}{\epsilon_0 c^2} A_1 A_2^* e^{i(k_1-k_2-k_T)z}. \tag{4.17}$$

The amplitude A_T varies slowly, so that the change is negligible for the propagation distance of a wavelength, then we can neglect the first term in Eq. 4.17. This approximation is called the *Slowly Varying Envelope Approximation* (SVEA). The wave equation is reduced to

$$\frac{dA_T}{dz} = \frac{2id_{\text{eff}}\omega_T}{\epsilon_0 n_T c} A_1 A_2^* e^{i\Delta k z}, \tag{4.18}$$

where $\Delta k = k_1 - k_2 - k_T$ is the momentum mismatch. The amplitudes of the optical waves also vary slowly and obey similar wave equations:

$$\frac{dA_1}{dz} = \frac{2id_{\text{eff}}\omega_1}{\epsilon_0 n_1 c} A_2 A_T e^{-i\Delta k z}, \tag{4.19}$$

and

$$\frac{dA_2}{dz} = \frac{2id_{\text{eff}}\omega_2}{\epsilon_0 n_2 c} A_1 A_T^* e^{i\Delta kz}. \tag{4.20}$$

The momentum mismatch is closely related to the velocity mismatch discussed in the section 3.3.3. When the two optical beams are polarized in the same direction and the dispersion is negligible near ω_1 and ω_2, we can define the optical refractive index as $n_O \equiv n_1 (= n_2)$. In this case, the momentum mismatch is proportional to the index mismatch, $\Delta n = n_O - n_T$, between the optical and THz waves:

$$\Delta k = n_O \frac{\omega_1}{c} - n_O \frac{\omega_2}{c} - n_T \frac{\omega_T}{c} = \Delta n \frac{\omega_T}{c}. \tag{4.21}$$

When the phase matching condition is satisfied, i.e., $\Delta k = 0$, the THz wave copropagates with the beat of the optical beams at the same velocity. Accordingly, the THz field undergoes a coherent amplification.

The ultimate upper limit of the optical-to-THz conversion efficiency is determined by the Manley-Rowe relations. The essence of the Manley-Rowe relations is that the creation and annihilation rates of photons should be equal at all frequencies involved in a nonlinear optical process. Imagine the initial intensities of the two optical beams and the THz wave are $I_1(0) = cN_1\hbar\omega_1/n_O$, $I_2(0) = cN_2\hbar\omega_2/n_O$, and $I_T(0) = 0$, where N_1 and N_2 are the initial photon number densities at ω_1 and ω_2. A downconversion with 100 % quantum efficiency yields $I_1(L) = 0$, $I_2(L) = c(N_1 + N_2)\hbar\omega_2/n_O$, and $I_T(L) = cN_1\hbar\omega_T/n_T$. Therefore, the optical-to-THz conversion efficiency obeys the inequality relation,

$$\frac{I_T(L)}{I_1(0)} \leq \frac{n_O \omega_T}{n_T \omega_1} \sim 10^{-3} - 10^{-2}. \tag{4.22}$$

4.2.2 Difference Frequency Generation with Two Pump Beams

When the two optical input beams have similar intensities, we can assume that the optical beams are non-depleted, i.e., A_1 and A_2 are constants. In practice, the optical-to-THz conversion efficiency is no more than 10^{-4} at best. With this non-depleted pump approximation, the integration of Eq. 4.18 for a propagation distance L yields

$$A_T(L) = \frac{2id_{\text{eff}}\omega_T}{\epsilon_0 n_T c} A_1 A_2^* \int_0^L e^{i\Delta kz} dz = \frac{2id_{\text{eff}}\omega_T A_1 A_2^*}{\epsilon_0 n_T c} \left(\frac{e^{i\Delta kL} - 1}{i\Delta k} \right), \tag{4.23}$$

which leads to the THz intensity,

$$I_T(L) = \frac{1}{2}\epsilon_0 c n_T |A_T(L)|^2 = \frac{8d_{\text{eff}}^2 \omega_T^2 I_1 I_2}{\epsilon_0^3 c^3 n_1 n_2 n_T} L^2 \text{sinc}^2 \left(\frac{\Delta kL}{2} \right). \tag{4.24}$$

The sinc function peaks at the origin and is negligible beyond $\left|\frac{\Delta kL}{2}\right| > \pi$. For a given Δk, $I_T(L) \propto \sin^2\left(\frac{\Delta kL}{2}\right)$. Thus, the maximum THz intensity is obtained when the crystal length is equal to the coherence length $l_c = \frac{\pi}{\Delta k}$.

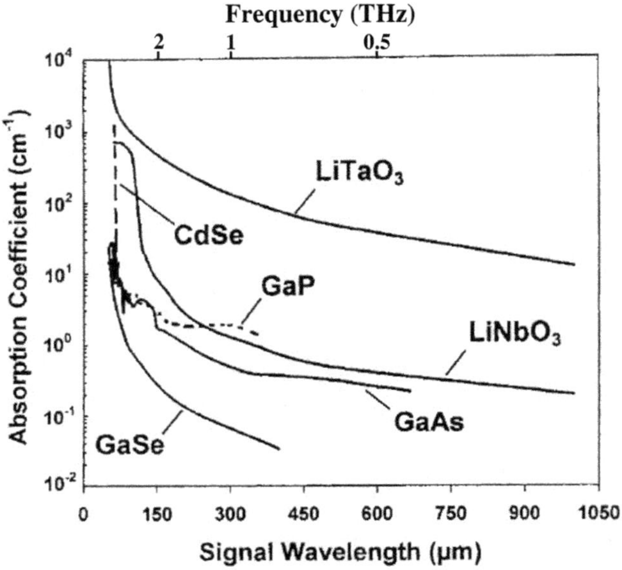

Fig. 4.6. Absorption spectra for an extraordinary wave in GaSe, GaAs, $LiNbO_3$, GaP, CdSe, and $LiTaO_3$. (Reprinted from [38].)

Among many nonlinear crystals examined for DFG, GaSe is the most efficient for THz generation. Quartz, $LiNbO_3$, GaP, and DAST (4-dimethylamino-N-methyl-4-stilbazolium-tosylate) are other nonlinear materials in which THz emission by DFG has been demonstrated. The generation efficiencies of these materials are substantially lower than GaSe, which has a few notable properties. First, its second-order nonlinear optical coefficient is very large: d_{22}=54 pm/V. Second, phase matching is attainable with optical pump beams in the infrared wavelength range. The output THz frequency is continuously tunable in the broad spectral range from 0.2 to 5.3 THz. Third, the linear absorption in GaSe is relatively low in the THz frequency range. Figure 4.6 shows the absorption coefficient of GaSe, compared with those of other nonlinear optical crystals [38]. The curves of GaSe, $LiNbO_3$, and $LiTaO_3$ are theoretical calculations. Due to defects and second-order phonon processes, experimental values are higher than the theoretical predictions. The absorption coefficient of GaSe is approximately 1 cm^{-1} in the sub-THz frequency range [85].

Figure 4.7 illustrates the geometry for angle-tuned phase matching of type-II DFG in GaSe. The phase-matching angle θ is the angle between the optic and the propagation axes. GaSe is birefringent due to its anisotropic crystal structure (hexagonal structure of $\bar{6}m2$ point group). The axis of anisotropy corresponds to the optic axis $\hat{c}$. GaSe is a uniaxial crystal because it has only one optic axis. The uniaxial birefringence is quantified by two refractive indices: n^o and n^e are the refractive indices for polarizations perpendicular and parallel to the optic axis. GaSe is negative uniaxial because $n^e < n^o$. Type-II

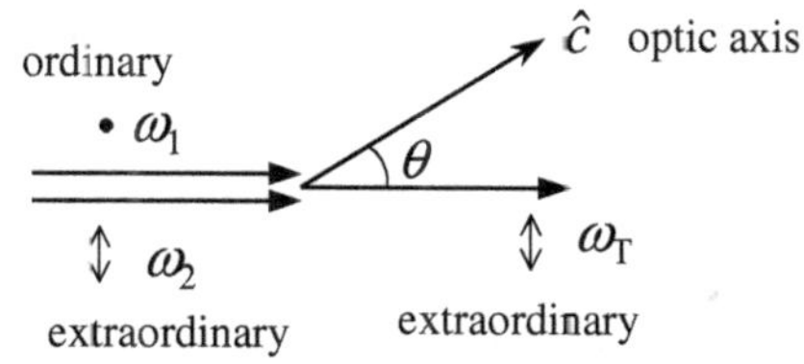

Fig. 4.7. Schematic diagram of angle-tuned phase matching of Type-II DFG in a negative uniaxial crystal

DFG in a negative uniaxial crystal is a nonlinear optical process where two orthogonally-polarized input beams, with ordinary and extraordinary polarizations, produce an output beam of extraordinary polarization. The ordinary polarization is normal to the optic axis, and the extraordinary polarization is in the plane containing the optic and propagation axes. Light of ordinary (extraordinary) polarization experiences the ordinary (extraordinary) refractive index, n^o ($n^e(\theta)$) while propagating. n^o is independent of the propagation direction. $n^e(\theta)$, however, is anisotropic:

$$\frac{1}{n^e(\theta)^2} = \frac{\sin^2\theta}{(n^e)^2} + \frac{\cos^2\theta}{(n^o)^2}. \tag{4.25}$$

n^e and n^o of GaSe are 2.46 and 2.81 at 1 μm, respectively [86].

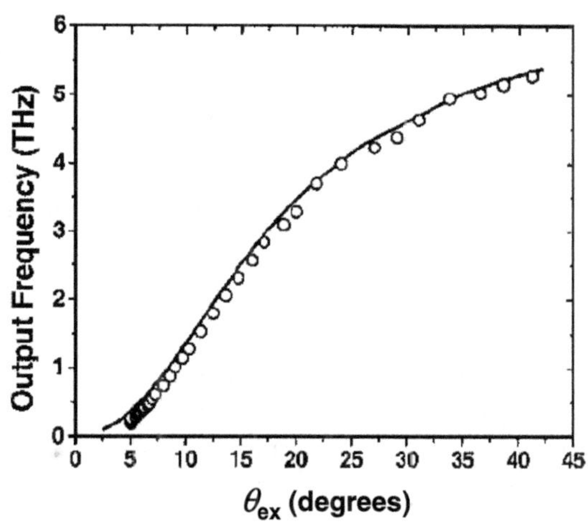

Fig. 4.8. Output frequency versus external phase-matching angle $\theta_{ex} = \sin^{-1}(n_O \sin\theta)$. Circles and solid curves, respectively, correspond to experimental and calculated results of refractive-index dispersion relations for GaSe. (Reprinted from [87].)

The phase matching condition, $k_T = k_1 - k_2$, is expressed as a function of the phase-matching angle θ,

$$n_T^e \omega_T = n_O^o \omega_1 - n_O^e(\theta)\omega_2. \tag{4.26}$$

Figure 4.8 shows the frequency tuning curve of GaSe as a function of the external phase-matching angle, $\theta_{ex} = \sin^{-1}(n_O \sin\theta)$, for the optical wavelength, $\lambda_1 = 2\pi c/\omega_1$=1.064 μm [87]. The experimental data points (open circles) are compared with theoretical calculation (solid line). The coherent THz radiation output has a broad tuning range from 0.2 to 5.3 THz.

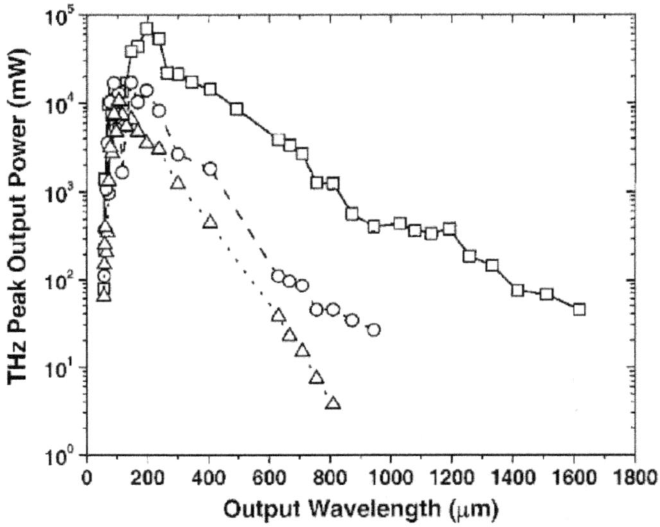

Fig. 4.9. Peak output power versus output wavelength for three GaSe crystals with thicknesses (along the z axis) of 4 mm (triangles), 7 mm (circles), and 15 mm (squares). (Reprinted from [87].)

Figure 4.9 shows the THz peak output power as a function of output wavelength for 4-mm, 7-mm, and 15-mm thick GaSe crystals [87]. The optical pump sources are a Q-switched Nd:YAG laser (wavelength, 1.064 μm; pulse duration, 10 ns; pulse energy, 6 mJ; repetition rate, 10 Hz) and the tunable idler output (duration, 5 ns; pulse energy, 3 mJ) of an optical parametric oscillator (OPO) pumped by the same Nd:YAG laser. Accordingly, the pulse duration and the repetition rate of the THz pulses are 5 ns and 10 Hz, respectively. The maximum THz peak power of the 15-mm crystal is 69.4 W at 1.53 THz, which corresponds to an optical-to-THz conversion efficiency of 1.8×10^{-4}, a pulse energy of 0.5 μJ, and an average power of 5 μW.

Sophisticated optical pumping schemes facilitate much higher THz output power. Figure 4.10 shows the schematic of the THz generation from a quasi-phase-matched (QPM) GaAs crystal placed inside the cavity of a synchronously pumped optical parametric oscillator (OPO) [88]. The average THz output power is 1 mW at 2.8 THz with a bandwidth of 0.3 THz. The OPO is pumped by a mode-locked laser at 1.064 μm (7 ps pulse duration, 50 MHz repetition rate, and 10 W average output power). The gain medium of the OPO is a periodically-poled lithium niobate (PPLN) crystal. The OPO converts a 1.064-μm photon to two-photons near the degeneracy wavelength 2.128-μm. The OPO output spectra are shown in Fig. 4.10(b). The frequency splitting, which is in the THz frequency range, is tunable via temperature control of the PPLN crystal.

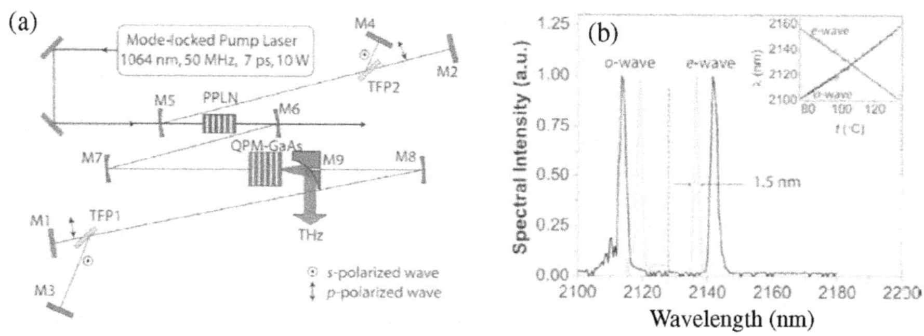

Fig. 4.10. (a) Schematic of a linear doubly resonant OPO with an "offset" cavity design. M1-M8, cavity mirrors; M9, off-axis parabolic mirror for THz outcoupling. (b) PPLN OPO line shapes near degeneracy for two different PPLN temperatures: T=90 °C (frequency splitting of 2.05 THz, black lines) and T=100 °C (frequency splitting of 0.96 THz, gray lines). The dotted line represents the degeneracy point 2.128 μm. The inset shows OPO tuning curves as a function of PPLN crystal temperature. (Reprinted from [88].)

4.2.3 Optical Parametric Amplification

Optical parametric generation is a second-order nonlinear optical process where the photon of a pump pulse is converted into two photons with lower energies. The sum of the two photon energies is equal to the pump photon energy: $\omega_p = \omega_i + \omega_T$, where pump and idler photons, ω_p and ω_i, are at optical frequencies and a signal photon, ω_T, is at THz frequency. The idler and THz waves are amplified when the phase-matching condition, $\mathbf{k}_p = \mathbf{k}_i + \mathbf{k}_T$, is satisfied.

The parametric process has been utilized to generate tunable narrowband THz waves in $LiNbO_3$ crystals. Figure 4.11 shows the momentum conservation of the pump, idler, and THz waves. The three wave vectors are noncollinearly phase matched. With a 1.064-μm optical pump, THz frequency is continuously tunable from 1 to 3 THz by changing the angle ϕ between the pump and the idler [89]. The angle is changed between 0.5° and 1.5° for this tuning range. The angle between the THz wave and the idler wave hardly changes at 65°.

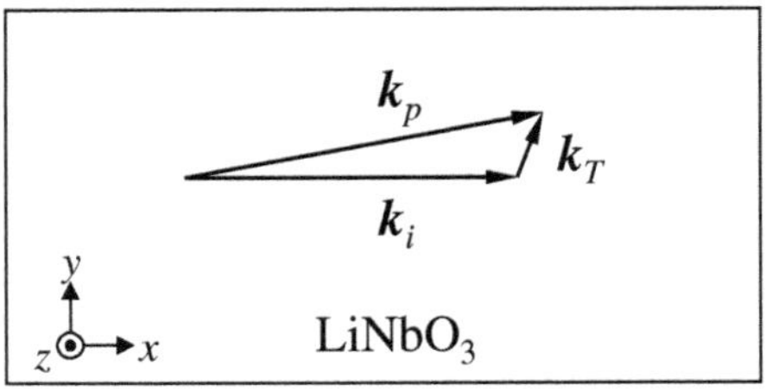

Fig. 4.11. Momentum conservation in $LiNbO_3$

When phase matching is perfect and the pump beam is nondepleted, the coupled wave equations 4.18 and 4.20 are given as

$$\frac{dA_T}{dz} = \frac{2id_{\text{eff}}\omega_T}{\epsilon_0 n_T c} A_p A_i^*, \tag{4.27}$$

and

$$\frac{dA_i}{dz} = \frac{2id_{\text{eff}}\omega_i}{\epsilon_0 n_i c} A_p A_T^*, \tag{4.28}$$

where A_p is a constant. Differentiation of Eq. 4.27 with respect to z and substitution of the A_i term into Eq. 4.28 leads to

$$\frac{d^2 A_T}{dz^2} = \frac{4d_{\text{eff}}^2\omega_T\omega_i}{\epsilon_0^2 n_T n_i c^2} |A_p|^2 A_T \equiv g_T^2 A_T, \tag{4.29}$$

where the exponential gain coefficient g_T is given as

$$g_T = \left(\frac{4d_{\text{eff}}^2\omega_T\omega_i}{\epsilon_0^2 n_T n_i c^2} \right)^{1/2} |A_p|. \tag{4.30}$$

The gain coefficient is proportional to the pump field amplitude.

Assuming the initial THz field $A_T(0) = 0$, we get the solution

$$A_T(z) = ie^{i\phi_p}\sqrt{\frac{n_i\omega_T}{n_T\omega_i}} A_i^*(0) \sinh g_T z, \tag{4.31}$$

where ϕ_p is the phase of the complex amplitude A_p. The intensity of the THz wave is

$$I_T(z) = \frac{1}{2}\epsilon_0 c n_T |A_T(z)|^2 = \frac{\omega_T}{\omega_i} I_i(0) \sinh^2 g_T z. \tag{4.32}$$

A notable implication of this equation is that the output THz intensity is proportional to the initial idler intensity.

We describe two methods to strengthen the idler intensity, shown in Fig. 4.12: (a) injection-seeded THz-wave parametric generator (TPG) and (b) THz-wave parametric oscillator (TPO). A CW Yb-fiber laser (wavelength, 1.070-μm) injects the seed beam of the idler into the TPG. The optical pump source is a Q-switched Nd:YAG laser (wavelength, 1.064 mm; pulse energy, 45 mJ/pulse; pulse duration, 15 ns; repetition rate, 10 Hz). The maximum THz pulse energy is 0.6 nJ with 45-mJ pump pulses, which corresponds to an optical-to-THz conversion efficiency of 1.3×10^{-8}. In a TPO, the idler beam is confined within an optical cavity, which gives rise to a significant enhancement of the idler intensity. For the specific design depicted in Fig 4.12(b), the $LiNbO_3$ crystal is placed inside the pump laser cavity. The maximum THz pulse energy is 5 nJ when the pump pulse energy is 1.3 mJ. Accordingly, the optical-to-THz conversion efficiency is 3×10^{-6}.

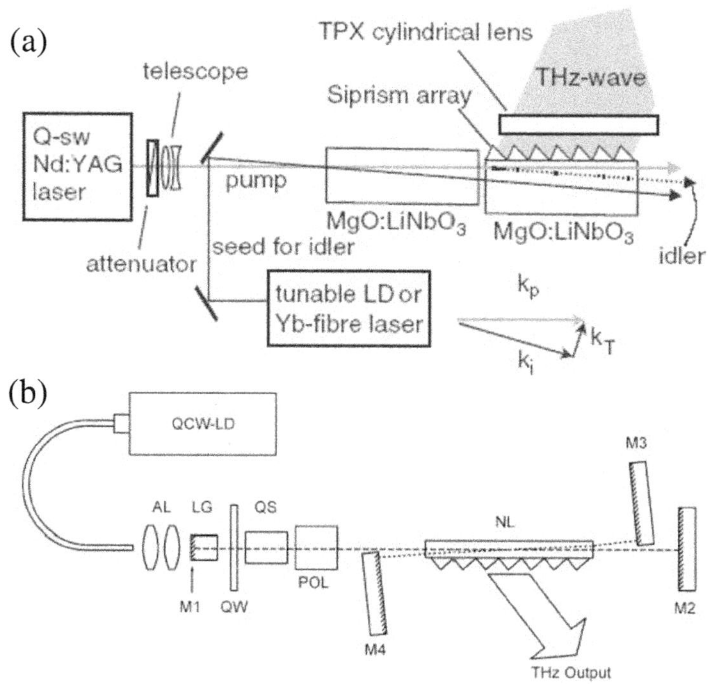

Fig. 4.12. Schematic diagrams of (a) the injection-seeded TPG (Reprinted from [90]. ©2001, American Institute of Physics) and (b) the non-collinear phase-matched TPO (Reprinted with permission from [91]. ©2006, American Institute of Physics).

4.3 Far-Infrared Gas Lasers

The basic design of THz gas lasers is similar to that of the typical laser system shown in Fig. 2.24. An extra component of importance is an intracavity waveguide used to confine the laser modes in the transverse direction. The gain media of THz gas lasers are molecular gases such as CH_3F, CH_3OH, NH_3 and CH_2F_2. The THz radiation originates from the rotational transitions of the molecules (see section 2.2.3). The molecules have permanent dipole moments, hence their rotational transitions are directly coupled to electromagnetic radiation via dipole interactions.

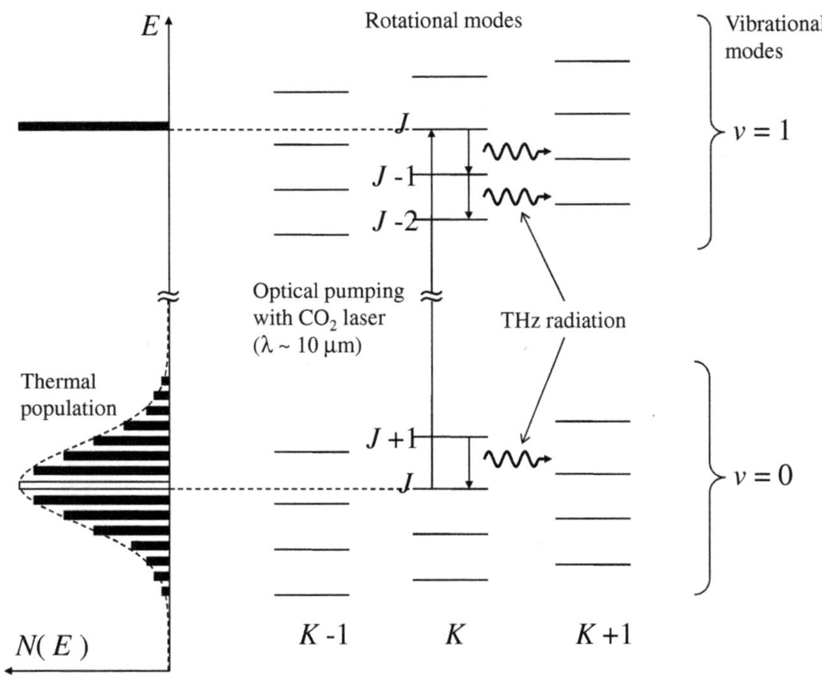

Fig. 4.13. Energy level diagram of optical excitation ($v = 0 \to 1$) and THz radiation ($J+1 \to J$ for $v = 0$, $J \to J-1$ and $J-1 \to J-2$ for $v = 1$) in an optically-pumped THz gas laser.

Figure 4.13 illustrates the lasing scheme of a typical THz gas laser. At room temperature, the molecules occupy the lowest vibrational mode ($v = 0$) with a thermal population

$$N(J,K) \propto g(J,K)e^{-E_{rot}(J,K)/k_BT} \tag{4.33}$$

of rotational states, where $E_{rot}(J,K)$ is the rotational energy eigenvalues (Eq. 2.177). Optical pumping with a CO_2 laser excites some of the molecules from the lowest to the first excited vibrational mode. For symmetric-top molecules, the vibrational-rotational transitions obey the selection rules $\Delta v = 1$, $\Delta J = 0$ or ± 1, and $\Delta K = 0$ (see section 2.2.3). The optically induced population inversions between $(J+1)$ and J-levels for $v = 0$ and between J and $(J-1)$-levels for $v = 1$ give rise to emissions at THz frequencies. The cascade transition from $(J-1)$ to $(J-2)$-level for $v = 1$ also contributes to the THz radiation.

Many chemical species have been examined for lasing in the THz region, and several hundred THz laser emission lines have been observed. Table 4.1 lists some of the stronger laser lines in the THz region [92].

Table 4.1. Laser lines of optically pumped THz gas lasers

Frequency (THz)	Molecule	Output Power (mW)
8.0	CH_3OH	~10
7.1	CH_3OH	~10
4.68	CH_3OH	>20
4.25	CH_3OH	~100
3.68	NH_3	~100
2.52	CH_3OH	>100
2.46	CH_2F_2	~10
1.96	$^{15}NH_3$	~200
1.81	CH_2F_2	<100
1.27	CH_2F_2	~10
0.86	CH_3Cl	~10
0.59	CH_3I	~10
0.525	CH_3OH	~40
0.245	CH_3OH	~10

Data from Ref. [92]

4.4 P-Type Germanium Lasers

P-type germanium THz lasers are electrically pumped all-solid-state lasers. The usual dopant is beryllium, which provides high optical gain. The lasing action is based on streaming motion and population inversion of hot-carriers in p-type Ge crystals submerged in crossed electric and magnetic fields.

THz photons are generated by stimulated transitions between two light-hole Landau levels. Classically, a charged particle in a magnetic field moves in circular motion with a cyclotron frequency proportional to the field strength. Because of spatial confinement, the quantum mechanical energy levels of the charged particle can only have discrete values. The discrete energy levels are called *Landau levels.*

A population inversion is accomplished by intricate intraband transitions among light hole and heavy hole states at low temperature (<40 K). In this cryogenic condition, the predominant mechanism of hole scattering in p-type Ge is spontaneous emission of optical phonons ($\hbar\omega_{OP}$=37 meV). If the applied electric field is strong enough, a hole is freely accelerated up to the optical phonon energy where it makes a transition to a lower energy state by emitting an optical phonon. This phenomenon is called *streaming motion*. A certain amount of heavy holes (HH) scatter into pseudo-stable Landau levels in the light-hole (LH) band. These pseudo-stable states are formed under the specific condition of crossed electric and magnetic fields. The accumulation of streaming heavy holes in the pseudo-stable LH states results in the population inversion between the stable Landau levels and lower energy levels. Figure 4.14 illustrates the process of population inversion in a p-type Ge crystal when crossed electric and magnetic fields are applied.

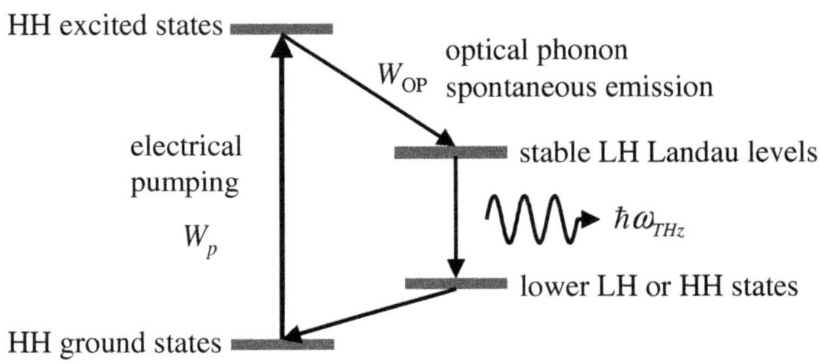

Fig. 4.14. Energy diagram of the population inversion in p-type Ge in crossed electric and magnetic fields.

The underlying process is well described by a semiclassical model [93]. We treat a hole in p-type Ge as a free carrier with a scalar effective mass m, the HH and LH bands in Ge being nearly parabolic and isotropic. The equation of motion of a free hole in crossed electric and magnetic fields is given as

$$\frac{d\mathbf{v}}{dt} = \frac{e}{m}\left(\mathbf{E} + \mathbf{v} \times \mathbf{B}\right), \tag{4.34}$$

where $\mathbf{v}$, e, and m are the hole velocity, charge, and mass, respectively. Considering two dimensional motion of the hole with the electric field $\mathbf{E} = (E, 0, 0)$ and magnetic field $\mathbf{B} = (0, 0, B)$, we obtain coupled equations for v_x and v_y,

$$\frac{dv_x}{dt} = \frac{e}{m}E + \omega_c v_y \quad \text{and} \quad \frac{dv_y}{dt} = -\omega_c v_x, \tag{4.35}$$

where $\omega_c = eB/m$ is the cyclotron frequency. The solution to these equations is a circular trajectory in the velocity space centered at $-v_d\,\mathbf{e}_y$:

$$\mathbf{v}(t) = v_d \sin\omega_c t \,\hat{\mathbf{e}}_x + v_d(\cos\omega_c t - 1)\,\hat{\mathbf{e}}_y, \tag{4.36}$$

where $v_d (\equiv E/B)$ is the drift velocity. We set the initial condition $\mathbf{v}(0) = 0$, which corresponds to the hole being in the ground state without external fields. The circular trajectory then passes through the origin.

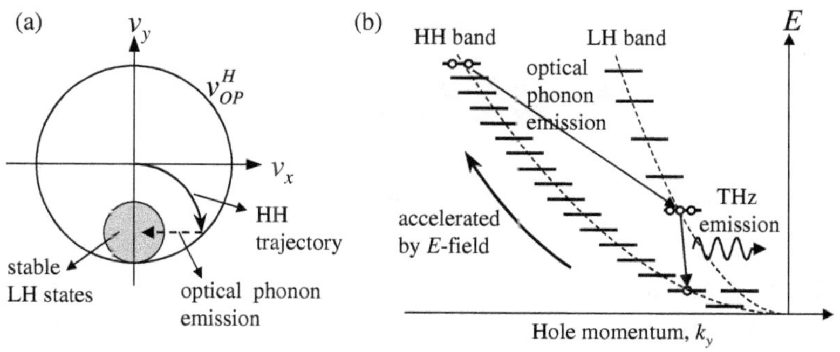

Fig. 4.15. (a) The heavy hole emits an optical phonon and scatters into a LH state with a long lifetime (shaded area) when the HH trajectory crosses the circle of $v = v^H_{OP}$ in the two dimensional velocity space. (b) Population inversion and stimulated THz emission from a p-type Ge crystal in cross electric and magnetic fields.

The hole energy oscillates between 0 and $W_H (\equiv 2mv_d^2)$ during the circular motion. If the maximum hole energy W_H surpasses the optical phonon energy, $\hbar\omega_{OP} \equiv \frac{1}{2}mv_{CP}^2$, as the hole is accelerated so that its velocity exceeds v_{OP}, the hole loses its kinetic energy by emitting an optical phonon, and scatters into a lower energy state. Suppose the drift velocity satisfies the inequality relation,

$$\frac{1}{2}v^H_{OP} < v_d < v^H_{OP}. \tag{4.37}$$

as depicted as Fig. 4.15(a). When the HH trajectory crosses the circle indicating $v = v^H_{OP}$, the heavy hole emits an optical phonon spontaneously. Some of the heavy holes jump into the shaded area indicating LH Landau levels with long lifetimes. Since the HH and LH effective masses are $m_H = 0.35m_e$ and $m_L = 0.043m_e$ in Ge, the drift velocity is smaller than $v^L_{OP} \approx \sqrt{8}v^H_{OP}$, i.e., the light hole energy cannot exceed the optical phonon energy. Consequently, the optical-phonon scattering is negligible for light holes in the shaded area, and the LH levels have long lifetimes. Figure 4.15(b) illustrates the intraband transitions among the HH and LH Landau levels to prepare the population inversion. THz photons are emitted by stimulated transitions from the long-lifetime LH Landau levels to lower HH or LH Landau levels.

The schematic diagram of a p-type Ge laser is shown in Fig. 4.16 [94]. The Ge crystal doped with acceptors such as beryllium, is placed between two permanent magnets which generate a horizontal magnetic field of ~1 T.

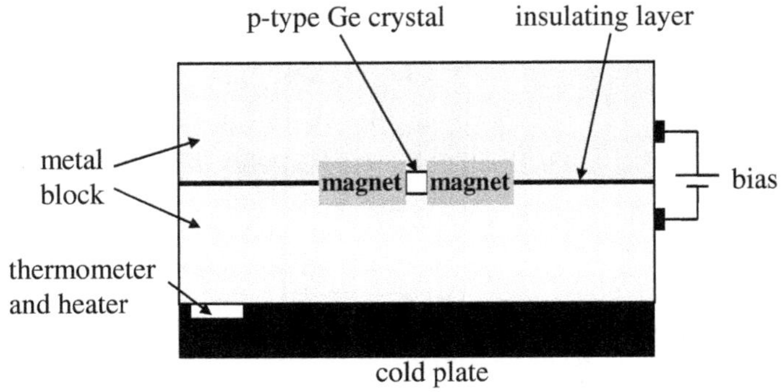

Fig. 4.16. Schematic diagram of a p-Ge THz laser [94].

The doping concentration is in the vicinity of 10^{14} cm^{-3}. The copper blocks separated by a thin insulating layer are biased to generate an electric field (~1 kV/cm) in the vertical direction. The cold plate, part of a closed cycle refrigerator system, keeps the crystal at cryogenic temperature (<40 K). The typical pulse energy is a few microjoules with a pulse duration of several microseconds. Due to electrical heating, the duty cycle is no more than a few per cent. The average output power reaches up to a few watts. The frequency is continuously tunable over the spectral range from 1 to 4 THz by changing the electric and magnetic field strength.

4.5 Frequency Multiplication of Microwaves

Another class of solid-state THz sources is founded on microwave technology. Taking the output of a microwave synthesizer—high-end models are capable of generating output signals in the frequency range of 10-100 GHz—as a seed, the solid-state THz sources multiply the frequency of the incoming microwave using Schottky barrier diodes. The microwave synthesizers and the frequency multipliers are the products of sophisticated engineering involving state-of-the-art electronics technologies. We shall look into the very basic mechanisms that the operation of these high-frequency electronic devices is based on.

Solid-state microwave sources include Gunn diodes and tunnel diodes. They are two-terminal negative resistance devices. When these devices are coupled to a resonator, they convert DC electric power into an AC signal of microwave frequencies. The key characteristic of the device for microwave generation is the negative resistance, as shown in the I-V curve of Fig. 4.17. Figure 4.17 also shows a simplified circuit diagram for a microwave oscillator. If the magnitude of the negative resistance of the diode is adjusted to cancel the resistance of the oscillator, the circuit oscillates without attenuation, and thus radiates a continuous electromagnetic wave.

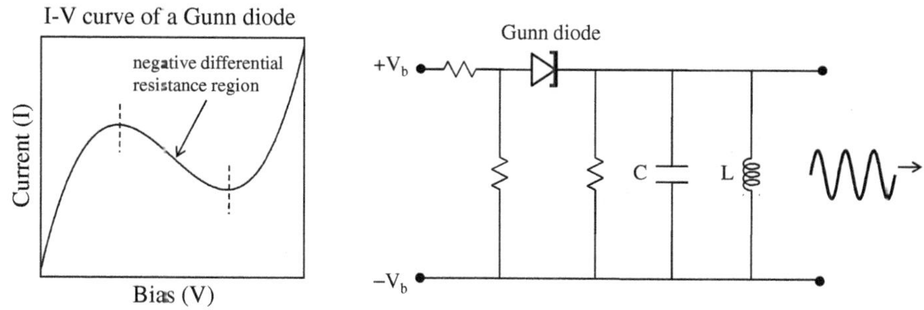

Fig. 4.17. Gunn diode I-V curve and oscillator circuit

Conceptually, the frequency multiplication of microwaves is analogous to the harmonic generation of optical waves in a nonlinear crystal. In a frequency multiplier, Schottky diodes function as nonlinear media, converting incoming microwaves into their harmonic waves. The Schottky diode employs a metal-semiconductor junction known as a Schottky barrier. The high conductivity of the metal contact gives rise to very high switching rates (>10 GHz). Figure 4.18 shows the schematic diagram of a frequency tripler [95]. The tripler, monolithically fabricated on a GaAs-based structure, is a split-block waveguide design including an array of Schottky diodes. The device consists of receiving and emitting antennas, input and output waveguides, and a circuit for harmonic-generation. Efficient third harmonic generation relies on the optimal design of the shape and dimension of the circuit elements.

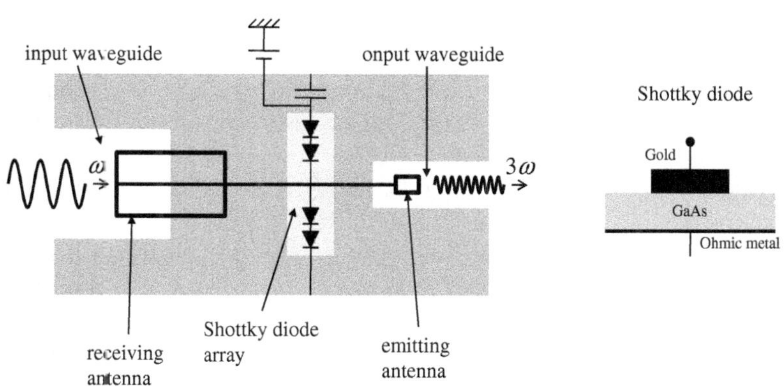

Fig. 4.18. Schematic diagram of a frequency tripler [95]

Solid-state THz emitters are compact, operate at room temperature, and have a narrow bandwidth, $\Delta\nu/\nu \sim 10^{-6}$. The output power of solid-state THz emitters falls off as $\sim 1/\nu^3$ as the frequency increases. The average output power around 1 THz is typically in the range of 10-100 μW. Table 4.2 lists nominal output powers of solid-state THz emitters at their representative frequencies.

Table 4.2. Output power of Schottky diode frequency multipliers

Frequency (THz)	Output Power (mW)
0.2 - 0.3	10 - 100
0.5 - 0.8	0.1 - 10
1.0 - 1.5	0.01 - 0.1

Schottky barrier diodes are also used as microwave detectors. A typical solid-state microwave detector includes an asymmetric Schottky diode assembly that downconverts the signal frequency by nonlinear mixing. The basic concept is analogous to optical rectification (see section 3.3). For incoming radiation given as $E_S \cos(\omega_S t)$, the output signal of the diode array is expressed as

$$V_O = \chi[E_S \cos(\omega_S t)]^2 = \frac{1}{2}\chi E_S^2 + \frac{1}{2}\chi E_S^2 \cos(\omega_S t), \tag{4.38}$$

where χ is the quadratic nonlinear coefficient of the device. The DC component is proportional to the radiation intensity.

4.6 Quantum Cascade Lasers

A quantum cascade laser (QCL) is a semiconductor heterostructure laser. The heterostructure forms a superlattice of periodically alternating layers of dissimilar semiconductors. An external bias induces transitions between two quasi-two-dimensional states in the superlattice, which gives rise to the emission of THz radiation. Introductory discussions about the elctronic structures of quasi-two-dimensional systems are presented in section 8.1.4.

The basic operation of QCLs relies on two essential processes: intersubband transitions and cascading. Unlike the radiative recombination of electron and hole pairs in conventional diode semiconductor lasers, an intersubband transition involves only electrons: it is therefore called a unipolar process. In a QCL, an electron undergoes an intersubband transition in one period of the superlattice, then the same electron is injected into the next period and subject to another intersubband transition. This process repeats until the electron reaches the end of the superlattice. The term "cascading" reflects the repetitive nature of the multiple intersubband transitions. It is noteworthy that a cascading process beats the conventional limitation of quantum efficiency, i.e., multiple photons can be generated by a single electron. We describe details of these processes in the following sections.

There are two major difficulties to overcome in the development of THz QCLs. First, the energy of a 1-THz photon, 4.1 meV, is so low that thermal excitations can easily disturb the electron configuration, most importantly the population inversion, necessary for lasing in a QCL. The output power of QCLs reaches up to 100 mW at liquid helium (L-He) temperature, but falls off quickly as the temperature increases. At present, the highest operating temperature of THz QCLs is in the vicinity of 180 K [96]. Second, it is difficult to confine laser modes in a small volume due to the long wavelength of THz radiation. Mode confinement is important because stimulated emission efficiency is proportional to the light intensity in the gain medium. The dielectric waveguides of conventional solid-state lasers cannot be used in THz QCLs because the penetration depth of evanescent waves is far longer than the size of the active regions. Metal waveguides are necessary components of a THz QCL despite their high loss.

4.6.1 Lasing and Cascading

A schematic presentation of lasing and cascading in a QCL is shown in Fig. 4.19. One period of the QCL superlattice structure consists of an injector, an injection barrier, and an active region. Intersubband transitions in the active region, comprising multiple QWs, give rise to radiative emission. Laser action in typical QCLs is based on a three-level system. A population inversion is created between levels 3 and 2. The superlattice structure is biased by a static electric field. For cascading, the bias voltage is adjusted so that level 1 of an active region is aligned with level 3 of the next period. Then, the used electrons in one active region are injected into and recycled in the next period.

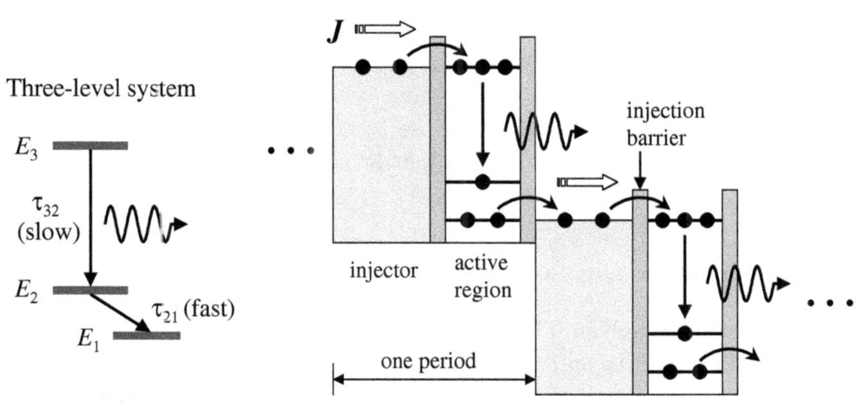

Fig. 4.19. Schematic illustration of lasing and cascading in a QCL superlattice structure.

A large population inversion is necessary for effective laser action. Thus, efficient injection into the upper level and fast depletion of the lower level

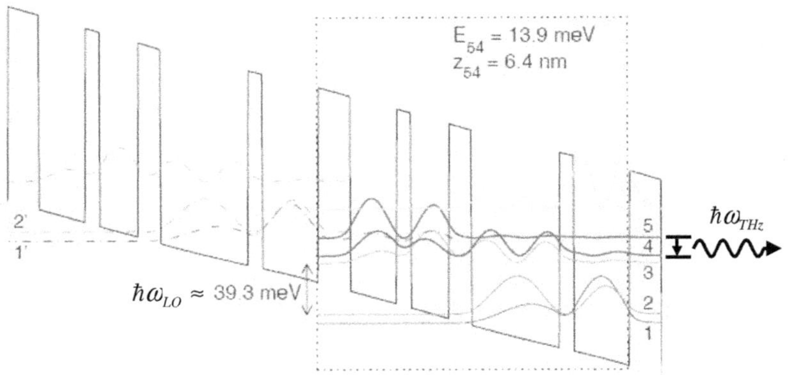

Fig. 4.20. Conduction band structure of a QCL biased at 64 mV/period, which corresponds to a field of 12.2 kV/cm. The four GaAs QW module grown in GaAs/Al_0.15$Ga_{0.85}$As is outlined. Beginning with the left injection barrier, the layer thicknesses in Å are 54/78/24/64/38/148/24/94. (Reprinted with permission from [97]. ©2003, American Institute of Physics.)

are highly desirable. Also, stimulated emission must be the dominant decay process between the two lasing levels. Several ingenious designs of superlattice structures have been developed to fulfill these requirements. Figure 4.20 shows an example [97]. One period of the superlattice structure consists of four QWs grown in GaAs/$Al_{0.15}Ga_{0.85}As$. Levels 5 and 4 are the upper and lower lasing levels, respectively. THz photons of 3.44 THz or 14.2 meV are generated through the transition. The transition via LO phonon scattering is forbidden due to the small energy separation, thus the upper level has a long lifetime. The ample overlap of the two wave functions indicates a high rate of radiative transition between the two levels. The device expedites the fast depletion of the lower level via LO-phonon emission, where the transition energy from levels 3 and 4 to levels 1 and 2 is close to the LO-phonon energy. Levels 1 and 2 are strongly coupled to the upper levels of the next period, and thus injection by tunneling is efficient.

4.6.2 Prospective Development

A major breakthrough in THz QCL technology came in 2002 when the first demonstration of THz emission from a QCL based on chirped superlattice active regions was reported [98]. The progress with this new technology has been remarkable since it was first demonstrated. Except p-type Ge lasers, THz QCLs are currently the only solid-state source whose coherent output power exceeds 10 mW above 1 THz. Radiation power reaches up to ~200 mW at L-He temperature (4 K), and a few milliwatts can be obtained near 100 K. The frequency range extends from 1.5 to 4.5 THz. This evolving technology faces two main challenges for broad device applications: room temperature operation and low frequency generation below 1.5 THz.

4.7 Backward Wave Oscillators

Backward wave oscillators (BWOs), or carcinotrons, are electron vacuum tubes in which an electron beam interacts with a travelling electromagnetic wave. The electrons are slowed down by a metal grating called a comb slow-wave structure, and the kinetic energy of the electrons is transferred to the electromagnetic wave. The device was named a BWO because the electron beam and the electromagnetic wave move in opposite directions.

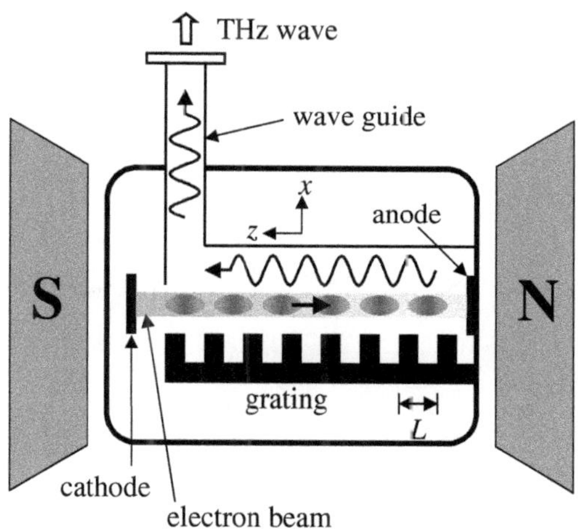

Fig. 4.21. Schematic diagram of backward wave oscillator.

Figure 4.21 shows a BWO schematic. Electrons are emitted from the heated cathode and accelerated by a DC electric field applied between the cathode and the anode. The magnets collimate the electron beam. The periodic structure of the grating induces a spatial modulation of the longitudinal electric field, which gives rise to an energy modulation in the electron beam. The periodic perturbation drives electrons into bunches. The bunched distribution of the electrons excites surface waves on the periodic structure. If the electron beam velocity matches the phase velocity of the surface wave, the kinetic energy of the electrons is transferred coherently to the electromagnetic wave. This implies that the frequency of the electromagnetic wave is determined by the electron velocity. Thus, the frequency is tunable by adjusting the bias voltage. Because the group velocity of the surface wave moves opposite to the phase velocity, the energy transferred to the field is transported and amplified in the backward direction. The radiation comes out through a waveguide coupled to the cavity.

Now we discuss how the backward wave propagates. Imagine a monochromatic electromagnetic wave in a periodic structure with a period L along the z axis. The electric field is written as

$$\mathbf{E}(\mathbf{r},t) = \mathbf{E}(\mathbf{r})e^{-i\omega t}, \tag{4.39}$$

where a modulus of the field is a periodic function with the same period L. Therefore, the field amplitude is given as

$$\mathbf{E}(x,y,z,t) = \mathbf{A}(x,y,z)e^{i(k_0 z-\omega t)}, \tag{4.40}$$

where $\mathbf{A}(x,y,z) = \mathbf{A}(x,y,z+L)$ and $0 \le k_0 L \le \pi$. The periodic function $\mathbf{A}(x,y,z)$ is expanded as a Fourier series,

$$\mathbf{A}(x,y,z) = \sum_{m=-\infty}^{\infty} \mathbf{A}_m(x,y)e^{\frac{2\pi m}{L}z}. \tag{4.41}$$

Substitution of the Eq. 4.41 into Eq. 4.40 yields

$$\mathbf{E}(x,y,z,t) = \sum_{m=-\infty}^{\infty} \mathbf{A}_m(x,y)e^{i(k_m z-\omega t)}, \tag{4.42}$$

where

$$k_m = k_0 + \frac{2\pi}{L}m. \tag{4.43}$$

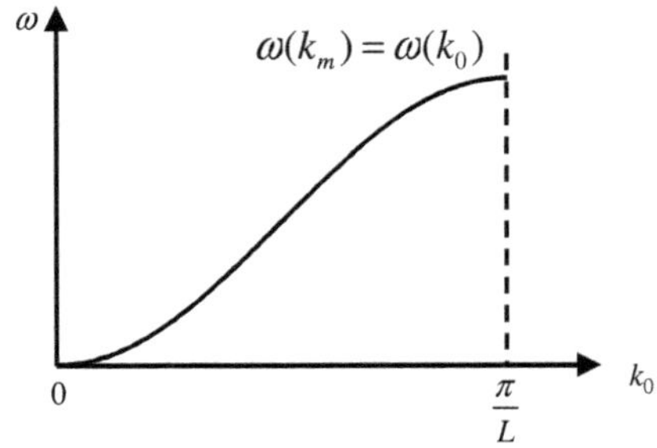

Fig. 4.22. Dispersion relation for an electromagnetic wave in a periodic structure

The bunched electron beam in a BWO interacts with negative modes ($m < 0$) of the electromagnetic wave [99]. To maximize the efficiency of the energy transfer, the electrons travel with the phase velocity of a negative mode. The phase velocity is expressed as

$$v_p = \frac{\omega}{k_m} = \frac{\omega}{k_0 + \frac{2\pi}{L}m} < 0. \tag{4.44}$$

On the other hand, the dispersion relation of the periodic system (see Fig. 4.22) indicates that the group velocity is independent of m and moves opposite to the phase velocity:

$$v_g = \frac{d\omega}{dk_m} = \frac{d\omega}{dk_0} \geq 0. \tag{4.45}$$

Operation of a BWO in the THz frequency range requires sophisticated engineering of the device [100]. The typical size of a BWO is about 30cm×30cm×30cm, and most of the volume is taken up by magnets. Typical magnetic fields generated by the magnets are ~1 T. The vacuum tube is just several centimeters in length and less than one centimeter in diameter. Inside the tube, a metal grating consists of a few hundred grooves with a period of ~10 μm. The miniature vacuum tube must endure harsh conditions: the bias voltage reaches up to 6.5 kV; the cathode is heated up to high temperature (1,200°C); and the high vacuum is kept around 10^{-8} Torr.

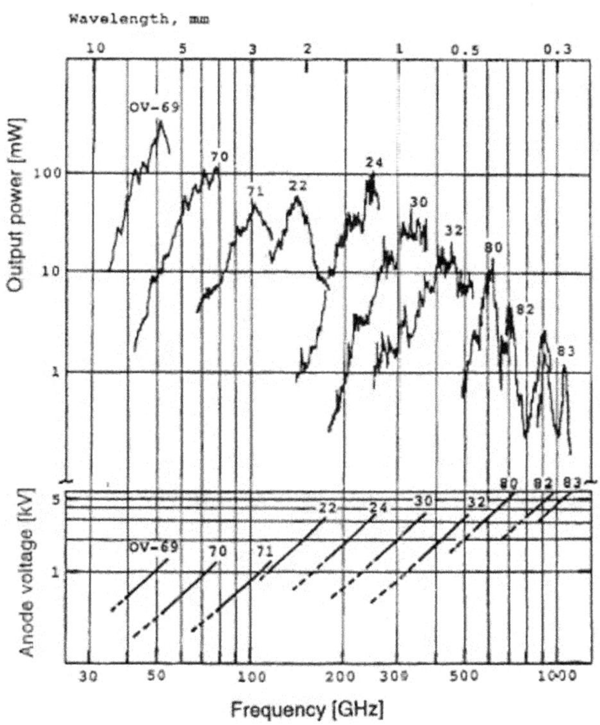

Fig. 4.23. BWO output power and anode voltage versus operating frequencies. (Reprinted from [100].)

Figure 4.23 shows the output power and the anode voltage of a BWO versus operating frequencies from 0.03 to 1 THz [100]. Ten different tubes cover the whole spectral range. Each tube is tunable in a range of ~10 % of its central frequency. The output power reaches up to 100 mW below 200 GHz. The power falls off rapidly with increasing frequency, e.g., ~1 mW at 1 THz.

4.8 Free-Electron Lasers

Free-electron lasers (FELs) use a relativistic electron beam passing through a wiggler, a series of magnets arranged to supply periodic, transverse magnetic field, to generate coherent electromagnetic radiation. The periodically alternating magnetic field forces a sinusoidal oscillation of the electrons, and hence the radiation is monochromatic.

The underlying physical process behind FELs is applicable to a wide spectral range from microwaves to x-rays. Their broad and continuous tunability is a prime advantage of FELs. The radiation wavelength is determined by a small number of parameters such as wiggler period, magnetic field strength, and electron beam energy. FELs can also produce high-power radiation because FELs, in which electrons are the only medium involved in lasing, are free of the conventional problems of high-power laser systems such as thermal lensing and material damaging.

4.8.1 Operational Principles

Figure 4.24 shows a FEL resonator in which a laser beam copropagates with an electron beam periodically modulated by a wiggler magnet array. The electrons travel at approximately the speed of light, $v \approx c$, thus a relativistic analysis is necessary to describe the electrodynamics. In the frame moving with the mean velocity of the electrons, the electron motion is described as a harmonic oscillation around the cavity axis, and the oscillating electrons emit radiation. In the rest frame, most of the radiation power is amassed in the forward direction near the electron beam axis. The relativistic Doppler effect asserts that, in the forward direction, the rest-frame wavelength undergoes a severe spectral shift to the short-wavelength side:

$$\lambda = \frac{\lambda'}{\gamma\,(1 + v/c)} \approx \frac{1}{2\gamma}\lambda', \tag{4.46}$$

where λ and λ' are the wavelengths in the rest and moving frames, respectively, and

$$\gamma = \frac{1}{\sqrt{1 - v^2/c^2}} \tag{4.47}$$

is the Lorentz factor. If the magnetic field is weak enough that the electron motion perpendicular to the beam axis is nonrelativistic, then λ is attained by simple reasoning. In the moving frame, the electrons see the contracted period of the wiggler array,

$$\lambda'_w = \lambda_w/\gamma, \tag{4.48}$$

where λ_w is the wiggler period. The resonance condition $\lambda' = \lambda'_w$, together with Eq.4.46 and 4.48, leads to

$$\lambda = \frac{1}{2\gamma^2}\lambda_w. \tag{4.49}$$

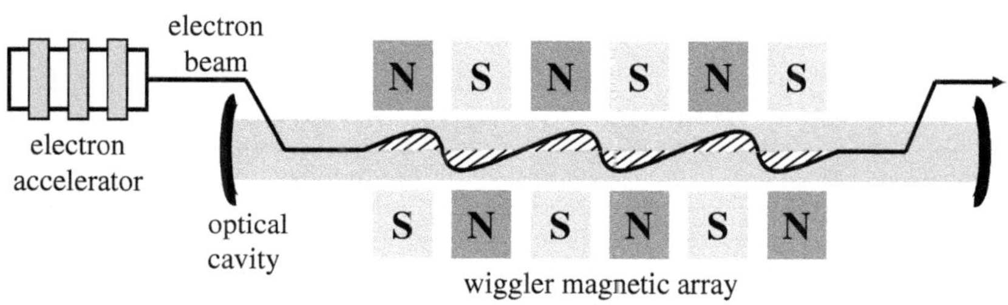

Fig. 4.24. Schematic diagram of a FEL. The relativistic electron beam passes through a periodically alternating magnetic array called a wiggler or an undulator. The alternating magnetic field forces the oscillation of electrons, which generates coherent electromagnetic radiation.

With the magnetic field effect being taken into account, the rest-frame wavelength is expressed as [101]

$$\lambda = \frac{\lambda_w}{2\gamma^2}\left[1+\kappa^2\right], \tag{4.50}$$

where

$$\kappa = \frac{eB\lambda_w}{2\pi m_e c} \tag{4.51}$$

is the wiggler parameter. In order to balance the need to maximize the radiation power and to constrain the wavelength, κ is usually close to unity. Given that the electron energy $\gamma m_e c^2$ is varied from 1 to 5 MeV and λ_w is 1 cm, the frequency $\nu = c/\lambda$ is tunable in the range of 0.5-3 THz.

Interacting with the wiggler field and the radiation, the electron beam is bunched, where the modulation period matches the laser wavelength. This bunching effect is vital to the radiation being coherently amplified. Making multiple round trips in the laser cavity, the laser beam is amplified through the synchronization between the radiation field and the electron oscillation.

4.8.2 Free Electron Laser Facilities

A typical wiggler array in a THz FEL consists of a few hundred magnets extending several meters. The total length of the resonator is on a similar scale. A typical beam diameter is a few milimeters. The bulkiest component of a FEL is the electron accelerator, whose size extends at least several tens of meters. In the THz regime, it is either a RF linac accelerator or an electrostatic accelerator. FELs operate in a pulsed mode: typical pulse duration, energy, and repetition rate are several microseconds, microjoules and hertz, respectively. Some new FELs are designed to produce a picosecond pulse train within a microsecond pulse. In this way, the peak power reaches the megawatt range, which is three orders of magnitude higher than that of pseudo-CW operation.

Table 4.3. THz Free-Electron Lasers

Location	Name	Frequency	Type
iFEL (Japan)	1, 4, 5	3-60 THz	linac
FOM (Netherlands)	FELIX 1,2	1.2-100 THz	linac
LURE-Orsay (France)	CLIO	2-100 THz	linac
FZ Rossendorf (Germany)	FELBE	1.5-20 THz	SC-linac
Stanford CA (USA)	FIREFLY	4-20 THz	SC-linac
UCSB CA (USA)	FIR-FEL	0.88-4.5 THz	electrostatic

http://sbfel3.ucsb.edu/www/fel_table.html

FELs are small scale particle accelerators, thus one takes up a huge space, and its construction and maintenance are expensive. All FELs are multi-user facilities maintained by sizable core groups. Table 4.3 lists some of the THz FELs open to outside researchers.

4.9 Thermal Detectors

Thermal detectors such as bolometers, Golay cells, and pyroelectric devices are commonly used for observation of CW THz radiation. A radiation absorber attached to a heat sink is the basic element of all thermal detectors. Radiation energy absorbed by the absorber is converted into heat, and a thermometer measures the temperature increase induced by the heat. The absorber has a low heat capacity so that the heat brings about acute temperature changes. Each type of thermal detector is distinguished by the specific scheme it uses to measure the temperature difference between the absorber and the heat sink. The absorbed radiation energy is determined by calibrating the measurement output.

We briefly describe the basic operational schemes of the most commonly used thermal detectors. A bolometer is equipped with an electrical resistance thermometer to measure the temperature of the radiation absorber. Usually, the thermometer is made of a heavily doped semiconductor such as Si or Ge, exploiting the fact that the resistance of such materials is susceptible to temperature. A bolometer is a cryogenic detector which operates at or below L-He temperature for high detection sensitivity. A pyroelectric detector exploits a pyroelectric material in which temperature changes give rise to spontaneous electric polarization, thus changing dielectric constant. A key component of this type of detector is a capacitor containing this pyroelectric material. A change in the detector temperature causes an electric charge to appear across the electrodes. The current flow neutralizing this bias is the measure used to determine the temperature variation. The radiation absorber of a Golay cell is a blackened thin metal film on a substrate. The heat is transferred to a small volume of gas in a sealed chamber behind the absorber so that pressure increases in the chamber. A reflective and flexible membrane is attached to the back side of the chamber, and an optical reflectivity measurement detects the membrane deformation induced by the pressure increase.

A favorable property of thermal detectors is that they respond to radiation over a very broad spectral range, which is unattainable for most photon detectors. On the other hand, thermal detectors are relatively slow compared to typical photon detectors, because the radiation absorber must reach thermal equilibrium before a temperature measurement can take place. The time constant of a typical bolometer is $\sim$0.1 ms at L-He temperature. Golay cells and pyroelectric detectors work at ambient temperature, but they are much slower than bolometers, with time constants on the order of 1 s.

Because of the strong influence of ambient temperature, detection of desired THz radiation requires special care to compensate for environmental effects. A common, yet powerful scheme to distinguish desired THz radiation from background signal is to modulate the intensity of the incident THz beam and measure the consequent changes in output signal.

4.9.1 Bolometers

A bolometer is a thermal detector that uses a material whose electric resistance is sensitive to temperature change [102]. In order to acquire high detection sensitivity, it is operated at or below L-He temperature. While a bolometer can measure radiation power over an extremely wide spectral range from THz to X-rays, this detection scheme is particularly important for measuring THz waves because it is the most sensitive one in their spectral range.

If a bolometer has separate components for key functions, the device is called a composite bolometer. Some bolometer designs use components performing multiple functions. Composite bolometers, however, show better performance in general, because each function of the components can be optimized independently. Figure 4.25 shows the key components and the operational schemes of a typical composite bolometer. The device consists of a

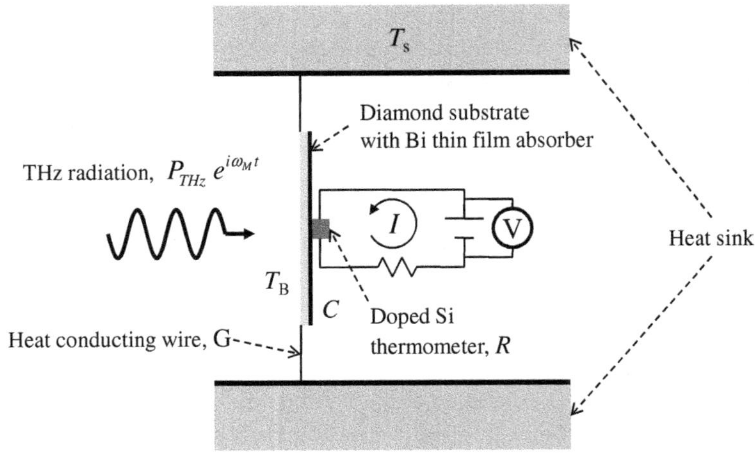

Fig. 4.25. Schematic diagram of a typical composite bolometer. Key components and their functions are illustrated. $P_{THz}e^{i\omega_M t}$ is the incident THz power modulated at the angular frequency ω_M. The thermometer is characterized by the temperature-dependent resistance R. C is the heat capacity of the absorber/substrate/thermometer unit. T_B is the substrate temperature. G is the dynamic thermal conductance of the heat conducting wire connecting the substrate to the heat sink. The bias V varies to keep the current I, flowing through the thermometer, constant.

radiation absorber deposited on a substrate and a thermometer whose electrical resistance varies according to the substrate temperature. The substrate is connected to a heat sink by heat conducting wires. The heat sink includes a L-He dewar to keep the detector at cryogenic temperature. Radiation energy absorbed by the absorber is converted into heat, which raises the temperature of the absorber and thermometer assembly. The temperature increase induces a change in the thermometer resistance, which is detected by measuring the change of the electrical signal across it.

Figure 4.26 shows a simple circuit used to measure the thermally induced current. The radiation absorber functions as one of the resistors of the balanced Wheatstone bridge. When the radiation induces a change in the absorber resistance, the bridge becomes out of balance and current flows through the galvanometer. Some bolometers include another matching absorber in the opposite arm of the bridge. It is shielded from the incident radiation to compensate for ambient effects.

An important characteristic of a bolometer is the detector responsivity R_V, which is defined as the voltage change across the thermometer upon incident radiation with one unit of power, thus its SI unit is V/W. The responsivity is derived by balancing the energy flow passing through the thermometer. The incident THz radiation $P_{THz}e^{i\omega_M t}$ is modulated at the angular frequency ω_M, thus the bolometer temperature T_B also oscillates at the same frequency: $T_B =$

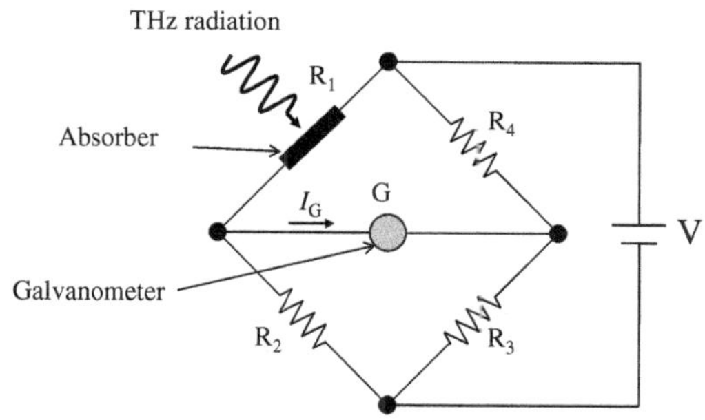

Fig. 4.26. Bolometer circuit with a balanced Wheatstone bridge and a galvanometer.

$T_S + T_1 e^{i\omega_M t}$, where T_S is the temperature of the heat sink. For simplicity, we assume that there is no background signal. The absorbed radiation power and the power stored in the absorber/substrate/thermometer unit is equal to the power loss through the conducting wire [102]:

$$P_{THz} e^{i\omega_M t} + \frac{d}{dt}\left(I^2 R\right) = G T_1 e^{i\omega_M t} + \frac{d}{dt}\left(C T_1 e^{i\omega_M t}\right). \tag{4.52}$$

The varying bias V keeps the current constant. Because the temperature change is small, we can approximate the resistance as

$$R(T) \simeq R(T_S) + \left(\frac{dR}{dT}\right)_{T=T_S} T_1 e^{i\omega_M t}. \tag{4.53}$$

From the Eqs. 4.52 and 4.53 we get

$$\frac{P_{THz}}{T_1} = G + i\omega_M C - I^2 \left(\frac{dR}{dT}\right)_{T=T_S}. \tag{4.54}$$

By definition the responsivity is written as

$$R_V = \frac{V}{P_{THz}} = \frac{I T_1}{P_{THz}} \left(\frac{dR}{dT}\right)_{T=T_S}. \tag{4.55}$$

Using Eq. 4.54 we rewrite Eq. 4.55 as

$$R_V = \frac{I \left(\frac{dR}{dT}\right)_{T=T_S}}{G - I^2 \left(\frac{dR}{dT}\right)_{T=T_S} + i\omega_M C}. \tag{4.56}$$

A useful parameter for characterizing the thermometer is

$$\alpha = \frac{1}{R}\left(\frac{dR}{dT}\right)_{T=T_S} \tag{4.57}$$

Incorporating the influence of the temperature dependence of R we define the effective thermal conductance as

$$G_e = G - I^2 \left(\frac{dR}{dT}\right)_{T=T_S} = G - \alpha I^2 R. \tag{4.58}$$

The temporal behavior of temperature rise and decay is characterized by the thermal time constant, $\tau_{Th} = C/G_e$. Using these parameters we rewrite Eq. 4.56 as

$$R_V = \frac{\alpha I R}{G_e(1 + i\omega_M \tau_{Th})} \tag{4.59}$$

and

$$|R_V| = \frac{\alpha I R}{G_e(1 + \omega_M^2 \tau_{Th}^2)^{1/2}}. \tag{4.60}$$

Responsivity of a typical Si bolometer is $\sim 10^7$ V/W at L-He temperature. The noise equivalent power (NEP), which is the radiant power needed to generate signal equivalent to noise, is $\sim 10^{-14}$ W·Hz$^{-1/2}$.

The theoretical analysis above sheds light on the favorable properties of the bolometer components to achieve high detection sensitivity. A radiation absorber of good quality must have nearly flat spectral responsivity over a broad spectral range with low reflectivity and high absorptivity. Low heat capacity is also an indispensable property. Thermal properties of the substrate are critical: high thermal conductivity and low heat capacity are preferable. A typical, and highly efficient, absorber/substrate assembly contains a thin metal film deposited on a thin crystalline dielectric substrate. Bi is widely used for the absorbing material because of its relatively low conductivity, high absorptivity, and very small heat capacity. The low conductivity is beneficial because of the practical reason that a Bi film used for efficient absorption is relatively thick, thus it is easy to keep the film thickness uniform. Diamond is considered to be the best material for the substrate because of its excellent thermal conductivity and high transparency in the THz region below 30 THz. Sapphire also has a very good thermal conductivity, and is transparent below 10 THz. The thermometer is thermally attached to the absorber/substrate assembly. Low heat capacity and electrical noise are preferable properties. Response of its electrical resistance to temperature change must be consistent and adequate for desired operating conditions. Heavily doped Si or Ge are the most widely used materials for the thermometer. The doping brings the material close to the metal-insulator transition with a majority impurity and a compensating minority impurity. Then, the electrical resistance is not only low enough at cryogenic temperature, but also sensitive to minute temperature changes. Resistance of doped semiconductors has the following temperature dependence [102]:

$$R = R_0 \exp\left(\sqrt{\frac{A}{T}}\right), \tag{4.61}$$

where R_0 and A are constants.

In order to maximize detection sensitivity, the absorber area needs to be minimized. In free space, it is hard to focus an incident THz beam on a small spot because of its relatively long wavelength compared to the size of the absorber. Most bolometers employ multi-mode waveguides to improve coupling to the incident THz waves. The Winston light concentrator is widely used because the solid angle is maximized ($\Omega = \pi$) at the exit aperture. Figure 4.27 illustrates a waveguide structure comprising a series of the Winston concentrator: cone a is used to accept the input and to reject background, cone b is used to increase the filtering efficiency, and cone c is used to collimate the beam onto the absorber [102].

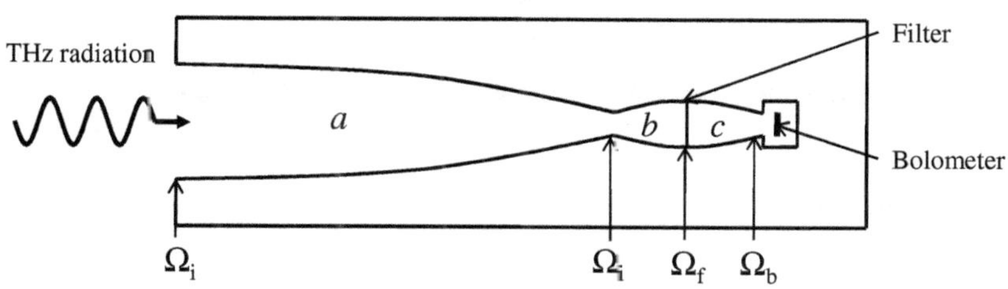

Fig. 4.27. Bolometer waveguide structure to collimate the incident radiation and limit background signal.

4.9.2 Pyroelectric Detectors

A pyroelectric crystal, also called a polar crystal, is spontaneously polarized since each unit cell of the crystal has a permanent electric dipole moment aligned with a specific crystal axis. Actually, spontaneous polarization is a fundamental property of many crytalline systems. Ten of the 32 crystal classes are pyroelectric. If an external electric field can reverse the dipole, the material is said to be ferroelectric. All ferroelectric materials are pyroelectric, but not the other way around. The spontaneous electric polarization in a pyroelectric material is susceptible to changes in temperature, and this physical phenomenon is called pyroelectricity.

Commonly used materials for pyroelectric detectors are triglycine sulfate (TGS), deuterated triglycine sulfate (DTGS), lithium tantalate ($LiTaO_3$), and barium titnate ($BaTiO_3$). Table 4.4 lists the pyroelectric coefficients and Curie temperatures of these materials. TGS and DTGS detectors have relatively high sensitivity at THz frequencies compared to other pyroelectric detectors.

The spontaneous polarization is accompanied by surface charge, which is neutralized by free carriers forming a steady state. Due to the charge balancing act, a piece of pyroelectric material normally shows no macroscopic dipole, unlike a bar magnet. The charge balancing, however, is the key process for

Table 4.4. Pyroelectric coefficients and Curie temperatures

Crystal	p (nC/cm^{2}K)	T_C (°C)
[a]TGS	33-40	49
[a]DTGS	25-30	57-62
[b]$LiTaO_3$	19	620
[b]$BaTiO_3$	20	135

[a]http://www.girmet.com/tgs_dtgs.htm
[b]Ref. [103]

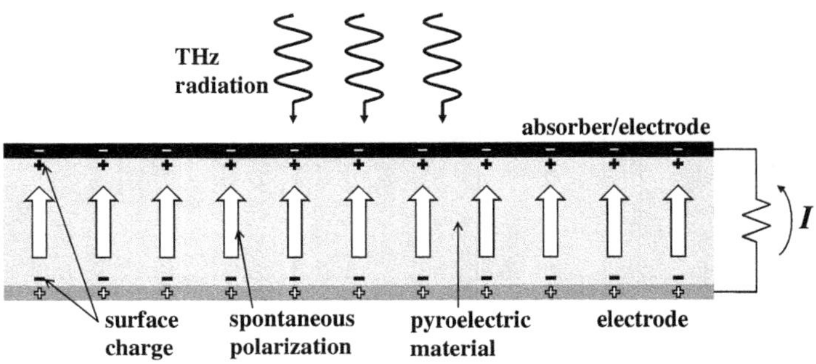

Fig. 4.28. Schematic diagram of a typical pyroelectric detector.

detecting THz radiation. Figure 4.28 illustrates the basic scheme of a pyroelectric detector. A pyroelectric crystal cut perpendicular to its polar axis is sandwiched between two electrodes [104]. The top one is usually blackened to be a radiation absorber. If the electrode is transparent in the desired spectral region, the crystal surface itself is treated to absorb incident radiation. The heat generated by incident radiation raises the temperature of the pyroelectric crystal. The increase in temperature induces a reduction of the spontaneous polarization and, simultaneously, the surface charge. The electrodes attached to the two opposite crystal surfaces form a capacitor. If the circuit is closed, a current flows through it to compensate the change in the surface charge. Circuits in real detectors are a little more complicated: the pyroelectric capacitor functions as either a voltage or current source. Figure 4.29 shows circuit diagrams for a voltage and a current signal. In order to get high responsivity The load resistance R is usually very large: 10^{10} - 10^{12} Ω.

The surface charge density is related to the polarization by

$$\sigma_b = \mathbf{P} \cdot \mathbf{n}, \tag{4.62}$$

where $\mathbf{P}$ is the polarization and $\mathbf{n}$ is a unit vector normal to the surface. On the detector surface, a net current I is induced by a change in the polarization,

$$I = \frac{dQ}{dt} = \frac{d(\sigma_b A)}{dt} = A\left(\frac{dP}{dt}\right), \tag{4.63}$$

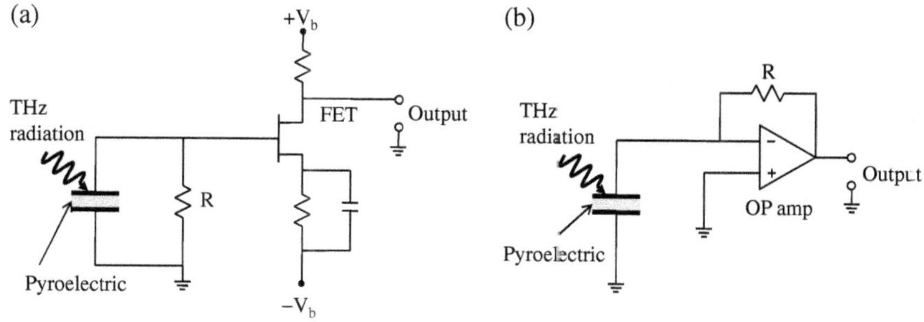

Fig. 4.29. Detection circuits for (a) voltage and (b) current signal. R is the load resistance. FET: field-effect transistor, OP amp: operational amplifier.

where A is the surface area. The polarization current density, therefore, is

$$\mathbf{J}_p = \frac{\partial \mathbf{P}}{\partial t} \tag{4.64}$$

When the change in polarization is produced by a change in temperature, we can rewrite Eq. 4.63 as

$$I = A\left(\frac{dP}{dT}\right)\left(\frac{dT}{dt}\right) = pA\left(\frac{dT}{dt}\right), \tag{4.65}$$

where $p = (dP/dT)$ is the pyroelectric coefficient, indicating how sensitive the material polarization is to a change in temperature [103, 105]. Since the pyroelectric current is too small for direct measurement, practical detectors use high impedance amplifiers.

Pyroelectric detectors are used in AC mode because no pyroelectric current is generated in a steady state. When the incident THz power is modulated at ω_M, thus undulating the corresponding temperature change at the same frequency, the thermodynamic equation balancing the power absorption and dissipation can be written as

$$\begin{aligned} P_{THz} e^{i\omega_M t} &= G T_1 e^{i\omega_M t} + \frac{d}{dt}\left(C T_1 e^{i\omega_M t}\right) \\ &= \left(G + i\omega_M C\right) T_1 e^{i\omega_M t}, \end{aligned} \tag{4.66}$$

which leads to the time dependent temperature

$$T(t) = T_1 e^{i\omega_M t} = \frac{1}{G(1 + i\omega_M \tau_{Th})} P_{THz} e^{i\omega_M t}, \tag{4.67}$$

where C and G are the heat capacity and the thermal conductance of the detector, and $\tau_{Th} = C/G$ is the thermal time constant. We assume perfect absorption of the incident power. From Eq. 4.65 and 4.67 we obtain the current modulation

$$I(t) = \frac{i\omega_M pA}{G(1+i\omega_M\tau_{Th})} P_{THz} e^{i\omega_M t}, \tag{4.68}$$

The current source is in parallel with its own capacitance and resistance. With the parallel capacitance C_E and resistance R of the detection circuit, the detector responsivity can be expressed as

$$\begin{aligned} R_V &= \frac{V(t)}{P_{THz}(t)} = \frac{I(t)R}{P_{THz}(t)\,(1+i\omega_M RC_E)} \\ &= \frac{ip\omega_M RA}{G(1+i\omega_M\tau_{Th})\,(1+i\omega_M\tau_E)}, \end{aligned} \tag{4.69}$$

thus

$$|R_V| = \frac{p\omega_M RA}{G\left(1+\omega_M^2\tau_{Th}^2\right)^{1/2}\left(1+\omega_M^2\tau_E^2\right)^{1/2}}, \tag{4.70}$$

where $\tau_E = RC_E$ is the RC constant of the circuit. Typical responsivity and NEP of a DTGS detector at a modulation frequency of ~10 Hz are ~1 kV/W and ~10^{-9} W·Hz$^{-1/2}$, respectively.

4.9.3 Golay Cells

A Golay cell is a sensitive pneumatic radiation detector for a broad radiation spectrum from millimeter waves to the near infrared [106]. It can sense radiation power levels down to 10 μW at THz frequencies. Figure 4.30 shows the basic components and operation of a typical Golay cell. Modulated THz radiation passes through the front window and is absorbed by the absorbing film. The absorbed energy heats up a small volume of gas enclosed in the pneumatic chamber, which induces thermal expansion of the gas. The resulting pressure rise deforms the flexible mirror attached to the backside of the chamber. An optical beam from a light emitting diode is focused on the flexible mirror. The reflected beam is collected and focused onto a photodetector. The deflection caused by the membrane deformation is sensed by the detection readout system.

High detection sensitivity of a Golay cell requires several conditions of its components. Ideally, the absorbing film should be the only medium experiencing heat exchange. To do this, materials for the window and the pneumatic chamber must be heat insulators of high quality, and the gas should be transparent over the whole detection spectrum. The heat conductivity of the gas should also be small. Xenon is a commonly used gas, because it is one of the least conductive gases available.

The Golay cell is the most sensitive detector among thermal radiation detectors which operates at room temperature. Responsivity of a Golay cell is in the range of kV/W when the modulation frequency is a few tens of hertz. Typically obtained NEP is 0.1-1 nW·Hz$^{-1/2}$. It is no surprise that the original design has changed little since its first development 60 years ago. A notable

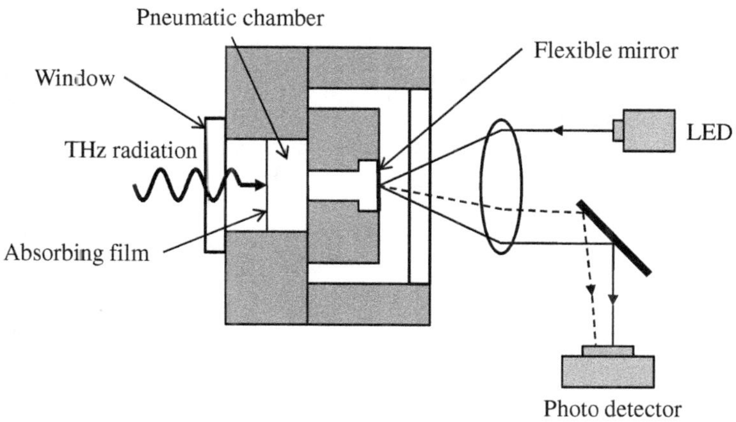

Fig. 4.30. Schematic diagram of a Golay Cell.

and practical disadvantage of the old design, however, is that the detector size is relatively large. Recent efforts to defeat this shortcoming have resulted in a few miniaturized Golay cells being fabricated by micromachining. Their detection schemes are based on either electric capacity variations [107] or tunneling displacement transducers [108, 109].

4.10 Heterodyne Receivers

Heterodyne detection is based on frequency downconversion in a nonlinear device, accomplished by mixing signal with reference radiation at a fixed frequency. The basic principle of this technique is analogous to that of photomixing (section 4.1) and difference frequency generation (section 4.2). We assume that the nonlinear device (a.k.a. "mixer") is characterized by a quadratic nonlinearity. For signal and reference radiation given as $E_S\cos(\omega_S t)$ and $E_{LO}\cos(\omega_{LO}t)$, the output signal contains the five frequency components:

$$\begin{aligned} V_O &= \chi\left[E_s\cos(\omega_S t) + E_{LO}\cos(\omega_{LO}t)\right]^2 \\ &= \frac{1}{2}\chi\left(E_s^2 + E_{LO}^2\right) \\ &\quad + \frac{1}{2}\chi E_S^2\cos(2\omega_S t) \\ &\quad + \frac{1}{2}\chi E_{LO}^2\cos(2\omega_{LO}t) \\ &\quad + \frac{1}{2}\chi E_S E_{LO}\cos[(\omega_S + \omega_{LO})t] \\ &\quad + \frac{1}{2}\chi E_S E_{LO}\cos[(\omega_S - \omega_{LO})t], \end{aligned} \tag{4.71}$$

where χ is the quadratic nonlinear coefficient. The heterodyne detector filters out the first four components and measures the last term of the difference frequency (a.k.a. "intermediate frequency"), $\omega_D = |\omega_S - \omega_{LO}|$. The rationale of the frequency downconversion technique is that the output signal at ω_D in the radio frequency or microwave region, much lower than ω_S and ω_{LO}, can be readily measured by electronic devices such as spectrum analyzers.

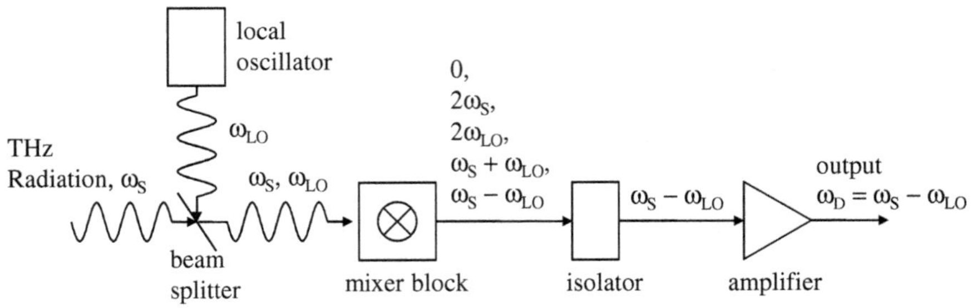

Fig. 4.31. Block diagram of signal flow in a heterodyne detector

Figure 4.31 illustrates the signal flow of typical heterodyne detection. One of the key components is the local oscillator (LO), which provides the reference radiation. The LO output power is a crucial factor in determining the detector performance, as V_O is proportional to E_{LO}. Solid-state emitters have commonly been used as LOs in the region of 0.1-1 THz, and gas lasers above 1 THz. Because of their compactness and high power, quantum cascade lasers are a promising THz LO for future applications. An advantage of heterodyne detection is that the spectrum of the incident light can be resolved by scanning the LO output frequency and measuring the signal with a spectrum analyzer. If the bandwidth of the incident radiation is within the LO tuning range, the whole spectrum can be measured with a spectral resolution largely determined by the LO linewidth.

Another key component of heterodyne detection is a mixer with nonlinear characteristics. Schottky diodes are commonly used as mixers in the spectral range below 1 THz. Schottky mixers have relatively low sensitivity and require at least several milliwatts of LO power. Above 1 THz, the most sensitive mixers are cryogenic detectors such as hot-electron bolometers [111, 112, 110]. Due to their high sensitivity, hot-electron bolometers require very low LO power ($\sim$10 nW). A commonly used hot-electron bolometer is an indium antimonide (InSb) detector, characterized by the thermal relaxation time of $\sim$1 ms and the noise-equivalent power (NEP) of $\sim 10^{-12}$ W$\cdot$Hz$^{-1/2}$ [111]. More recently, niobium (Nb) and niobium nitride (NbN) in a superconducting phase have been utilized as detection materials for hot-electron bolometers [112, 110]. Figure 4.32 shows an integrated hot-electron bolometer mixer consisting of a superconducting NbN microbridge. For it to have a large detection bandwidth,

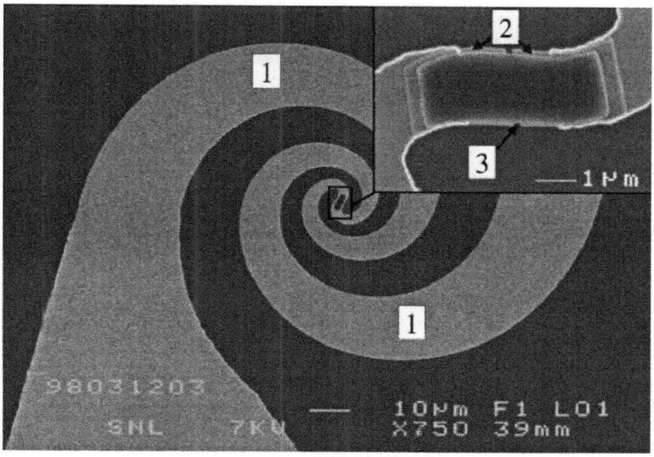

Fig. 4.32. Micrograph of an integrated mixer based on a NbN superconducting hot-electron bolometer. The insert shows the magnified image of the detection area: the NbN microbridge (2) between contact pads (3) of the Au spiral antenna (1). (Reprinted with permission from [110]. ©1998 American Institute of Physics.)

the thermal relaxation time of the detector must be short. The bolometers using Nb and NbN utilize different cooling mechanisms. A thin NbN film is characterized by a very short thermal relaxation time (<10 ps), thus NbN bolometers utilize an intrinsic phonon cooling mechanism. Since the thermal relaxation time of Nb is much longer, Nb bolometers use a very short and thin Nb strip to facilitate diffusion cooling. For both of the superconducting bolometers, the thermal response time reaches down to several 10's of picoseconds. Consequently, the detection bandwidth of the superconducting hot-electron bolometers is in the range of several gigahertz.

5

Terahertz Optics

Having learned how to generate and detect THz radiation, we are ready to discuss how to harness and manipulate THz waves. This chapter is devoted to materials, devices, and physical phenomena, which can be utilized in various THz applications.

5.1 Dielectric Properties of Solids in the Terahertz Region

THz optoelectronic devices bridge the technological gap inaccessible by RF and microwave techniques and infrared spectroscopy. The THz gap is also an intermediate spectral region for the optical properties of dielectrics. The Drude mechanism and Debye relaxation are dominant processes governing dielectric properties of solids in the low frequency region below the microwave band. These physical processes are probed by RF and microwave devices which measure the dielectric constants of materials. The low-frequency effects diminish in the THz region as frequency increases. On the higher frequency side, the mid-IR range, optical properties of dielectrics are overwhelmed by lattice vibrations, or more precisely, their IR-active quantized normal modes, optical phonons. Studying lattice vibrations is a major subject of infrared and Raman spectroscopy. The THz region is usually out of optical phonon resonances, but the low-energy tail of the spectrum is a major source of absorption in this spectral range. The dielectric response of vibrational modes decays as frequency decreases in the THz region.

The Drude mechanism accounts for the transport properties of free carriers in a material. The electrical conductivity derived from the Drude model is expressed as

$$\sigma(\omega) = \frac{\sigma_0}{1 - i\omega\tau} \tag{5.1}$$

with the static conductivity

$$\sigma_0 = \frac{n_q q^2 \tau}{m_q}, \tag{5.2}$$

where n_q is the charge density, m_q is the particle mass, and τ is the relaxation time. Thus, the electric permittivity of the material is given as

$$\epsilon(\omega) = \epsilon_0 + i\frac{\sigma(\omega)}{\omega} = \epsilon_0 + \frac{i\sigma_0}{\omega(1 - i\omega\tau)}. \tag{5.3}$$

From this equation we can see that the relaxation time is the sole parameter governing the frequency-dependent optical properties. The relaxation time of a high-quality intrinsic semiconductor is in the vicinity of a picosecond at room temperature. The time scale is a few orders of magnitude shorter in most dielectric media. It is noteworthy that the inverse of the time scale falls into the THz range.

Dielectric relaxation, also called Debye relaxation, refers to the delayed response of a dielectric medium to applied electric fields. The momentary delay of the response is accounted for by random thermal fluctuations which slow down the reorientations of the dipole moments in the material. The simplest model to describe the relaxation process is expressed as the Debye equation

$$\epsilon(\omega) = \epsilon(\infty) + \frac{\epsilon(0) - \epsilon(\infty)}{1 - i\omega\tau_D}, \tag{5.4}$$

where the frequency-dependent electric permittivity is determined by the static permittivity $\epsilon(0)$, the high-frequency permittivity $\epsilon(\infty)$, and the Debye relaxation time τ_D. The Debye relaxation time varies widely depending on the material system. Typical time scales range from microseconds to nanoseconds at room temperature.

Usually the lowest optical phonon resonance of a dielectric crystal is in the vicinity of 10 THz. A simple approximation of the dielectric response of lattice vibrations is the harmonic oscillator model described in section 2.1.4. The electric permittivity derived from the model is expressed as

$$\epsilon(\omega) = \epsilon_L(0) + \frac{f_L}{\omega_L^2 - \omega^2 - i\omega\gamma_L}, \tag{5.5}$$

where $\epsilon_L(0)$ is the static permittivity of lattice vibrations, ω_L is the resonance frequency, f_L is the oscillator strength, and γ_L is the damping constant.

These three microscopic processes, the Drude mechanism, dielectric relaxation, and lattice vibrations, are dominant absorption processes of solids in the THz region. Figure 5.1 depicts the typical dielectric response at THz frequencies. The effects of the three absorption mechanisms are shown in the form of the refractive index

$$n(\omega) = \Re[\sqrt{\epsilon_r(\omega)}] \tag{5.6}$$

and the absorption coefficient

$$\alpha(\omega) = \frac{2\omega}{c}\Im[\sqrt{\epsilon_r(\omega)}] \tag{5.7}$$

in a log scale as functions of frequency.

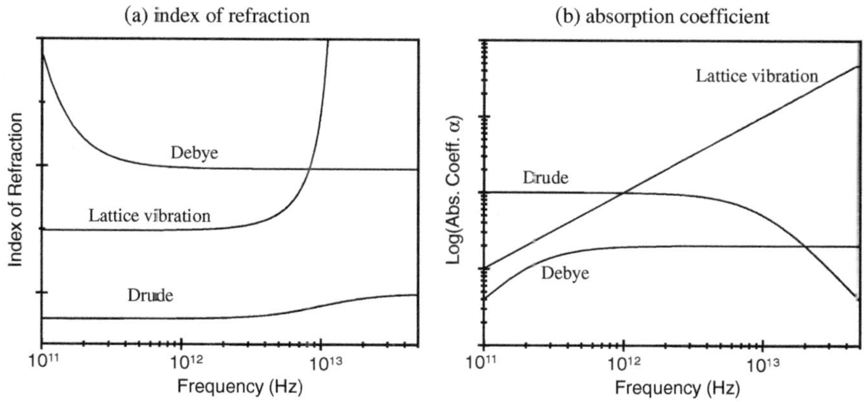

Fig. 5.1. Frequency-dependent refractive index and absorption coefficient of a typical dielectric medium in the THz region

5.2 Materials for Terahertz Optics

The characteristic optical properties of solids in the THz region depend on different physical mechanisms from those in other spectral ranges. Free-carrier effects, in particular, are relatively strong; phonon resonances make materials opaque in this spectral range. Ordinary glasses commonly used in optical regions are useless for THz applications because extrinsic dielectric losses from charged defects are too high. Some material types, however, are highly transparent at THz frequencies. Transmissive THz materials include polymers, dielectrics, and semiconductors. Polymers such as polyethylene, Teflon (PTFE), and TPX are transparent and almost dispersionless at THz frequencies. Their absorption coefficients are less than 0.5 cm^{-1} at 1 THz, and show nearly quadratic increases with frequency. The average refractive indices vary little among the polymers ranging from 1.4 to 1.5. Commonly used dielectrics and semiconductors are silicon, germanium, gallium arsenide, quartz, fused silica, and sapphire. Silicon is the most transparent and the least dispersive material among the dielectrics as well as the polymers. The absorption coefficient of a high-purity crystal is less than 0.1 cm^{-1} below 3 THz, and the variation of the refractive index, 3.4175, is less than 0.0001 in the same range. The effects of free carriers and lattice vibrations in other dielectrics and semiconductors are far greater than those in silicon. Basic optical components such as windows and lenses are made of transmissive THz materials. Polished metal surfaces or metal coating mirrors are commonly used as THz reflectors. Typical reflectivity is 98-99% in the THz region.

5.2.1 Polymers

The polyethylene family, including high-density polyethylene (HDPE) and low-density polyethylene (LDPE), is the most widely used polymer group because of its excellent clarity. HDPE and LDPE also have other favorable properties: they are isotropic, easily machinable, and chemically stable. High-quality polyethylene has a sharp lattice mode at 2.2 THz with a bandwidth of 0.2 THz. Although this shows its high degree of ordered structure, the absorption line is problematic for applications near that frequency. The refractive index of HDPE, 1.526, is slightly larger than that of LDPE, 1.513. PTFE is highly transparent on the lower frequency side of the spectrum, yet its absorption is a little higher than that of polyethylene. It becomes significantly lossy above 3 THz. The refractive index is 1.432 below 3 THz. Its high resistance to corrosion and adsorption makes this material suitable for chemical and biological studies. Polypropylene (PP) is another transmissive material, though it is not as widely used as other polymers. Its absorption spectrum is similar to that of PTFE below 3 THz. It has multiple lattice modes above that frequency. The average refractive index is 1.498 on the low frequency side. TPX, polyolefine based on poly 4 methyl pentene-1, is transparent not only at THz frequencies, but also in the visible spectral range. Its THz refractive index, 1.457, is also close to its optical counterpart. This distinctive advantage has been exploited in many quasi-optical systems. In the THz region its absorption coefficient is slightly higher than that of polyethylene, but

Table 5.1. Optical Constants of Polymers

Polymer	n	α (cm^{-1}) at 1 THz
LDPE[a,c]	1.51	0.2
HDPE[a,c,f,g]	1.53	0.3
PTFE[a,d,g]	1.43	0.6
PP[a,d,g]	1.50	0.6
TPX[a,b]	1.46	0.4
Tsurupica[h]	1.52	0.4

[a]reference [113]
[b]reference [114]
[c]reference [115]
[e]reference [116]
[f]reference [117]
[g]reference [118]
[h]http://www.mtinstruments.com/thzlenses/index.htm

lower than that of PTFE. It is more mechanically rigid than other polymers. Tsurupica, formerly known as picarin, is a relatively new material developed at RIKEN in Japan. Tsurupica is also highly transparent in both the THz and visible spectral ranges. The THz refractive index is 1.52, which is almost the same as for visible light. Mechanically, it is strong enough to withstand optical polishing.

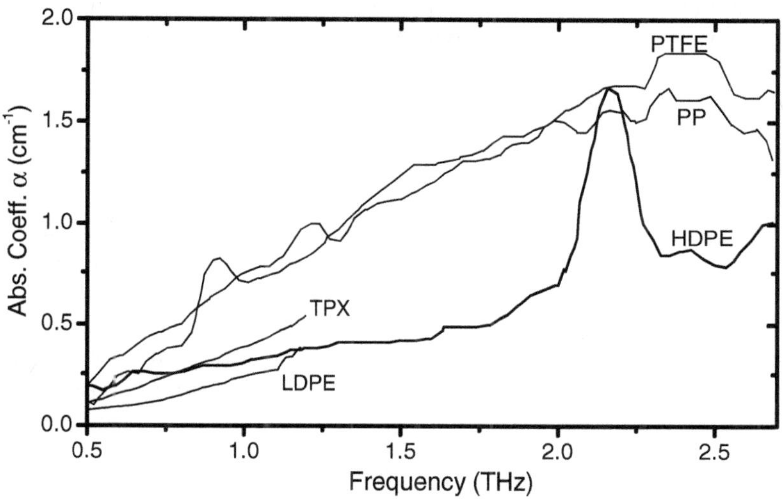

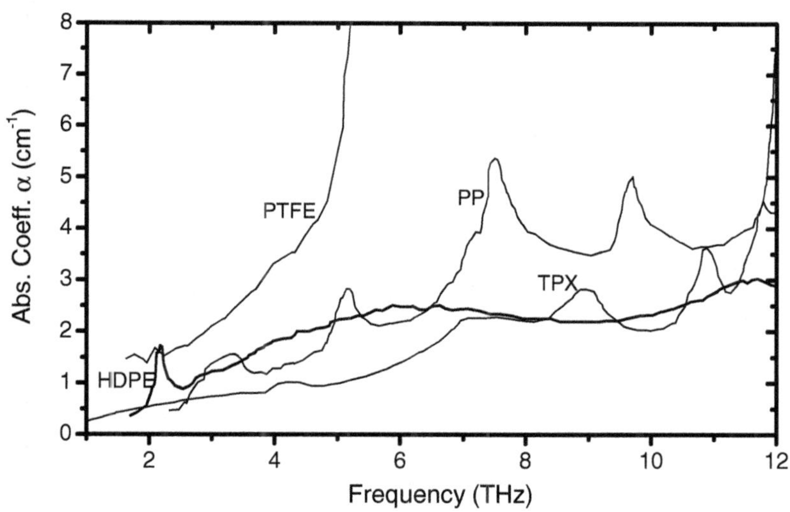

Fig. 5.2. Absorption coefficients of the polymers versus frequency in the two spectral ranges of (a) 0.5-2.5 THz and (b) 2-12 THz. (Data from Refs. [113, 114, 115, 116, 117, 118])

The average THz refractive indices over the range 0.5-3 THz and the absorption coefficients at 1 THz are given in Table 5.1 for the aforementioned polymers. The details of the frequency-dependent absorption are shown in Fig. 5.2 for two spectral windows, (a) the low frequency side, 0.5-2.5 THz, and (b) the high frequency side, 2-12 THz.

5.2.2 Dielectrics and Semiconductors

Crystalline silicon, whose optical properties in the THz region haven been studied more extensively than any other material, is not only the most transparent but also the least dispersive medium in that spectral range. Moreover, its mechanical and electrical properties are known in great detail, its price is low, high-quality crystals are readily available, and there are numerous fabrication techniques specified for the material. Silicon is arguably the most important raw material for THz device development.

Crystals of high symmetry are inactive in the infrared region because first-order transitions between lattice vibration modes are forbidden by symmetric selection rules. The lack of first-order absorption in silicon can be simply understood by the symmetric charge distribution in the crystal structure. Crystalline silicon, composed of one element, has no dipole moment to be coupled with external electric fields. Thus, absorption involving lattice vibrations

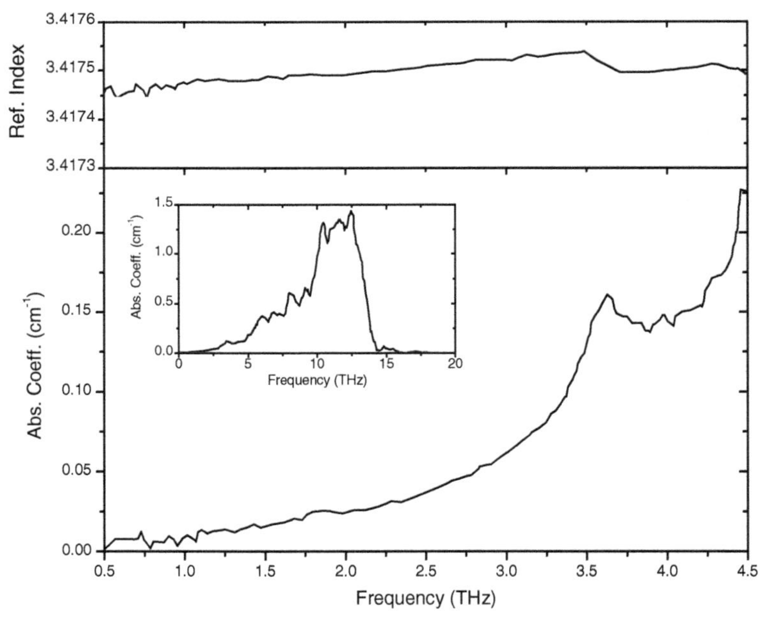

Fig. 5.3. Dispersion and absorption spectra of float-zone, high-resistivity silicon. Calculated two-phonon absorption in the infrared region is shown in inset. (Data from Refs. [119, 120])

should be dominated by second-order (two-phonon) processes. Its two-phonon absorption coefficient is estimated to be less than 0.1 cm^{-1} below 3 THz [120]. On the low frequency side, the experimental data on absorption of commonly used silicon wafers cannot be explained by two-phonon contributions alone. The main culprit causing the additional absorption is free carriers from defect sites. Measurements on high-resistivity silicon samples show that absorption is proportional to conductivity, which is linearly dependent on carrier concentration, in the spectral range where two-phonon absorption is negligible.

High-purity silicon, produced using the float-zone crystal growth method, has a remarkably low carrier concentration ($< 4 \times 10^{11}$ cm^{-3} for n-type) and high resistivity (>10 kΩ-cm). Figure 5.3 shows the refractive index and the absorption spectrum of float-zone, high-resistivity silicon in the spectral range 0.5-4.5 THz [119]. The spectral feature, including the peak near 3.6 THz, is consistent with the calculated two-phonon absorption shown in inset [120].

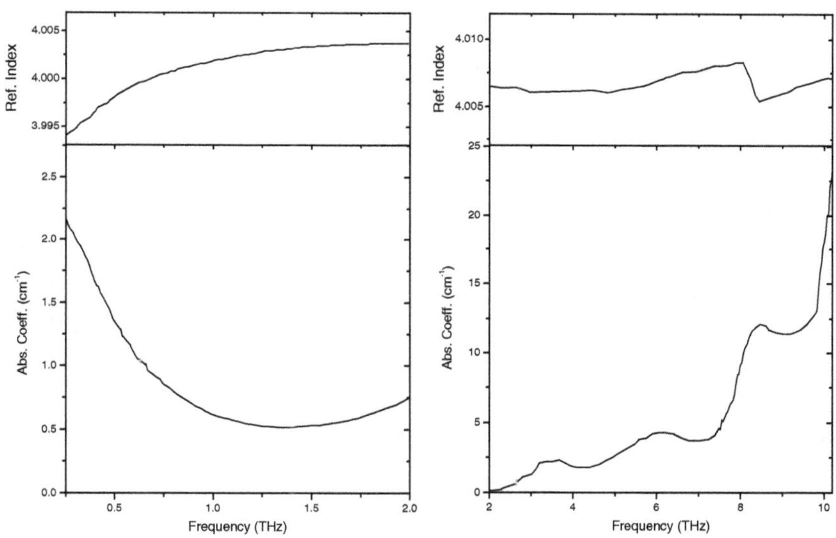

Fig. 5.4. Dispersion and absorption spectra of intrinsic germanium in the spectral ranges 0.2-2 THz and 2-10 THz. (Data from Refs. [23, 121].

As germanium has the same crystal structure as silicon, no first-order absorption is allowed. A reasonable yet casual guesswork may speculate that crystalline germanium should have similar optical properties as silicon in the THz region. A closer look, however, defies this prediction. Intrinsic germanium has a relatively small bandgap energy (0.66 eV). Consequently, the intrinsic carrier concentration at room temperature, 2×10^{13}, is significantly higher than that of silicon, and the resistivity, 46 Ω-cm, is relatively low. The carrier concentration is high enough that the Drude mechanism dominates the absorption of germanium in the THz region: the relaxation times for the electrons and the holes are 0.6 and 0.7 THz, respectively. The left panel of Fig. 5.4 shows the dispersion and absorption spectra of an intrinsic germanium crystal [23]. The spectral features are consistent with the Drude model discussed in section 5.1. The optical constants in the spectral range 2-10 THz are shown in the right panel of Fig. 5.4 [121]. Multi-phonon resonances are discernible at 3.5, 6.0, and 8.5 THz. The Drude mechanism is negligible in this spectral region.

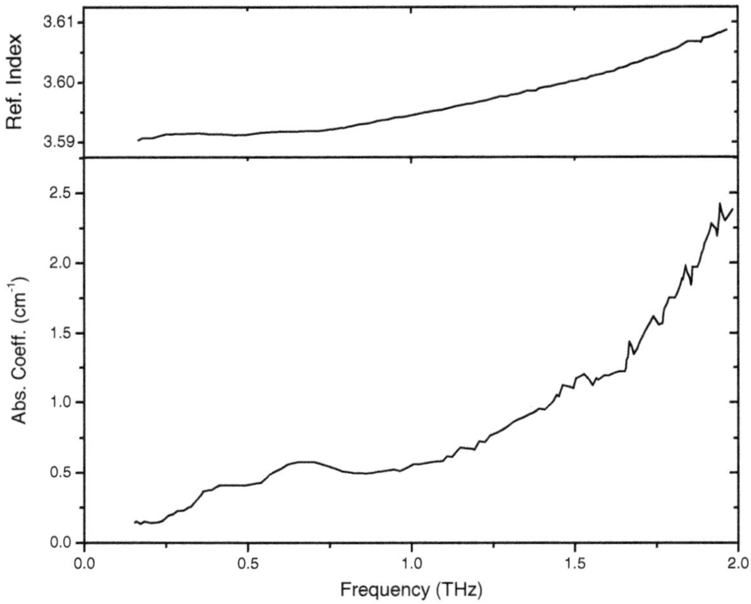

Fig. 5.5. Dispersion and absorption spectra of crystalline GaAs. (Data from Ref. [23])

The refractive index and absorption spectrum of gallium arsenide (GaAs) are shown in Fig. 5.5 [23]. The highly refined growth techniques of crystalline GaAs routinely produce high-resistivity (>10 MΩ-cm) crystals with very low carrier concentrations. The primary THz absorption mechanism in GaAs involves the optical phonon mode at 8.1 THz which has a large density of states due to the flat band structure. First-order absorption is allowed in the III-V material because the charge distribution between the two elements of the crystal is asymmetric, and the consequent dipole moments are directly coupled with an applied electric field. The phonon resonance is strong enough to give rise to the gradual increase of absorption as shown in Fig. 5.5, even though the spectral range is far from the resonant frequency. The two small peaks featured at 0.4 and 0.7 THz are accounted for by a two-phonon process in which an optical phonon is created and a LA phonon is annihilated [122].

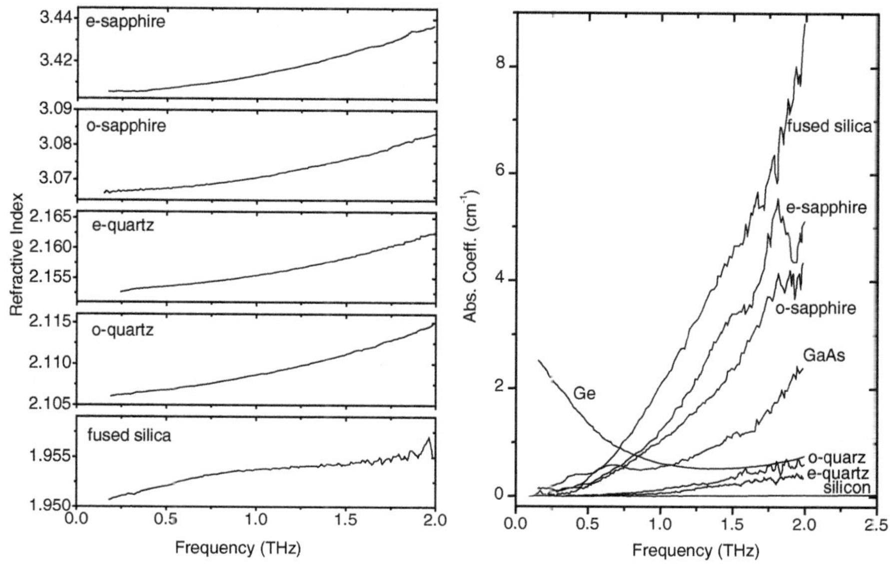

Fig. 5.6. Dispersion and absorption spectra of sapphire, quartz, and fused silica in the spectral range 0.2-2 THz. (Data from Ref. [23])

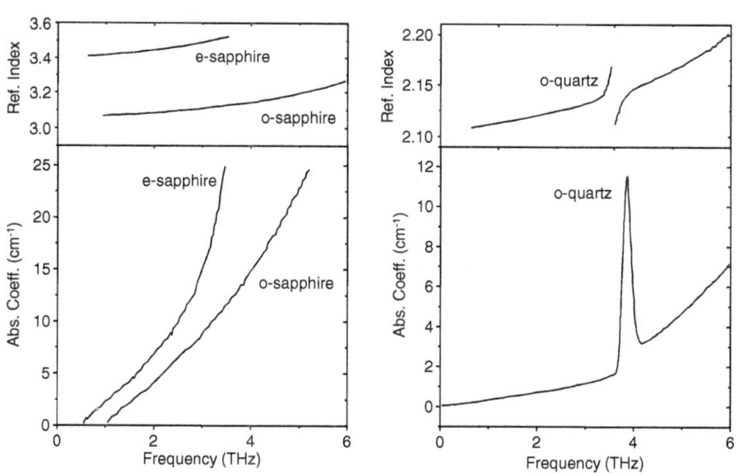

Fig. 5.7. Dispersion and absorption spectra of sapphire and quartz in the spectral range 1-6 THz. (Data from Ref. [121])

Absorption of crystalline sapphire, crystalline quartz, and fused silica is substantially higher than that of silicon in the THz region, but their high transmittance in the visible range is a useful property for some applications. The dispersion and absorption spectra for the dielectrics in the spectral range 0.2-2 THz are shown in Fig. 5.6. Sapphire and quartz are birefringent, thus their optical constants have different values for ordinary and extraordinary rays. The lack of long-range order in fused silica facilitates the coupling of THz fields with multi-mode lattice vibrations, which gives rise to the substantially stronger absorption compared with that of crystalline quartz, while its refractive index is slightly smaller. Figure 5.7 shows the optical constants of sapphire and quartz in the spectral range 1-6 THz [121]. A strong ordinary ray absorption line is featured at 3.87 THz.

5.2.3 Conductors

The reflectivity of a metal surface is near unity in the THz region, and hence metal-coated mirrors are widely used as reflectors for THz applications. Common metals such as copper, silver, gold, and aluminium are adequate for this purpose. The optical properties of a metal surface at THz frequencies are well accounted for by the Drude model. Moreover, the Drude conductivity is further simplified as

$$\sigma(\omega) = \frac{\sigma_0}{1 - i\omega\tau} \cong \sigma_0, \tag{5.8}$$

because $\omega\tau \ll 1$, where the relaxation times of these common metals are on the order of 10^{-14} s. The electrical conductivities of common metals are given in Table 5.2. The penetration depth, $\delta = \sqrt{2/\omega\mu_0\sigma_0}$, is less than 100 nm at 1 THz for common metals, thus a few-micron-thick layer is sufficient for a reflector.

Table 5.2. Electrical Conductivity and Penetration Depth

	Cu	Ag	Au	Al
σ_0 (10^6 S·m^{-1})	59.6	63.0	45.2	37.8
δ at 1 THz (nm)	65.2	63.4	74.9	81.9

When an electromagnetic wave bounces off the interface of air and metal, the reflectivity at normal incidence has the form

$$R(\omega) = \left| \frac{\sqrt{\epsilon_r(\omega)} - 1}{\sqrt{\epsilon_r(\omega)} + 1} \right|^2, \tag{5.9}$$

where the complex dielectric constant is expressed as

$$\epsilon_r(\omega) = \epsilon_b + i\frac{\sigma(\omega)}{\epsilon_0\omega} \cong i\frac{\sigma_0}{\epsilon_0\omega}, \tag{5.10}$$

where ϵ_b is the contribution from bound electrons. Since $\sigma_0/\epsilon_0\omega \gg \epsilon_b$ in the THz region, the equation for the reflectivity is reduced to

$$R(\omega) \cong 1 - \sqrt{\frac{8\epsilon_0\omega}{\sigma_0}}. \tag{5.11}$$

Figure 5.8 shows the reflectance spectra of silver, copper, gold, and aluminium at normal incidence. The solid lines depict calculations using Eq. 5.11, which agree well with the experimental data of references [123] and [124].

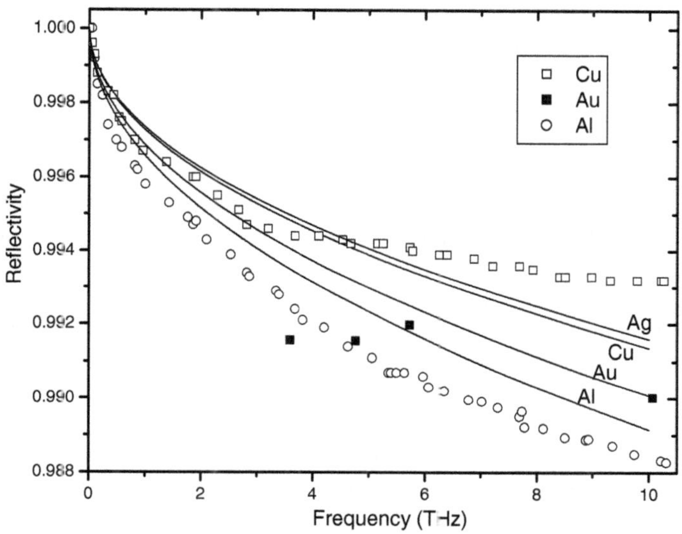

Fig. 5.8. Reflectivity versus frequency for Cu, Ag, Au, and Al. Solid lines indicate calculations using Eq. 5.11. Solid squares, open squares, and open circles are experimental data for gold, copper, and aluminium, respectively. (Data from Refs. [123, 124])

So far, we have neglected the real part of ϵ_r because it is negligible, compared with the imaginary part. It is negative and independent of frequency in the THz region. Its absolute value is much greater than unity. Including the real part, we can approximate Eq. 5.10 as

$$\epsilon_r(\omega) \cong -\frac{\sigma_0\tau}{\epsilon_0} + i\frac{\sigma_0}{\epsilon_0\omega}. \tag{5.12}$$

Given that τ=7.5 and 25 fs for Al and Cu [124], we obtain

$$\text{Al}: \ \epsilon_r(\nu) = -3.2 \times 10^4 + i\,6.7 \times 10^5\nu^{-1} \tag{5.13}$$

$$\text{Cu}: \ \epsilon_r(\nu) = -1.7 \times 10^5 + i\,1.1 \times 10^6\nu^{-1} \tag{5.14}$$

where $\nu = \omega/2\pi$ in THz.

An interesting material system for THz applications is transparent conductors such as tin doped indium oxide (ITO). Optical transmittance of ITO is reported as high as 95% while the conductivity reaches up to 10^6 S·m^{-1}. A conductivity of 10^6 S·m^{-1} corresponds to 98% reflectivity at 1 THz and 500 nm penetration depth. This unique property provides a special use for THz applications. A thin layer of ITO coated on a glass substrate can be utilized as a dichroic mirror reflecting THz radiation while transmitting an optical beam.

5.3 Optical Components

5.3.1 Focusing Elements

Off-axis parabolic mirrors are broadly used to focus or collimate THz beams. The reflective surface of an off-axis parabolic mirror, formed when a partial section is cut out from a paraboloid, as shown in Fig. 5.9(a), is usually coated with common metals such as aluminium and gold, whose reflectivity is in the vicinity of 99 % in the THz region (see Fig. 5.8). A clear advantage of reflective optics over conventional lenses is that there is little loss by reflection and absorption. It also works over a broad spectral range, including the optical region, without spectral aberration. Furthermore, parabolic mirrors are free from spherical aberration, focusing a parallel beam to a point or forcing the radiation from a point source highly collimated. This is an important trait, especially for a quasi-optical THz system dealing with diffraction-limited beams. Because the alignment of an off-axis parabolic mirror is highly sensitive to astigmatism and other collimation errors, a high-precision procedure is necessary in practical applications to avoid them.

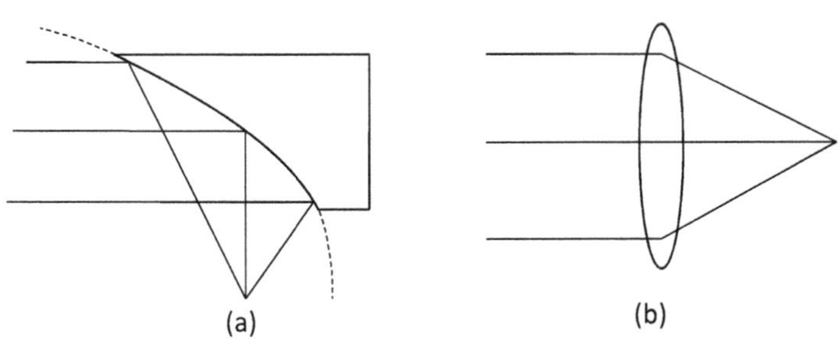

Fig. 5.9. (a) Off-axis parabolic mirror and (b) singlet lens

As discussed in section 3.2.3, substrate lenses, collimating lenses and hyper-hemispherical Lenses, are being used for collecting radiation from photoconductive emitters and coupling it to receivers. The low-loss polymers, dielectrics, and semiconductors (see section 5.2) potentially make good materials for THz lenses. THz singlet lenses for general purposes made of silicon, polyethylene, teflon, and Tsurupica, are commercially available.

Diffractive optics, such as Fresnel zone-plates, have also been fabricated and tested [125, 126], but are not broadly available at present.

5.3.2 Antireflection Coatings

Fresnel, or reflection, loss is one of the main loss mechanisms in THz optical systems, because the majority of low-loss dielectrics and semiconductors used for THz components have relatively large refractive indices in the THz region. Antireflection (AR) coatings can greatly reduce Fresnel losses.

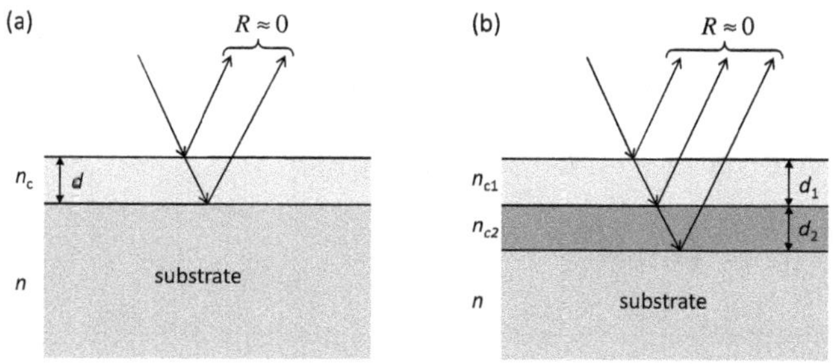

Fig. 5.10. Schematic diagrams for (a) a single-layer and (b) multilayer antireflection coatings.

Figure 5.10 illustrates a schematic representation of single-layer and multilayer AR coatings. The AR coating schemes exploit the destructive interference between the reflective waves from multiple interfaces. Single-layer AR coatings require two conditions. First, the reflection coefficients of the two interfaces at normal incidence must be equal,

$$\frac{1-n_c}{1+n_c} = \frac{n_c - n}{n_c + n}, \tag{5.15}$$

which yields

$$n_c = \sqrt{n}. \tag{5.16}$$

Second, in order for the two reflected waves to destructively interfere the effective path length in the coating layer must be a half wavelength of the incoming wave, i.e., the coating layer thickness is

$$d = \frac{\lambda}{4n_c}. \tag{5.17}$$

A small number of THz-AR coating techniques have been developed so far. Polyethylene ($n \approx 1.5$) AR coatings on quartz, sapphire, and CaF_2 have been constructed by a thermal bonding technique in which a polyethylene film is placed in physical contact with a substrate and heated to a temperature just below its melting point [127]. Another promising coating material is silicon oxide (SiO_2), whose refractive index ($n = 2$) is close to the square root of Ge ($n = 4.0$) and GaAs ($n = 3.6$) refractive indices. An AR coating technique used with this material is to glue a SiO_2 plate onto Ge and GaAs wafers and control the coating thickness by mechanical polishing [128]. Epitaxial growth techniques control film thicknesses with high accuracy, but a technical impediment for growing THz AR coatings is that the required film thickness, in the range of 10-100 μm, exceeds the capabilities of typical epitaxial techniques. This technical difficulty was circumvented when a plasma-enhanced chemical-vapor deposition (CVD) method made a huge improvement on the film growth speed. It has been used to grow SiO_x AR coatings on Ge wafers [129]. A drawback of these single-layer AR coatings is that the AR bandwidth is narrow. A broadband AR bandwidth can be obtained by using a multilayer interference film. The plasma-enhanced CVD technique has been applied to grow multilayer AR coatings on Ge substrates [130].

5.3.3 Bandpass Filters

Thin metallic meshes have been used as bandpass filters in the THz region. The optical properties of mesh filters are accounted for by the dynamics of surface plasmon polaritons in metal-dielectric interfaces. We will review this subject from a broader perspective in section 5.5.3. In the present section, we will approach this problem from a phenomenological point of view. A typical structure of mesh filters is the inductive grid shown in Fig. 5.11(a). An electromagnetic wave incident on the metal grid structure gives rise to electromagnetic induction as the field induced surface currents flow in the closed loops of the grid. At the same time, charge distribution varies in time depending on the field amplitude, phase, and polarization. Elaborating on this picture systematically, we can obtain an equivalent circuit representation based on the transmission line theory as a simple model to describe the optical properties of the metal grid [131]. The inset of Fig. 5.11(b) shows an equivalent circuit to the inductive grid. The transmission of this circuit is expressed as

$$T(\omega) = \frac{(\omega^2 - \omega_0^2)^2 R^2 + \omega^2 \omega_0^2 Z_0^2}{(\omega^2 - \omega_0^2)^2 (1 + R^2) + \omega^2 \omega_0^2 Z_0^2} \tag{5.18}$$

$$\approx \frac{\omega^2 \omega_0^2 Z_0^2}{(\omega^2 - \omega_0^2)^2 + \omega^2 \omega_0^2 Z_0^2} \quad \text{for } R \ll Z_0, \tag{5.19}$$

where ω_0 is the resonant frequency, R is the loss resistance, and

$$Z_0 = \omega_0 L = \frac{1}{\omega_0 C} \tag{5.20}$$

is the normalized impedance of L and C at resonance. Fig. 5.11(b) shows the transmission spectrum for $R \ll Z_0$. The resonant wavelength $\lambda_0 = 2\pi c/\omega_0$ is on the same scale as the grid period g.

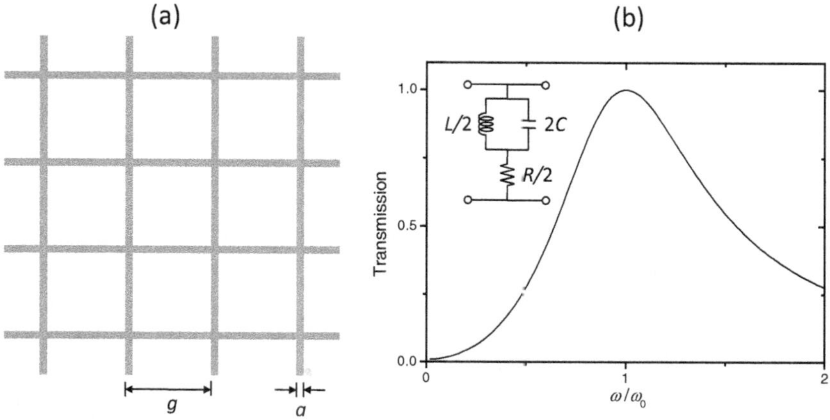

Fig. 5.11. Metal-mesh filter: (a) inductive grid structure and (b) transmission spectrum. Inset shows the equivalent circuit.

Several different types of grid structures have been examined in an effort to better control the transmission band structure. The array of cross-shaped apertures shown in Fig 5.12(a) [132] is one of them which has some useful properties. The transmission spectra of four filters with different dimensions ($g/a/b$ in μm are 402/251/66, 201/126/33, 154/98/28, and 113/71/19) are shown in Fig 5.12(b). The peak transmissions at the central frequencies are almost unity for all of the filters. The central frequency of the passband is mainly determined by the the length of the cross a. The bandwidth tends to get narrower as the ratios of g/a and g/b are increased.

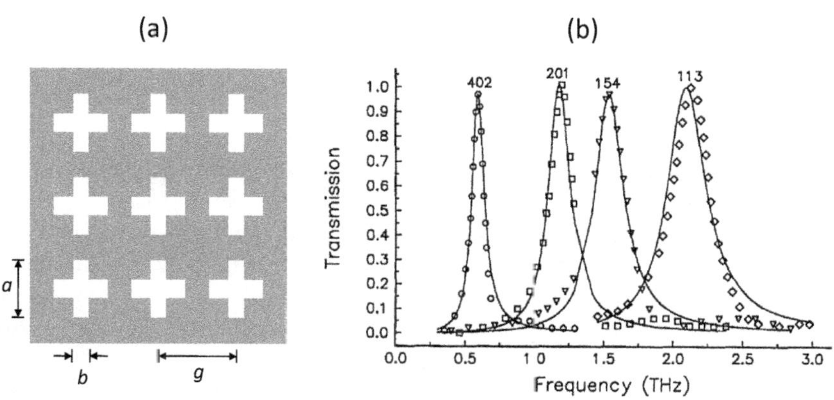

Fig. 5.12. Resonant bandpass filter with an array of cross-shaped apertures: (a) schematic diagram of the structure and (b) transmission spectra for the four filters whose dimensions $g/a/b$ in μm are 402/251/66, 201/126/33, 154/98/28, and 113/71/19. (Reprinted from [132].)

5.3.4 Polarizers

Free-standing metal wire grids are commonly used as polarizers in the THz region. Figure 5.13 illustrates the structure of a typical wire-grid polarizer with a circular frame: the thin metal wires, each with diameter a, form the regular array of the grid with period g, placed in a flat plane. The underlying mechanism of its polarization selectivity is quite simple. Imagine an electromagnetic wave is incident on a wire-grid polarizer. If the electric field is parallel to the wires, the electrons in the wires can move freely along the wire direction responding to the incident field. In this case, the polarizer behaves much like a typical metal surface, thus most of the incident beam is reflected by the polarizer. On the other hand, if the field is perpendicular to the wires, the wave does not see much of the wires and passes through the polarizer, because the movements of the electrons in the direction perpendicular to the wires are highly restricted. In general, the transmission at an angle θ between the grid direction and the polarization has the following relation: $T(\theta) = \sin^2 \theta$.

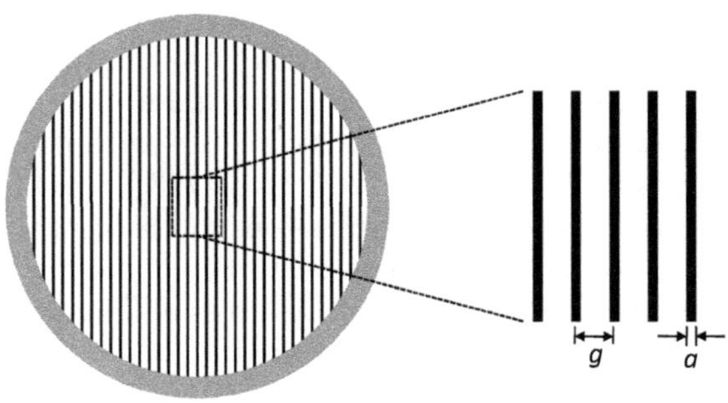

Fig. 5.13. Wire-grid polarizer. Transmission and extinction ratio are determined by the grid period g and the wire diameter a.

Wire grids are usually made of tungsten because it has the highest tensile strength among metals and excellent corrosion resistance. The wire diameter is $\sim$10 μm, and the grid period is in the range of 20-200 μm. In practical applications, the transmission does not disappear completely when the field is parallel to the grid direction. The distinction ratio $T_{\perp}/T_{\parallel}$ is enhanced as the grid period g is decreased, while the cutoff frequency of $T_{\perp}$ is reduced at the same time. For a polarizer with a=10 μm and g=25 μm, it is measured that $T_{\perp}$ is about 0.98 at 1 THz and drops to 0.95 at 3 THz, and $T_{\perp}/T_{\parallel}$ is $\sim$1000 at 1 THz and $\sim$200 at 3 THz. The reflectivity of a grid polarizer is usually higher than 0.95 in a broad spectral range for a field parallel to the grid direction, thus wire-grid polarizers are often used as beam splitters.

5.3.5 Wave Plates

A wave plate is an optical component used to control the polarization state of light. A birefringent crystal has different refractive indices for different polarizations: ordinary and extraordinary refractive indices n_o and n_e. Using the birefringence of crystals, we can modify the waves polarization state. For two monochromatic waves with their polarizations parallel to the ordinary and extraordinary axes, propagating a distance d in a birefringent crystal, the phase delay $\Delta\phi$ between them is given as

$$\Delta\phi = \frac{\omega}{c}(n_e - n_o)d. \tag{5.21}$$

The thickness of a half-wave plate is chosen to produce a π phase delay so that the polarization of linearly polarized light can be rotated from 0 to 90° by adjusting the relative angle between the optic axis and the incident polarization. The $\pi/2$ phase delay of a quarter-wave plate changes linearly polarized light to a circularly polarized light and vice versa when the linear polarization is aligned to dissect the ordinary and extraordinary axes. In between, any arbitrary elliptical polarization state can be obtained.

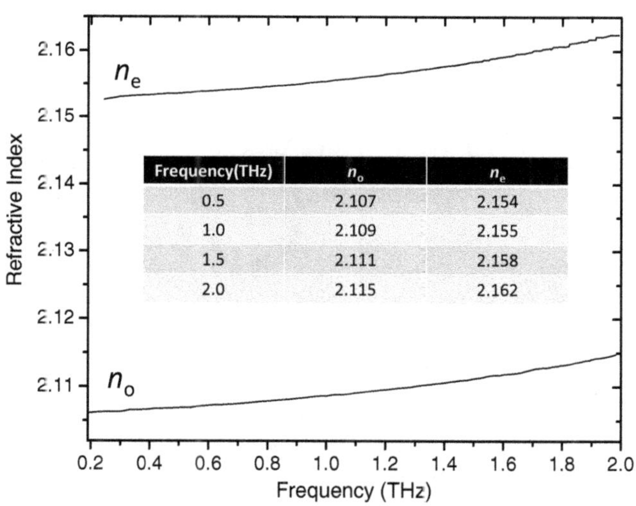

Fig. 5.14. Ordinary and extraordinary refractive indices of crystalline quartz in the THz region. (Data from Ref. [23])

Crystalline quartz is a birefringent crystal with some excellent properties for use as a wave plate. As shown in Figs. 5.6 and 5.7, quartz is strongly birefringent as well as highly transparent in the THz region. Figure 5.14 shows the ordinary and extraordinary refractive indices of quartz in the THz region. For $\Delta n = n_e - n_o \simeq 0.047$ in this spectral range, the thickness of a quarter wave plate is determined to be

$$d_{\frac{\lambda}{4}} = \frac{c}{4\nu \Delta n} \simeq \frac{1.6}{\nu(\text{in THz})} \quad (\text{in mm}), \tag{5.22}$$

where $\nu = \omega/2\pi$ is the frequency.

A significant drawback of a conventional birefringent wave plate is that it can only be used at a single wavelength. One approach to broaden the bandwidth is to fabricate a wave plate by stacking up multiple quartz plates. The THz achromatic quarter-wave plate made of six quartz plates has an almost flat phase retardation from 0.3 to 1.7 THz [133]. Another technique is to electrically control the birefringence of liquid crystal. With this technique, the phase delay at 1 THz is continuously tunable from 0 to $\pi/2$ by adjusting the bias voltage [134].

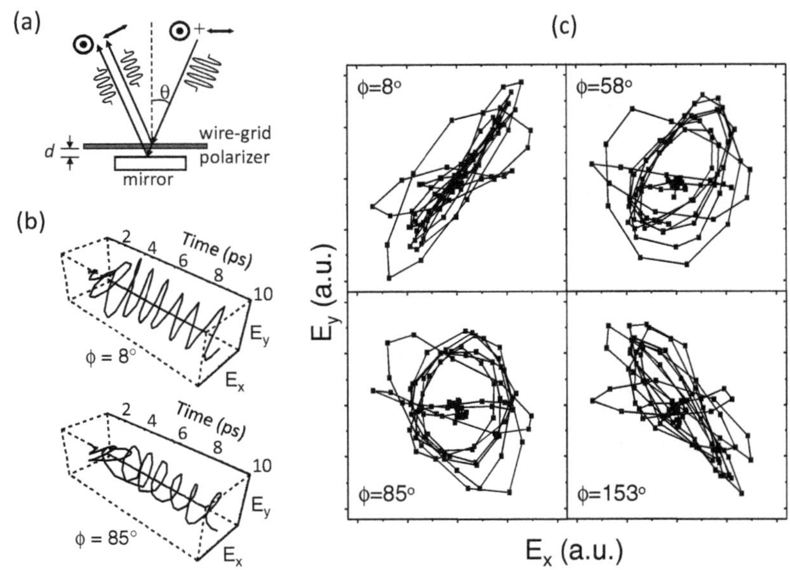

Fig. 5.15. (a) Schematic diagram of the broadband THz wave plate consisting of a wire-grid polarizer and a mirror. (b) Three-dimensional plots of THz E-field vectors for $\Delta\phi$ =8° and 85°. (c) Polarization trajectories in the x-y plane for $\Delta\phi$=8°, 58°, 85°, and 153°. (Data from Ref. [135])

An alternative method to control THz polarization is illustrated in Fig. 5.15(a) [135]. Elliptical polarization can be obtained from the reflection of a linearly polarized THz wave by a combination of a wire-grid polarizer and a mirror with variable spacing d. The mirror and the polarizer reflect two perpendicularly polarized THz waves, and the phase delay

$$\Delta\phi = \frac{\omega}{c}\frac{2d}{\cos\theta}, \tag{5.23}$$

where θ is the incident angle, is controlled by adjusting the spacing d. Figure 5.15(b) shows the THz E-field vectors of the linearly ($\Delta\phi$ =8°) and circularly ($\Delta\phi$ =85°) polarized THz pulses in the x-y plane. The phase delay is continuously tunable, and several snap shots of the phase control are shown in Fig. 5.15(c): the polarization trajectories in the x-y plane for $\Delta\phi$=8°, 58°, 85°, and 153°.

5.4 Terahertz Waveguides

A waveguide is a device used to carry electromagnetic waves from one place to another without significant loss in intensity while confining them near the propagation axis. The most common type of waveguides for radio waves and microwaves is a hollow metal pipe. Waves propagate through the waveguide, being confined to the interior of the pipe. A representative waveguide in the optical region is an optical fiber. Fiber-optic communication and a variety of other applications exploit the extremely low attenuation and dispersion of silica-based optical fibers in the optical communication band of 1.3-1.6 μm. Several microwave and optical waveguide technologies have been examined in the THz region. The major challenge of THz waveguide technologies is the relatively strong absorption in most of the conventional waveguide structures, which prevents THz wave transmission over long distances.

5.4.1 Theory of Rectangular Waveguides

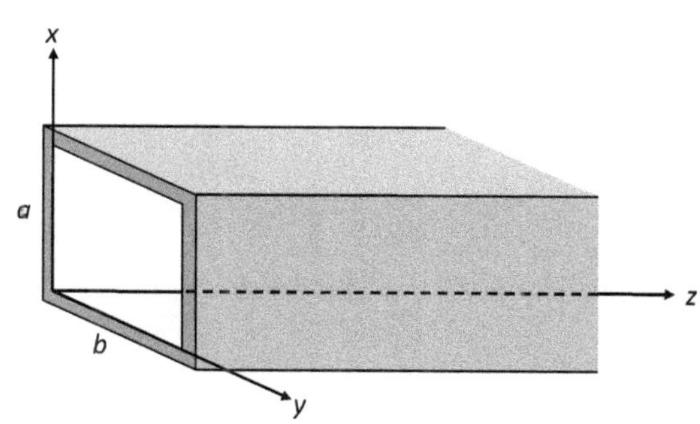

Fig. 5.16. Rectangular waveguide

To get a sense of how waves propagate in a waveguide, we look into a metal tube of rectangular shape (Fig. 5.16). The electric and magnetic fields of a monochromatic wave travelling through the waveguide in the positive direction of the z-axis have the generic form

$$\mathbf{E}(x, y, z, t) = \mathbf{E}_0(x, y)e^{i(kz-\omega t)}, \tag{5.24}$$

$$\mathbf{B}(x, y, z, t) = \mathbf{B}_0(x, y)e^{i(kz-\omega t)}, \tag{5.25}$$

where $\mathbf{E}_0 = E_x\mathbf{e}_x + E_y\mathbf{e}_y + E_z\mathbf{e}_z$ and $\mathbf{B}_0 = E_x\mathbf{e}_x + B_y\mathbf{e}_y + B_z\mathbf{e}_z$. Unlike propagation in free space, guided waves, in general, are not transverse, i.e., the longitudinal components E_z and B_z do not vanish. Inserting Eqs. 5.24 and 5.25 into Maxwell's equations (Eqs. 2.1, 2.2, 2.3, and 2.4) and manipulating the equations, we obtain the wave equations for E_z and B_z:

$$\left[\frac{\partial^2}{\partial x^2} + \frac{\partial^2}{\partial y^2} + \frac{\omega^2}{c^2} - k^2\right] E_z(x, y) = 0, \tag{5.26}$$

$$\left[\frac{\partial^2}{\partial x^2} + \frac{\partial^2}{\partial y^2} + \frac{\omega^2}{c^2} - k^2\right] B_z(x, y) = 0. \tag{5.27}$$

Noting that these two equations are independent of each other, we can classify guided waves into different types of modes. If $E_z = 0$ we call the waves transverse electric (TE) modes. Similarly, transverse magnetic (TM) modes have no longitudinal component of the magnetic field, $B_z = 0$. A TEM mode has neither electric nor magnetic field in the longitudinal direction. A hollow waveguide, however, does not support TEM modes.

Suppose we are interested in TE modes. We obtain $B_z(x, y)$ by solving Eq. 5.27, then determine the other components of electric and magnetic fields using the following relations obtained from Maxwell's equations:

$$E_x = \frac{i\omega}{(\omega/c)^2 - k^2}\frac{\partial B_z}{\partial y}, \tag{5.28}$$

$$E_y = \frac{-i\omega}{(\omega/c)^2 - k^2}\frac{\partial B_z}{\partial x}, \tag{5.29}$$

$$B_x = \frac{ik}{(\omega/c)^2 - k^2}\frac{\partial B_z}{\partial x}, \tag{5.30}$$

$$B_y = \frac{i\omega}{(\omega/c)^2 - k^2}\frac{\partial B_z}{\partial y}. \tag{5.31}$$

The general solution of Eq. 5.27 has the form,

$$B_z(x, y) = [A\sin(k_x x) + B\cos(k_x x)] \cdot [C\sin(k_y y) + D\cos(k_y y)], \tag{5.32}$$

where the coefficients A, B, C, and D, and the wavenumbers, k_x and k_y, are determined by boundary conditions. Under the assumption that the metal is a perfect conductor, electromagnetic waves vanish inside the material. Accordingly, the electric and the magnetic fields satisfy the boundary conditions that the parallel components of the electric field and the normal component of the magnetic field vanish at the interior surface, $\mathbf{E}_{||} = 0$ and $\mathbf{B}_{\perp} = 0$. Applying the boundary condition $\mathbf{E}_{||} = 0$ to Eqs. 5.28 and 5.29, we obtain

$$B_z(x, y) = B_0\cos\left(\frac{m\pi}{a}x\right)\cos\left(\frac{n\pi}{b}y\right), \quad (m, n = 0, 1, 2, \cdots). \tag{5.33}$$

Figure 5.17 shows the spatial profile of the field intensity for the low-order TE modes.

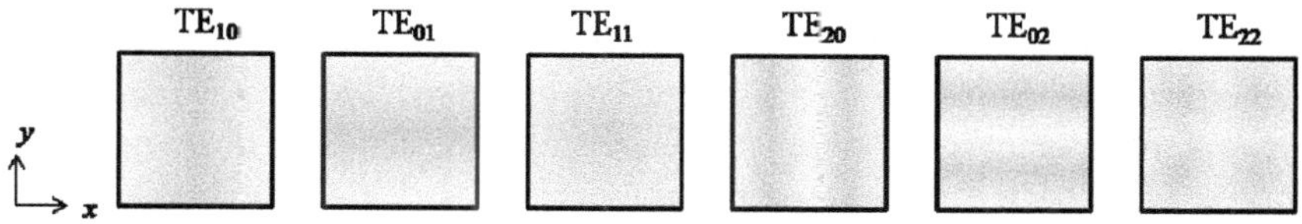

Fig. 5.17. Field distribution of TE modes in a cross section of a rectangular waveguide

Inserting Eq. 5.33 into the wave equation, we get the dispersion relation,

$$k = \frac{1}{c}\sqrt{\omega^2 - \omega_{mn}^2}, \tag{5.34}$$

where the cutoff frequency,

$$\omega_{mn} = \pi c \sqrt{\frac{m^2}{a^2} + \frac{n^2}{b^2}}. \tag{5.35}$$

If $\omega < \omega_{mn}$ the wavenumber k is imaginary, the wave attenuates exponentially as $e^{-|k|z}$. Therefore, the frequency of a travelling wave must be higher than the cutoff frequency. The phase and the group velocities,

$$v_{ph} = \frac{\omega}{k} = \frac{c}{\sqrt{1 - \omega_{mn}^2/\omega^2}}, \tag{5.36}$$

$$v_{gr} = \frac{\partial \omega}{\partial k} = c\sqrt{1 - \omega_{mn}^2/\omega^2}, \tag{5.37}$$

indicate that the waveguide is highly dispersive, especially, near the cutoff frequency.

5.4.2 Hollow Metallic Tubes

A comprehensive study has been conducted on rectangular and circular metallic waveguides [136]. THz time-domain spectroscopy (THz-TDS) is employed to measure broadband THz pulses propagating through the 25-mm-long metal tubes of various cross-sectional dimensions. Figure 5.18 shows the waveforms and amplitude spectra of the THz pulses transmitted through (b) a 250 μm×125 μm rectangular and (c) 280 μm-diameter circular brass waveguide, with the incoming waves being linearly-polarized single-cycle pulses.

The incoming Gaussian beam is coupled into the rectangular waveguide most efficiently with the lowest-order mode TE_{10} when the polarization is parallel to the y-axis. Consequently, THz wave transmission through this waveguide is effectively single-mode propagation. The cutoff frequency of the measured spectrum is consistent with the calculated value, ω_{10}=0.6 THz. The transmitted THz pulse is stretched to ~15 ps due to the strong group-velocity dispersion.

Group-velocity dispersion is even stronger in the circular waveguide, through which the pulses are stretched to ~40 ps. The interference fringes in the spectrum indicate that the waveguide propagation involves multiple modes. The sharp cutoff at 0.67 THz is close to the calculated cutoff frequency, ω_{11}=0.65 THz.

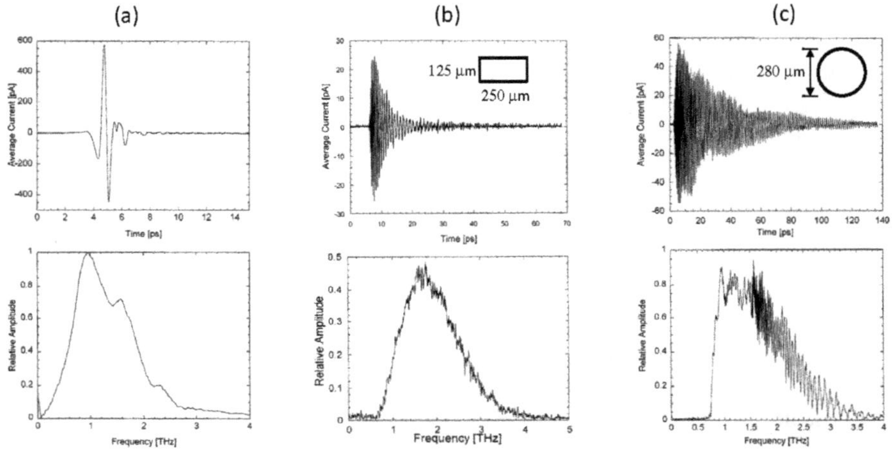

Fig. 5.18. THz waveforms and amplitude spectra of (a) the reference, (b) the transmitted pulses through a 25-mm long, 250 μm×125 μm rectangular brass waveguide, and (c) the transmitted pulses through a 25-mm long, 280 μm-diameter circular brass waveguide. (Reprinted with permission from [136]. ©2000, American Institute of Physics.)

Before we move on to other waveguide structures, it is worthwhile to review the intrinsic properties of metal-tube THz waveguides pertinent to practical applications. They provide the criteria against which other waveguides are assessed. First, there is a cutoff frequency below which no wave is allowed to propagate. Second, a broadband pulse undergoes a severe waveform distortion due to strong group-velocity dispersion. Third, long-distance propagation is limited by absorption: the absorption coefficient is in the range of ~1 cm^{-1}. Fourth, incident THz waves efficiently couple into only a few low-order modes: TE_{10} and TM_{12} modes in a rectangular guide and TE_{11}, TE_{12}, and TM_{11} modes in a circular guide.

5.4.3 Dielectric Fibers

A dielectric fiber is a cylindrical waveguide which confines light within the core of the fiber by total internal reflection. The core is surrounded by cladding, layers of material having lower refractive index than that of the core. Most of the dielectric waveguide modes are hybrid, i.e., neither E_z nor B_z vanishes. The fundamental mode of a single-mode fiber is the hybrid electric mode HE_{11}. The HE_{11} mode travelling in the core of refractive index n with an infinite cladding is expressed as a linearly polarized wave, $\mathbf{E}_{11}(\rho, \phi)e^{i(kz-\omega t)}$, and the components of $\mathbf{E}_{11}(\rho, \phi)$ are given as

$$E_x = E_0 J_0(\beta r), \tag{5.38}$$

$$E_z = E_0 \frac{i\beta}{k} J_1(\beta r) \cos\phi \tag{5.39}$$

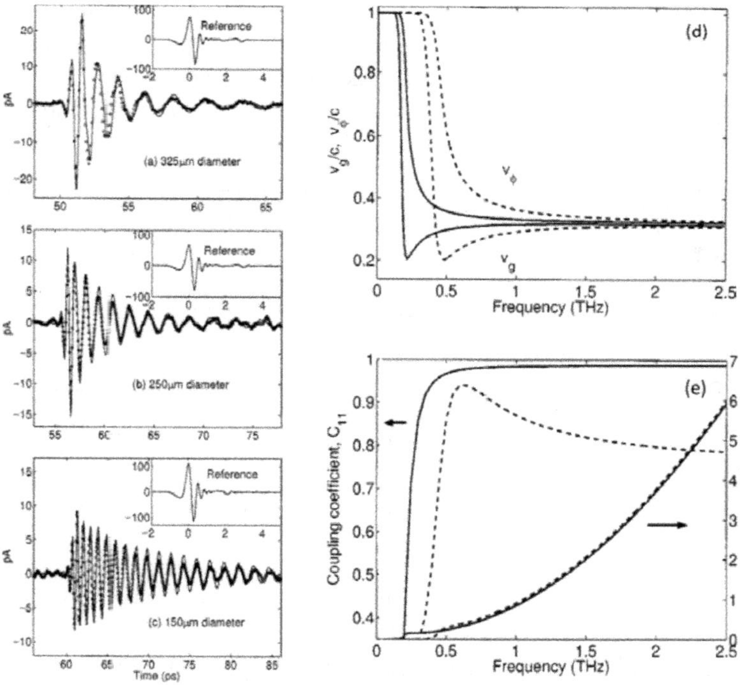

Fig. 5.19. Measured (circle) and calculated (solid line) THz pulses transmitted through sapphire fibers of diameter a and length l: (a) a=325 μm and l=7.3 mm, (b) a=250 μm and l=7.8 mm, and (c) a=150 μm and l=8.3 mm. The incoming single-cycle pulses are shown in the insets. (d) Calculated group and phase velocities in 325 μm-diameter (solid line) and 150 μm-diameter (dashed line) fibers for HE_{11} mode. (e) Coupling coefficient (left axis) and absorption coefficient (right axis) of 325 μm-diameter (solid line) and 150 μm-diameter (dashed line) fibers for HE_{11} mode. (Reprinted with permission from [137]. ©2000, American Institute of Physics.)

in terms of cylindrical coordinates, where the dispersion relation,

$$\beta = \sqrt{\frac{n^2\omega^2}{c^2} - k^2}, \tag{5.40}$$

is determined by boundary conditions. Since the HE_{11} mode has a linear polarization, an azimuthal symmetry, and a profile very close to a Gaussian distribution, coupling between a free-space beam and the mode can be highly efficient if the mode profile of the incoming beam matches well with that of the HE_{11} mode.

Figure 5.19(a)-(c) show broadband THz pulses transmitted through unclad sapphire fibers of three different diameters, 325, 250, and 150 μm, which are comparable to the central wavelength of the pulses [137]. The pulses undergo significant reshaping because the waveguides are strongly dispersive in the spectral range of interest, as shown in Fig. 5.19(d). The frequency-dependent coupling and absorption (Fig. 5.19(e)) also contribute to the pulse reshaping. The solid lines of Fig. 5.19(a)-(c) are calculations obtained by analyzing HE_{11} mode propagation in the fibers. The dominance of the HE_{11} mode is accounted for by its superior coupling efficiency from free space to waveguide.

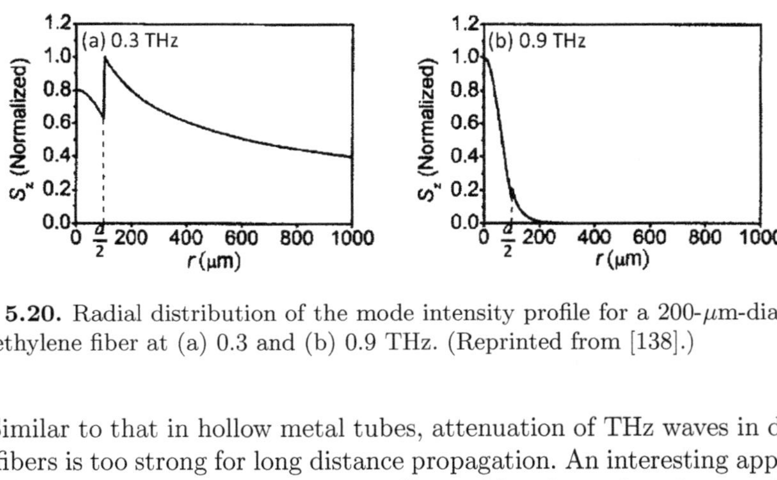

Fig. 5.20. Radial distribution of the mode intensity profile for a 200-μm-diameter polyethylene fiber at (a) 0.3 and (b) 0.9 THz. (Reprinted from [138].)

Similar to that in hollow metal tubes, attenuation of THz waves in dielectric fibers is too strong for long distance propagation. An interesting approach to reduce the attenuation is to use a fiber with subwavelength core diameter [138]. As shown in Fig. 5.20, if the fiber diameter is significantly smaller than the mode wavelength, only a small fraction of the wave propagates in the core of high loss while the greater part lies in free space. Consequently, the effective attenuation coefficient can be much lower than that of the core material. The attenuation constant of a 200-μm-diameter polyethylene fiber shown in Fig. 5.21(a) is significantly lower than that of polyethylene, in particular, at low frequencies. A drawback of this scheme is that the coupling is relatively inefficient due to substantial mode mismatch (see Fig. 5.21(b)).

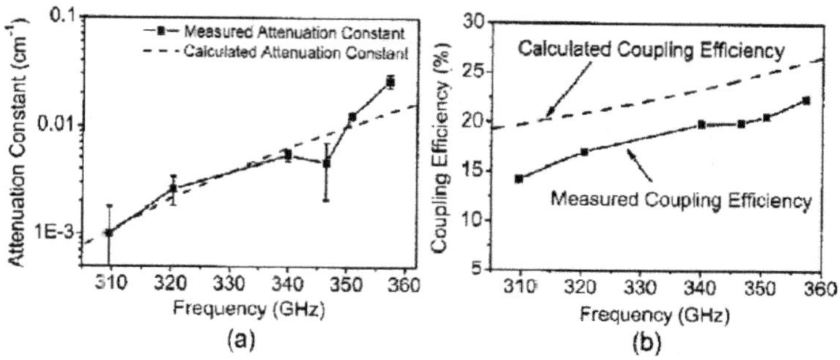

Fig. 5.21. Measured and calculated (a) absorption coefficient and (b) coupling efficiency of a 200-μm-diameter polyethylene fiber versus frequency. (Reprinted from [138].)

5.4.4 Parallel Metal Plates

An outstanding property of a parallel-plate metal waveguide is that its TEM mode is virtually dispersionless. Figure 5.22 illustrates the TEM mode travelling along the z-axis within the gap of two metal plates parallel to the y-z plane. The electric field is linearly polarized in the direction normal to the metal surfaces. The TEM mode is represented in the form of a plane wave, $\mathbf{E}(z,t) = \mathbf{e}_x E_0 e^{i(kz-\omega t)}$, inside the waveguide, vanishing elsewhere. The dispersion relation, $k = \omega/c$, is identical with that of free space, thus there is no cutoff frequency, and the group and the phase velocities are equal to the speed of light.

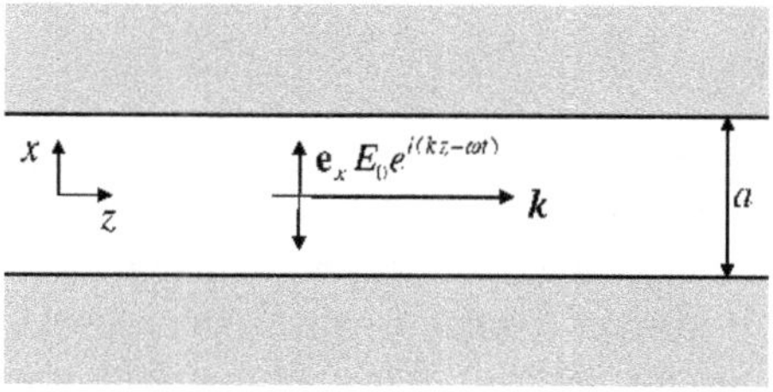

Fig. 5.22. TEM mode in parallel metal plates

Figure 5.23(a) shows that single-cycle THz pulses transmitted through a 12.6- and a 24.4-mm-long copper parallel-plate waveguide with a 108-μm gap retain their pulse shapes with little stretching [139]. Plano-cylindrical lenses are attached to the entrance and exit slits to couple the THz beam into and out of the waveguides. The power loss is relatively low compared with metal tubes and dielectric fibers: the amplitude absorption coefficient $\alpha/2$ is less than 0.2 cm^{-1} in the broad spectral range from 0.1 to 4.5 THz.

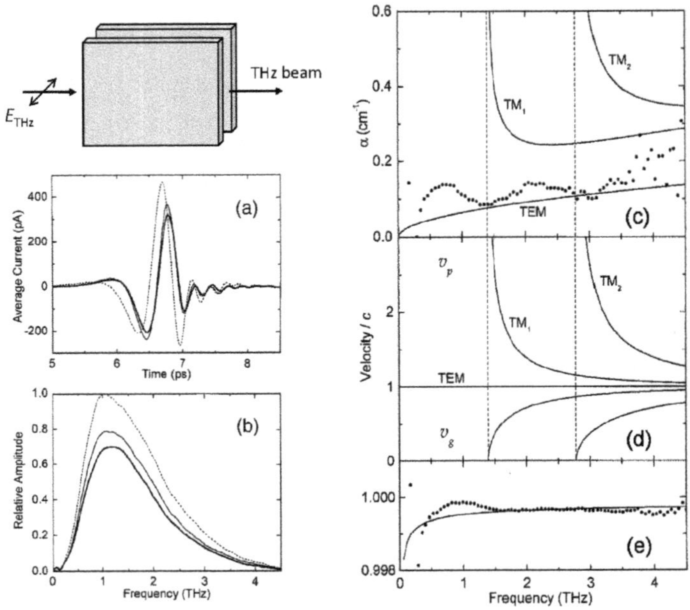

Fig. 5.23. (a) Waveforms and (b) spectra of broadband THz pulses transmitted through a 12.6- and a 24.4-mm-long copper parallel-plate waveguide with a 108-μm gap. The dashed lines represent the reference pulses. (c) Amplitude absorption coefficient, (d) phase and group velocity scaled with the speed of light, and (e) the scaled velocity of the TEM mode in the vicinity of unity. Solid lines are the calculations for the first three modes, and the filled circles indicate experimental data. (Reprinted from [139].)

5.4.5 Metal Wires

A fascinating development in guiding THz waves is that a bare metal wire turned out to be an effective waveguide with virtually no dispersion and little attenuation [140]. Electromagnetic waves propagating on a long cylindrical conductor are called surface plasmon waves, whose existence was first predicted by A. Sommerfeld in 1899. The Sommerfeld wave is a radially symmetric TM mode travelling along the cylinder axis. THz surface plasmon polaritons at a metal-dielectric interface are of great interest because of their unique properties applicable to near-field optics and subwavelength optics. A comprehensive review of this subject is featured in section 5.5.3. In the present section, we will focus on guided THz waves on a metal wire.

In the cylindrical coordinate system shown in Fig. 5.24, the electric and magnetic fields of the Sommerfeld wave propagating along the z-axis are expressed as

$$\mathbf{E}(\mathbf{r},t) = \left[\mathbf{e}_r E_r(r) + \mathbf{e}_z E_z(r)\right] e^{i(kz-\omega t)}, \tag{5.41}$$

$$\mathbf{B}(\mathbf{r},z) = \mathbf{e}_\phi B_\phi(r) e^{i(kz-\omega t)}. \tag{5.42}$$

E and **B**, having azimuthal symmetry, are independent of the azimuthal angle ϕ. In the transverse plane, **E** has only a radial component $E_r(r)$ and **B** has only azimuthal component $B_\phi(r)$.

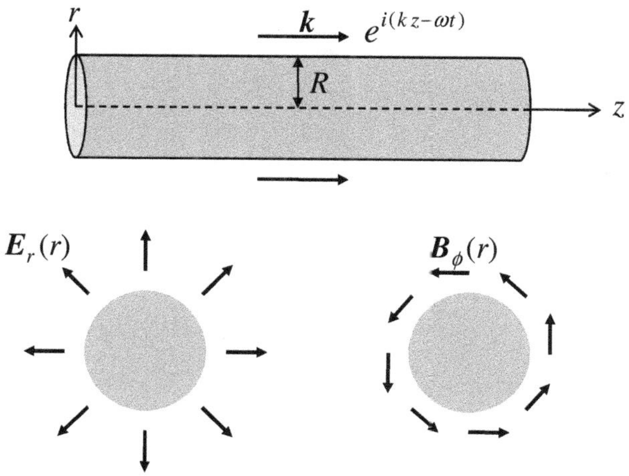

Fig. 5.24. Surface plasmon wave travelling on a cylindrical metal wire

The universal wave equation, Eq. 2.15, leads to the wave equation for the longitudinal electric field $E_z(r)$:

$$r^2 \frac{d^2}{dr^2} E_z(r) + r \frac{d}{dr} E_z(r) - \beta^2 r^2 E_z(r) = 0, \tag{5.43}$$

where the radial parameter β is associated with the dispersion relations

$$\beta_a^2 = k^2 - \frac{\omega^2}{c^2} \tag{5.44}$$

in free space and

$$\beta_c^2 = k^2 - \epsilon_r \mu_r \frac{\omega^2}{c^2} \cong k^2 - i\mu\sigma\omega \tag{5.45}$$

in a conductor with dielectric constant ϵ_r, relative permeability $\mu_r = \mu/\mu_0$, and conductivity σ. Solutions of Eq. 5.43 are modified Bessel functions, $I_0(\beta r)$ and $K_0(\beta r)$. Conforming to the fact that surface plasmon waves fade away from the metal surface, a proper expression of $E_z(r)$ takes the form

$$E_z(r) = \begin{cases} E_c\, I_0(\beta_c r) & \text{for } r < R \\ E_a\, K_0(\beta_a r) & \text{for } r > R \end{cases}. \tag{5.46}$$

Using Maxwell's curl equations, 2.3 and 2.4, we obtain the relations of E_r and B_ϕ with E_z:

$$E_r(r) = \frac{k}{i\beta^2}\frac{dE_z(r)}{dr}, \tag{5.47}$$

$$B_\phi(r) = \frac{k^2 - \beta^2}{i\omega\beta^2}\frac{dE_z(r)}{dr}, \tag{5.48}$$

where

$$\frac{dE_z(r)}{dr} = \begin{cases} \beta_c E_c I_1(\beta_c r) & \text{for } r < R \\ -\beta_a E_a K_1(\beta_a r) & \text{for } r > R \end{cases}. \tag{5.49}$$

The recurrence relations of the modified Bessel functions, $I_0'(\xi) = I_1(\xi)$ and $K_0'(\xi) = -K_1(\xi)$, are applied to attain Eq. 5.49.

The boundary conditions that E_z and B_ϕ are continuous at $r = R$ lead to the transcendental equation,

$$\frac{\epsilon_r}{\beta_c}\frac{I_1(\beta_c R)}{I_0(\beta_c R)} = -\frac{1}{\beta_a}\frac{K_1(\beta_a R)}{K_0(\beta_a R)}, \tag{5.50}$$

from which, and together with

$$\beta_a^2 - \beta_c^2 = (\epsilon_r\mu_r - 1)\,\frac{\omega^2}{c^2}, \tag{5.51}$$

we can determine β_a and β_c.

Relying on theoretical background, we shall attempt to gain some physical insight into the Sommerfeld wave. The asymptotic expansions of $I_1(\xi)$ for $\xi \gg 1$,

$$I_1(\xi) \sim \sqrt{\frac{\pi}{2\xi}}\, e^{\xi}, \tag{5.52}$$

and the series expansion of $K_1(\xi)$,

$$K_1(\xi) = \frac{1}{\xi} + \cdots, \tag{5.53}$$

are useful for understanding its properties. As discussed in section 5.2.3, a THz wave undergoes excessive attenuation in a metal with a typical penetration depth less than 0.1 μm. Consequently, the THz surface plasmon wave propagates almost entirely in free space. A rational ramification of this is that the longitudinal wavenumber k must be very close to ω/c. Eventually, we find $\beta_a \ll k$ according to Eq. 5.44. As Eq. 5.53 indicates that the mode amplitude is roughly proportional to $\frac{1}{\beta_a r}$ in free space, the mode is extended into a region where its dimension is much greater than the mode wavelength, $\lambda = 2\pi/k$. On the other hand, inside the conductor, the radial parameter β_c has the relation, $|\beta_c| \sim |\epsilon_r\mu_r|\frac{\omega}{c} \gg k$, because $|\epsilon_r\mu_r| \gg 1$. Assuming the wire radius R is comparable to or larger than λ, we obtain $\beta_c R \gg 1$. Then, the mode amplitude near the metal surface, proportional to $e^{\beta_c r}$, attenuates exponentially with the penetration depth, $\left|\sqrt{2/\omega\mu\sigma}\right| \ll \lambda$. As shown in Fig. 5.25, $B_\phi(r)$, continuous at the boundary, extends far and gradually attenuates as $\sim 1/\beta_a r$

in free space while decaying exponentially within a very short distance $\sim 1/\beta_c$ inside the conductor. The amplitude of the electric field inside the conductor is significantly smaller than that in free space. Since $E_z(r)$ is continuous at the boundary, the radial component $E_r(r)$ is dominant in free space. Virtually, the guided mode is a transverse wave in free space.

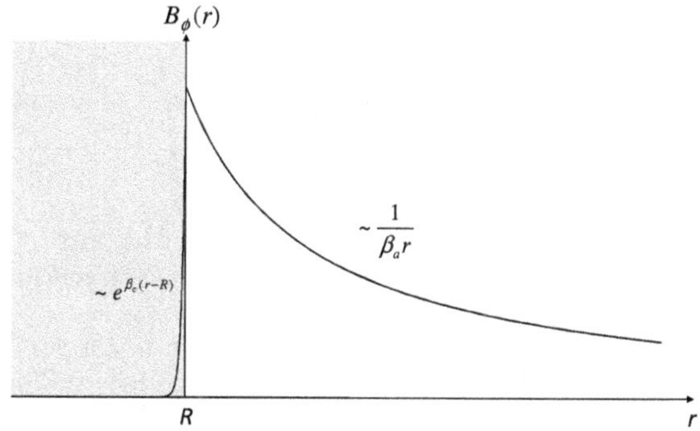

Fig. 5.25. Radial distribution of $B_\phi(r)$

An important question is how the surface plasmon wave attenuates in the propagation direction. We make an estimation of the wavenumber k assuming the wire is sufficiently thick so that R is at least several times larger than λ. Under this condition, it is valid that $I_1(\beta_c R)/I_0(\beta_c R) \approx K_1(\beta_a R)/K_0(\beta_a R) \approx 1$, then the transcendental equation (Eq. 5.50) is reduced to

$$\beta_c^2 \approx \epsilon_r^2 \mu_r^2 \beta_a^2. \tag{5.54}$$

Inserting Eqs.5.44 and 5.45 into Eq. 5.54, we obtain

$$k \approx \left(\frac{\epsilon_r \mu_r}{\epsilon_r \mu_r + 1}\right)^{1/2} \frac{\omega}{c} \approx \frac{\omega}{c} + i\frac{\omega^2}{2c^3 \mu \sigma}. \tag{5.55}$$

Here k is indeed very close to the free space wavenumber $\frac{\omega}{c}$ since $\left|\frac{\omega}{2c^2 \mu \sigma}\right| \ll 1$. The surface wave attenuates in the propagation direction as $\sim e^{-\alpha z}$ with the absorption coefficient, $\alpha = \Re\left\{\frac{\omega^2}{c^3 \mu \sigma}\right\} \ll k$. For example, it is estimated that $\alpha \sim 2 \times 10^{-4}$ cm^{-1} for a thick copper wire (σ_0=5.96 $\times 10^7$ S·m^{-1}).

How are these wave parameters affected as the wire radius decreases? The transcendental equation is helpful for a qualitative analysis of this matter. The ratio $K_1(\beta_c R)/K_0(\beta_a R)$, bigger than unity for a finite R, increases as R is decreased. Hence, β_a and α increase accordingly. For example, if $R \sim \lambda \sim$ 1 mm, β_a and α are roughly one order larger than the values for an infinitely thick wire. Figure 5.26 shows the experimental measurements on THz pulses transmitted through a copper wire waveguide with radius R=0.26 mm [141].

The mode extends several millimeters from the wire axis, and decays inversely with radial distance r. The absorption coefficient is in the range of 10^{-3} cm^{-1}, significantly lower than any other THz waveguide structures.

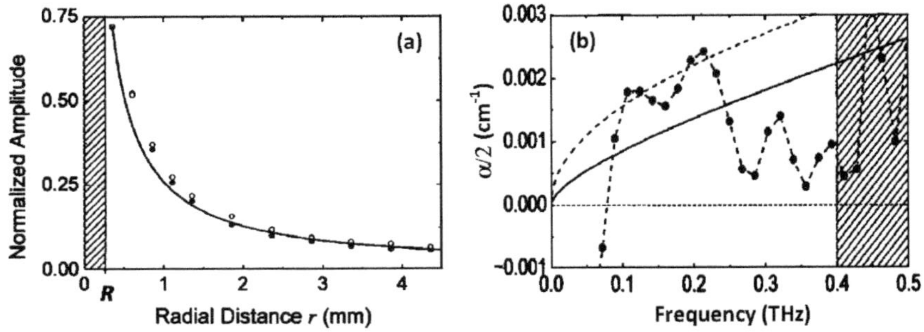

Fig. 5.26. Transmitted broadband THz pulses for a copper wire with radius R=0.26 mm. (a) Normalized peak amplitude (open circles) and spectral amplitude (filled circles) at 0.15 THz. The data are consistent with the solid line, a/r. (b) Measured (filled circle) and calculated (solid line) amplitude absorption coefficients. (Reprinted with permission from [141]. ©2005, American Institute of Physics.)

So far, we have discussed guided-wave propagation on a straight wire. What if the wire is bent? How efficiently does the wire waveguide bend and interconnect THz waves? In general, waveguides can transfer waves along straight lines with high efficiency yet even a moderate bending drastically reduces the efficiency due to a strong radiation loss. Metal wires are not an exception in this respect. The experimental data shown in Fig. 5.27 demonstrates that the radiation loss escalates sharply as the radius of curvature , R_c, is decreased [142]. The amplitude absorption coefficient increases from 0.03 cm^{-1} for a straight wire to 0.15 cm^{-1} for R_c=20 cm. In a phenomenological model incorporating the radiation loss into the overall attenuation, the attenuation coefficient of a bent wire is expressed as

$$\alpha_{R_c} = \alpha_0 \, e^{-\gamma R_c}. \tag{5.56}$$

The solid line in Fig. 5.27 fits the data to this model, which suggests that radiation loss is a dominant attenuation mechanism for bent wires.

A fundamental challenge for the wire waveguide scheme is that it is problematic to couple linearly-polarized THz waves into a wire, because the radially-polarized guided mode has a poor spatial overlap with the waves in free space. The coupling method shown in Fig. 5.28 employs a scattering mechanism which focuses linearly-polarized THz radiation onto a wire waveguide [140]. The coupling efficiency of this method, however, is discouragingly low: the upper limit is estimated as ~0.4% [143]. As a better alternative, Fig. 5.28(b) shows an ingenious technique to improve the coupling efficiency: THz waves are launched directly from a specially designed photoconductive

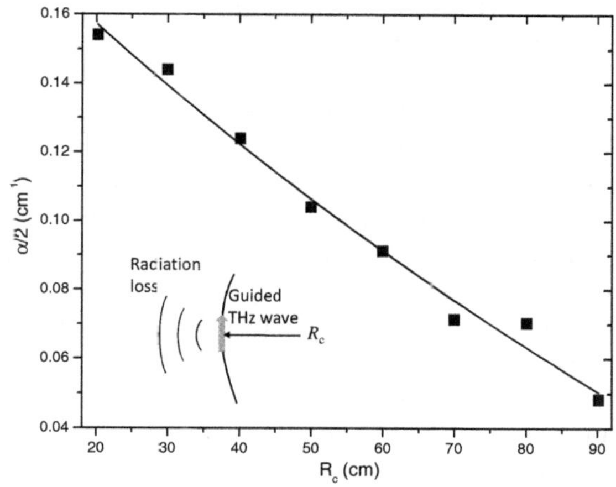

Fig. 5.27. Amplitude attenuation coefficient of a 0.9-mm-diameter, 21-cm-long stainless-steel wire as a function of the radius of curvature R_c. (Data from Ref. [142].)

emitter into the metal wire [141, 143]. The radially symmetric photocurrents flowing between the two concentric electrodes produce radially-polarized THz radiation which matches well with the guided-wave mode. A numerical simulation predicts that the coupling efficiency can reach up to 60% [143].

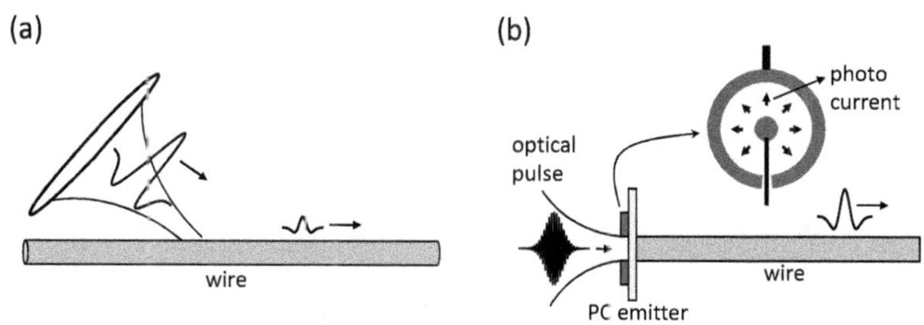

Fig. 5.28. (a) Scattering method used to couple linearly-polarized THz waves into a wire waveguide. (b) Direct launching of THz waves into a wire waveguide using a photoconductive emitter with a radially symmetric antenna structure.

5.5 Artificial Materials at Terahertz Frequencies

5.5.1 Metamaterials

A metamaterial refers to an artificially structured composite that exhibits exotic electromagnetic properties unattainable with naturally occurring materials. Figure 5.29 illustrates the basic concept behind electromagnetic metamaterials. The metamaterial consists of artificially structured elements embedded in a uniform matrix. As an electromagnetic wave interacts with the metamaterial, whose structural elements are smaller than the wavelength, the material system responds to the the wave like a homogeneous medium. That is, the electromagnetic properties of the metamaterial are characterized by an effective permittivity $\epsilon(\omega)$, and an effective permeability $\mu(\omega)$, independent of space. It is of great interest that some structural elements of particular design can give rise to extraordinary optical properties that are markedly different from those of the constituent materials. In particular, negative refractive index metamaterials have attracted a great deal of scientific and technological interests. While the peculiar properties of negative refraction were first predicted by Victor Veselago several decades ago [144], these material systems had remained hypothetical until 2000 when a composite medium showed simultaneously negative ϵ and μ over a band in the microwave range [145]. Since then, metamaterials have been demonstrated over a wide range of the electromagnetic spectrum from radio frequencies [146] to the near-infrared [147], including THz metamaterials [148].

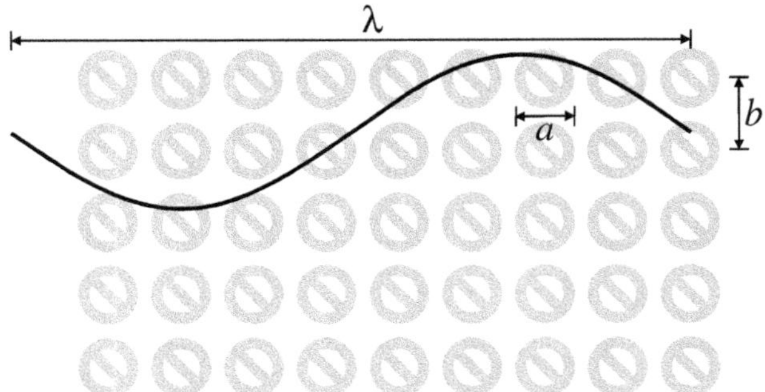

Fig. 5.29. Electromagnetic metamaterial

In order to understand the effects of negative refraction, it is necessary to review the basic concepts and theory of the interaction between electromagnetic waves and matter (see section 2.1). The refractive index of an isotropic medium is defined as

$$n(\omega)^2 = \epsilon_r(\omega)\mu_r(\omega), \tag{5.57}$$

where $\epsilon_r(\omega) = \epsilon(\omega)/\epsilon_0$ and $\mu_r(\omega) = \mu(\omega)/\mu_0$, and the refraction of an electromagnetic wave at an interface between vacuum and the medium is governed by Snell's law,

$$\sin\theta = n\sin\theta_r, \tag{5.58}$$

where θ is the angle of incidence and θ_r is the angle of refraction. All naturally occurring dielectric media have positive ϵ_r and μ_r, thus $n^2 > 0$. Although

Eq. 5.57 implies that n can be either $+\sqrt{\epsilon_r \mu_r}$ (> 0) or $-\sqrt{\epsilon_r \mu_r}$ (< 0), the boundary conditions at the interface warrant that θ_r is positive, and hence n is positive, too (Fig. 5.30(a)). Because the directions of the electric field $\mathbf{E}$, the magnetic field $\mathbf{B}$, and the wave vector $\mathbf{k}$ of an electromagnetic wave in this medium follow the so-called right-hand rule (e.g., $\mathbf{E} \parallel \mathbf{e}_x$, $\mathbf{B} \parallel \mathbf{e}_y$, and $\mathbf{k} \parallel \mathbf{e}_z$), a medium with positive n is called a right-handed material (RHM). On the other hand, if $\epsilon_r < 0$ and $\mu_r < 0$, θ_r has a negative value which satisfies the the boundary conditions, and thus the medium has a negative refractive index ($n < 0$), and is called a left-handed material (LHM) (Fig. 5.30(a)).

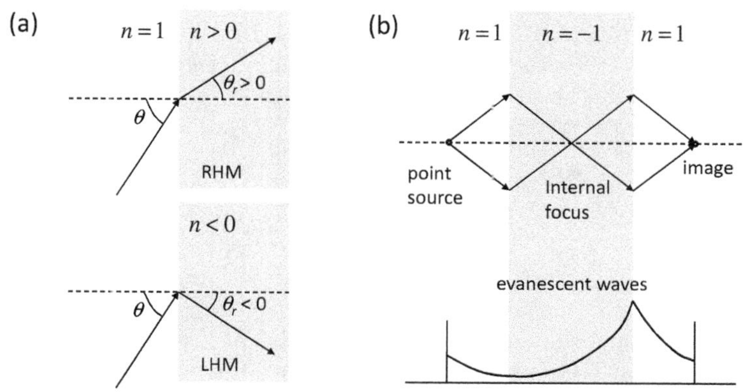

Fig. 5.30. (a) Refraction by right-handed and left-handed materials (b) Perfect focusing by a slab of a negative refractive index material

Arguably, the most important application of negative refraction is high-resolution imaging by use of perfect focusing, as shown in Fig. 5.30(b) [149]. The slab of LHM functions as a perfect lens, focusing the near-field as well as the far-field components of the point source. Consequently, the image can have spatial resolution beyond the diffraction limit. Perfect focusing is attainable because the near-field or evanescent components, which rapidly decay in RHMs, grow exponentially in LHMs.[1] Other notable ramifications of negative refraction include the phase velocity being antiparallel to the Poynting vector, the Doppler shift is reversed, and Cherenkov radiation is in the backward direction.

Split-ring resonators (SRRs) are the most commonly used magnetic elements to form a metamaterial. Figure 5.31(a) shows the schematic of an SRR consisting of two concentric bands of nonmagnetic conductor. When an oscillating magnetic field applied to the SRR has a nonvanishing component in

[1] The evanescent waves are characterized by large transverse wave vectors, and thus carry information of the subwavelength-scale variation. The growth of evanescent waves in LHMs does not violate the conservation of energy because no energy is transported by evanescent waves.

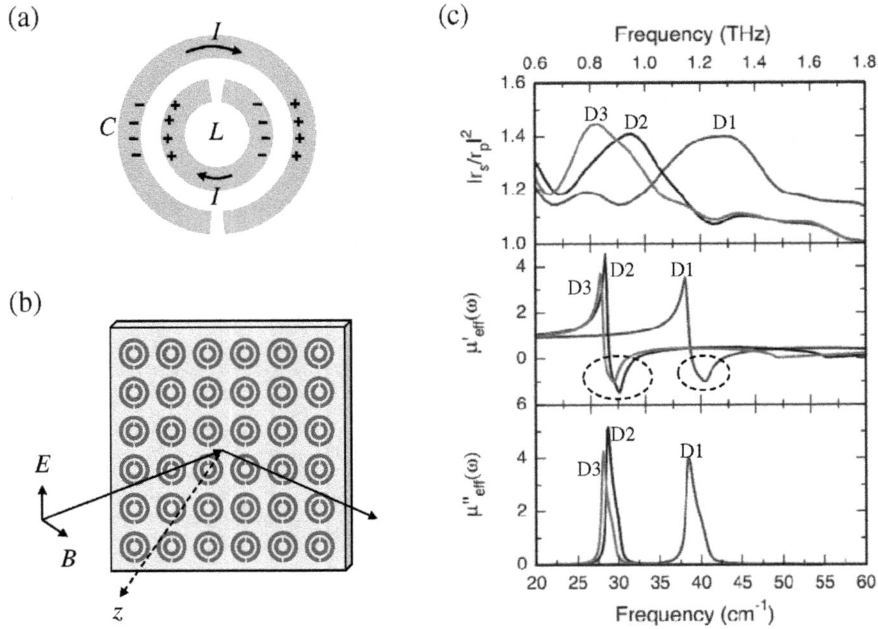

Fig. 5.31. Magnetic response of conducting split-ring resonators (SRRs) at THz frequencies. (a) SRR Geometry: equivalent to an LC resonator. (b) Ellipsometry measurements on a two-dimensional SRR array deposited on a dielectric substrate. (c) The ratio of the magnetic to electric response ($|r_s/r_p|^2$) (experiment) and the real (μ'_{eff}) and imaginary (μ''_{eff}) magnetic permeability (numerical simulation) for three samples of different sizes (D1<D2<D3). (From [148]. Reprinted with permission from AAAS.)

the direction normal to the surface, the changing magnetic field induces currents around the rings and, simultaneously, displacement currents $\left(J_D \propto \frac{\partial E}{\partial t}\right)$ produced by the accumulating charges flow through the gaps between the rings. Simply put, the SRR functions as an LC resonator of two capacitors and a double-ring inductor in series, driven by magnetic fields. Accordingly, the frequency-dependent effective permeability has the form

$$\mu_{\text{eff}}(\omega) = \mu'_{\text{eff}}(\omega) + i\mu''_{\text{eff}}(\omega) = 1 - \frac{F\omega^2}{\omega^2 - \omega_0^2 + i\gamma\omega}, \tag{5.59}$$

where μ' and μ'' are the real and imaginary part of the permeability, F is the oscillator strength depending on the SRR geometry, $\omega_0 \sim 1/\sqrt{LC}$ is the resonant frequency, and γ is the decay rate of the resonator due to resistive loss. The magnetic response of SRRs can be observed by the ellipsometry measurements shown in Fig. 5.31(b). As $B_z \neq 0$, the reflectivity coefficient r_s of s-polarization depends on the magnetic permeability of the SRR, while r_p exhibits no magnetic response because $B_z = 0$ for p-polarization. The top panel of Fig. 5.31(c) shows the ratio of the s- and p-pol reflectivities of three SRR samples as a function of frequency [148]. The broad peaks of the spectra are centered at the resonant frequencies scaling with sample dimensions: $\omega_1 > \omega_2 > \omega_3$ as D1<D2<D3. A numerical simulation, using the sample parameters as input, exhibits a strong diamagnetic behavior where $\mu'_{\text{eff}}(\omega)$ has negative values near and above the resonant frequencies.

Electrical response ca also be harnessed by artificial structures. In fact, it has been known for a long time that an array of conducting wires exhibit negative permittivity in a specific spectral range determined by its geometry. The frequency-dependent permittivity has the form

$$\epsilon_{\text{eff}}(\omega) = 1 - \frac{\omega_p^2}{\omega^2 - \omega_0^2 + i\gamma\omega}, \tag{5.60}$$

with the plasma frequency,

$$\omega_p^2 = \frac{n_{\text{eff}} e^2}{m^*}. \tag{5.61}$$

It gives the flexibility of controlling the electrical response, in that the effective electron density n_{eff} and mass m^*, as well as the resonant frequency ω_0, are governed by the geometry of the wire array. A three-dimensional structure of negative refractive index metamaterial can be constructed by alternatingly stacking layers of SRR arrays and conducting wire arrays [150].

An interesting application of metamaterials is to control and manipulate THz waves by use of active metamaterial devices. Figure 5.32(a) depicts the electric resonator element and the equivalent circuit of an active THz device. The parallel strips at the center function as capacitors, and the loops on the left- and right-hand sides as inductors. Because the loops are wound in opposite directions, there is no magnetic response from this element. The element is deposited on a n-GaAs layer on top of a SI-GaAs substrate. When no bias is applied between the element and the substrate, the split gap is shorted because of the relatively conductive substrate. Lacking capacitance, the element does not exhibit any resonant behavior. A reverse bias, however, depletes electrons in the n-GaAs layer, and hence the resistivity R_d between the split gap shoots up sharply, which makes the element act like an electric resonator. Figure 5.32(b) sketches the experimental arrangement for transmission measurements by THz-TDS. The active device consists of a planar array of metamaterial elements. All the elements are connected to a common electrode to serve as a Schottky gate. The active control of THz waves is shown in Figure 5.32(c). The polarization of the incident THz radiation is perpendicular to the connecting wires. In the transmission spectra of Fig. 5.32(c-1), the resonance dips at 0.72 and 1.65 THz grow as the gate bias is increased. The corresponding permittivity shown in Fig. 5.32(c-2) demonstrates significant modulations of the electric response of the device. As a comparison, the transmission spectra of a blank device with no metamaterial element (Fig. 5.32(c-3)) are featureless and independent of the gate bias.

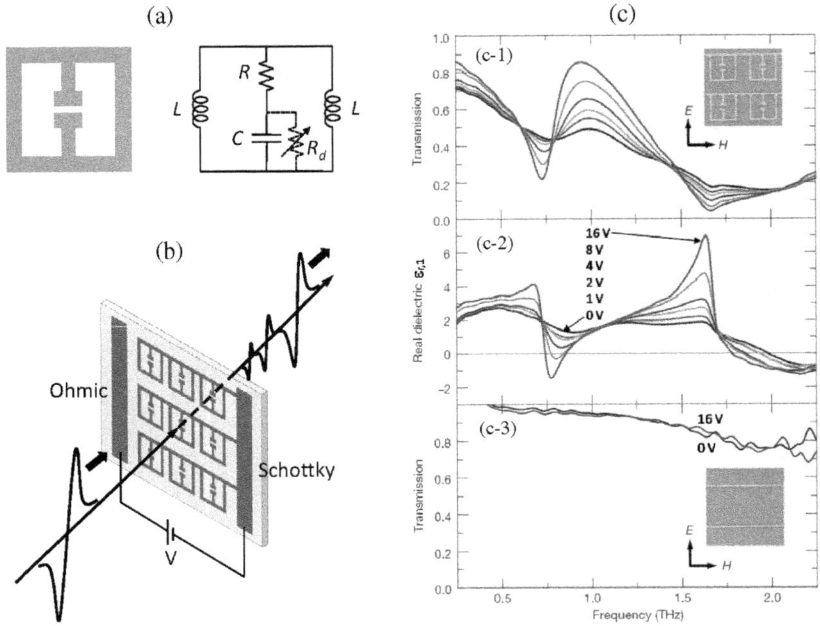

Fig. 5.32. Active control of THz waves by metamaterial devices. (a) Schematic of the structural element and the equivalent circuit. (b) Transmission measurement by THz-TDS (c) Switching performance for various gate biases. (c-1) Frequency dependent transmission and (c-2) the corresponding permittivity for gate biases from 0 to 16 V. (c-3) Transmission spectra of a blank device with no metamaterial element for gate biases of 0 and 16 V. (Reprinted by permission from Macmillan Publishers Ltd: Nature [151], ©2006.)

5.5.2 Photonic Crystals

A crystal is formed as a periodic array of atoms. In a crystalline conductor, electrons move freely, experiencing no scattering with atoms. An insulator has an energy band gap in which no electronic state is allowed. These unique electrical properties of crystals are governed by the wave nature of electrons subject to a periodic potential. Analogous to electron systems, a photonic crystal has a periodic lattice structure whose constituent media have distinctive dielectric constants. Figure 5.33 illustrates examples of one-, two-, and three-dimensional photonic crystals.

When an electromagnetic wave interacts with a photonic crystal whose lattice constant is comparable to the wavelength, the refracted and reflected waves from the lattice elements undergo strong interference, and ultimately form a standing wave called an electromagnetic mode. A group of modes form a continuous energy band. If the interference is completely destructive for

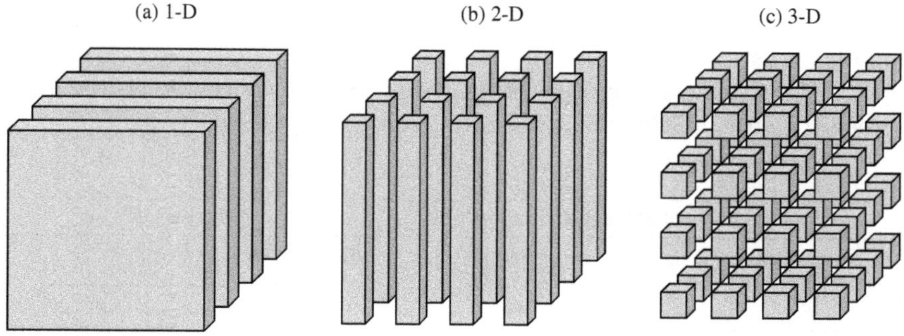

Fig. 5.33. Representative structures of 1-D, 2-D, and 3-D photonic crystals.

some frequency range, corresponding modes are forbidden in the photonic crystal, and the forbidden energy range is called a photonic band gap.

Naturally, we can ask what the photonic bandgap structures are good for. From a practical point of view, photonic crystals are an excellent base system upon which we can build compact, integrated optical circuits. As we discussed briefly in section 5.4.5, radiation losses from even a slight bend in nominal waveguides are significant as well as unavoidable. Waveguides built in a photonic crystal, however, can transport light with little or no loss at sharp bends. Therefore, together with other optical components, switching, mixing, and modulating optical signals can be accomplished in a small scale device based on photonic crystals.

A photonic crystal of dielectric media is characterized solely by a periodic dielectric function conforming to the relation

$$\epsilon_r(\mathbf{r}) = \epsilon_r(\mathbf{r} + \mathbf{u}) \tag{5.62}$$

for all Bravais lattice vectors $\mathbf{u} = n_1\mathbf{a_1} + n_2\mathbf{a_2} + n_3\mathbf{a_3}$, where $\mathbf{a_1}$, $\mathbf{a_2}$, and $\mathbf{a_3}$ are the lattice vectors, and n_1, n_2, and n_3 are arbitrary integers. The electromagnetic modes in the photonic crystal take the generic form (see section 2.1)

$$\mathbf{H}(\mathbf{r}, t) = \mathbf{H}(\mathbf{r})e^{-i\omega t}, \tag{5.63}$$

$$\mathbf{E}(\mathbf{r}, t) = \frac{i}{\omega\epsilon_0\epsilon_r(\mathbf{r})}\nabla \times \mathbf{H}(\mathbf{r}, t), \tag{5.64}$$

where the spatial mode function $\mathbf{H}(\mathbf{r})$ is determined by the wave equation

$$\nabla \times \left[\frac{1}{\epsilon_r(\mathbf{r})}\nabla \times \mathbf{H}(\mathbf{r})\right] = \frac{\omega^2}{c^2}\mathbf{H}(\mathbf{r}). \tag{5.65}$$

Solving the wave equation together with the periodic boundary condition, we can obtain eigenvalues $\omega(\mathbf{k})$ and eigenfunctions $\mathbf{H_k}(\mathbf{r})$, where $\mathbf{k}$ is a Bloch wave vector lying in the Brillouin zone. The periodic boundary condition warrants an infinite number of discrete eigenvalues $\omega_n(\mathbf{k})$ for each $\mathbf{k}$, and $\omega_n(\mathbf{k})$ is a continuous function for a given band index number n. The band structure of a photonic crystal refers to the information contained in the dispersion relation $\omega_n(\mathbf{k})$. The wave equation is scalable, thus the basic concepts and theoretical framework are applicable to the entire electromagnetic spectrum.

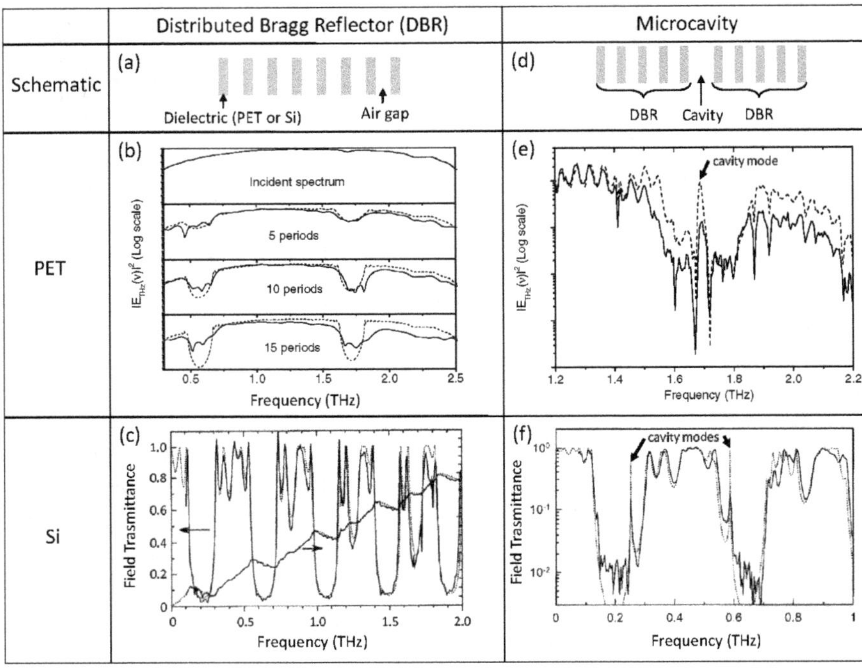

Fig. 5.34. 1-D photonic crystal devices at THz frequencies: distributed Bragg reflectors and microcavities. (a), (d) Schematics (b), (e) Spectra of broadband THz pulses transmitted through the components made of polyethylene terephthalate (PET). The dotted-lines depict theoretical calculations. One period of the structures consists of a 75-μm PET film (n=1.65) and a 125-μm air gap. The three DBR samples have 5, 10, and 15 periods. The 1-D resonant cavity consists of two DBRs with 2.5 periods of PET/air separated by a 250-μm air gap. (Data from Ref. [152]) (c), (f) Field transmittance of Si (n=3.42) components. The dotted-lines depict theoretical calculations. The DBR has three 100-μm Si wafers separated by 350-μm air layers. The air gap in the resonant cavity is 548-μm thick. (Reprinted from [153].)

Although they have not been called by this name, 1-D photonic crystals have been used as basic optical components such as mirrors, filters, and resonant cavities for a long time. Figure 5.34 shows a few examples of 1-D photonic crystal devices at THz frequencies. The devices are made of either polyethylene terephthalate (PET) or high-resistive silicon (Si). A distributed Bragg reflector (DBR) consists of multiple layers of alternating dielectric media, and has stopbands in which propagation of light is forbidden. Figure 5.34(b) and

(c) show stopbands of the PET and the Si DBRs in their transmission spectra measured by THz-TDS. The resonant cavities are formed by two DBRs separated by an air gap, and have cavity modes in the middle of the DBR stopbands. The cavity modes of the PET and Si microcavities are shown in Figure 5.34(e) and (f).

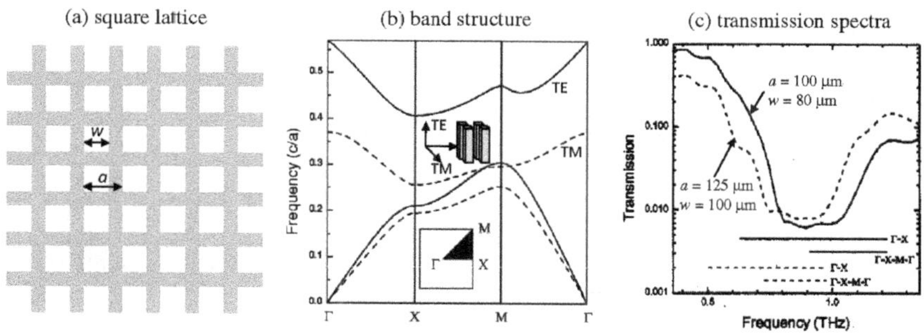

Fig. 5.35. (a) 2-D photonic crystal with square lattice. w is the hole width and a is the lattice constant. (b) Photonic band structure of Si (n=3.41) for w/a=0.8. Solid and dashed lines indicate TE and TM bands, respectively. The unit of frequency is c/a, where c is the speed of light in free space. The inset shows the irreducible Brillouin zone of a 2-D square lattice. (c) Transmission spectra for TE-polarized light propagating in the Γ-X direction: solid lines for a=100 μm ,w=80 μm and dotted lines for a=125 μm, w=100 μm. The horizontal lines correspond to the calculated 2-D TE band gap and the TE band gap in the Γ-X direction. (Reprinted with permission from [154]. ©2003, American Institute of Physics.)

Microfabrication techniques used for semiconductor devices have been employed to fabricate photonic crystals of high precision. 2-D photonic crystals with a square lattice, shown in Fig. 5.35(a), were made of high-resistivity Si using deep reactive ion etching [154]. Figure 5.35(b) depicts the photonic band structure of the square lattice ($w/a = 0.8$) for TE and TM polarizations. The TE polarization is perpendicular to the plane of the lattice, and the TM polarization is parallel to the plane. The TE and TM band gaps are $0.10c/a$ and $0.004c/a$, respectively. Figure 5.35(c) shows the spectra of TE-polarized THz radiation for the samples of a=125 μm and a=100 μm. The stopbands of the two samples are clearly shown and consistent with the numerical calculations of the TE band gap.

In general, no confinement is imposed on light in the vertical direction of a 2-D photonic crystal. A parallel-plate waveguide can be employed to accomplish vertical confinement. As we discussed in section 5.4.4, a TEM mode in a parallel-plate waveguide is not only bound in a narrow gap of conducting plates, but also is dispersionless. Figure 5.36(a) illustrates the schematic of a 2-D photonic bandgap structure embedded in a parallel-plate waveguide. The square lattice consists of SU-8 polymer (n = 1.7) cylinders.

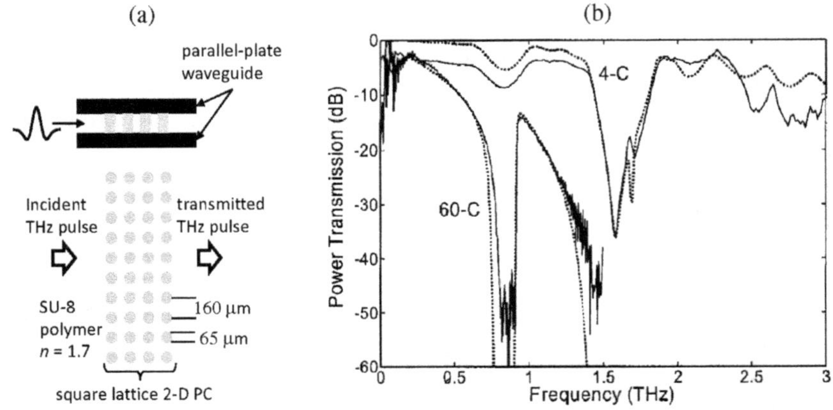

Fig. 5.36. (a) Schematic diagram of the THz transmission measurements of a 2-D photonic crystal, a square lattice of SU-8 polymer cylinders (n=1.7), embedded in a parallel-plate waveguide. The dielectric cylinders are 65-μm diameter and 70-μm high, and the lattice constant is 160 μm. (b) Transmission spectra for the 4-column and 60-column samples. The dotted lines show theoretical calculations. (Reprinted from [155].)

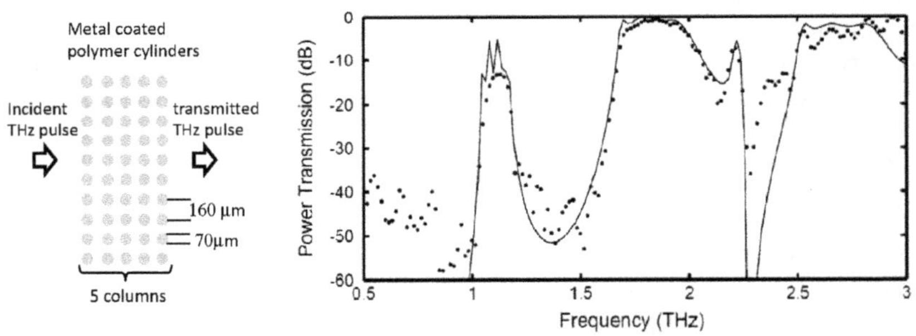

Fig. 5.37. Measured (dots) and calculated (solid line) transmission spectra of a 2-D metallic photonic crystal of square lattice (a=160 μm). The sample has five columns of gold coated cylinders (70-μm diameter and 80=μm height). (Reprinted from [156]. ©2007 IEEE)

Figure 5.36(b) shows the transmission spectra of 4-column and 60-column samples. SU-8 is relatively lossy at THz frequencies ($\alpha \cong 18\nu_{THz}$ cm^{-1}, where ν_{THz} is in THz), and hence absorption is discernible. It becomes stronger at higher frequencies and/or with more columns. The spectral features of the stopbands are consistent with the theoretical calculations depicted as dotted lines.

The transmission loss can be reduced by coating the dielectric cylinders with metal. Metallic photonic crystals are impractical in the optical region because of large ohmic losses, but the effects are much smaller in the THz regime.

Figure 5.37 shows the transmission spectrum of a 2-D metallic photonic crystal. The gold coating on the cylinders (~2.5-μm thick) is considerably thicker than the penetration depth of gold at THz frequencies ($\delta < 0.1$ μm at 1 THz), thus the photonic bandgap structure is considered to be purely metallic.

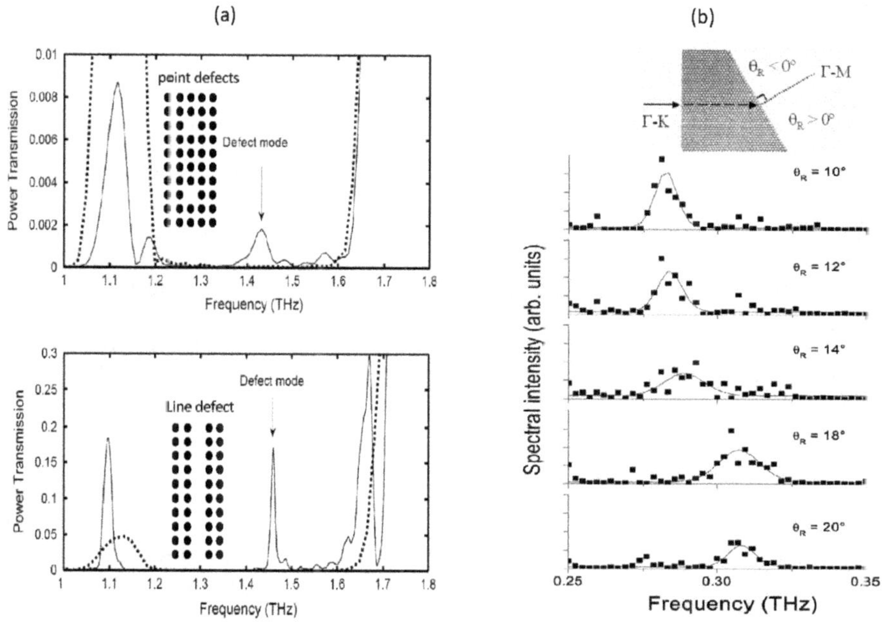

Fig. 5.38. THz devices based on 2-D metallic photonic crystal (a) Defect modes of two samples with point defects and a line defect. The square lattice (a=160 μm) consists of gold coated cylinders (70-μm diameter and 80=μm height). (Reprinted from [156]. ©2007, IEEE) (b) Anomalous refraction from a superprism containing a hexagonal array (a=400 μm) of circular holes (360-μm diameter) in a 305-μm thick Si wafer. (Reprinted from [157].)

A few examples of THz devices based on 2-D photonic crystals in a parallel-plate waveguide are shown in Fig. 5.38. Defects in a photonic crystal give rise to defect modes. Figure 5.38(a) shows the sharp features of defect modes of two square lattice samples containing point defects and a line defect, respectively. An experimental demonstration of the superprism effect is shown in Fig. 5.38(b). Since a photonic crystal is, in general, highly dispersive due to its complex band structure, a prism made of photonic crystal can exhibit a drastic change of refraction angle with a slight change of incident angle. This phenomenon is called the superprism effect. Certain superprism structures even allow negative refraction. The spectrum of diffracted THz radiation shown in Fig. 5.38(b) displays an abrupt spectral shift as the angle is varied over a small range, the characteristic superprism effect.

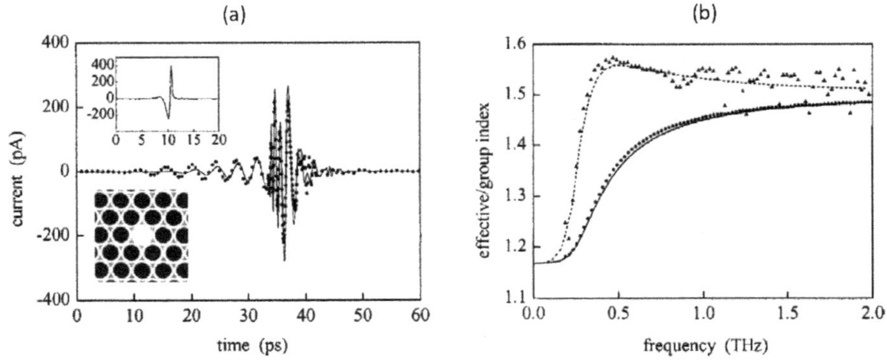

Fig. 5.39. Guided-wave propagation in a THz photonic crystal fiber. (a) Measured (dots) and calculated (solid line) THz waveform after propagating through a 2-cm long plastic photonic crystal fiber. The insets show the measured input waveform and an optical micrograph of the cross-section of the photonic crystal fiber with a high index defect. (b) Effective (dots: measured, solid line: calculated) and group (triangles: measured, dashed: calculated) indices. (Reprinted with permission from [158]. ©2002, American Institute of Physics.)

Another type of THz devices utilizing 2-D photonic crystals is photonic crystal fibers. The inset at the bottom of Fig. 5.39 shows the cross-section of a photonic crystal fiber made of HDPE tubes [158]. A high-index defect of HDPE is placed at the center of the structure. THz waves propagate along the defect in the vertical direction of the photonic crystal. The lattice constant is 500 μm, and the tube thickness is 50 μm. Figure 5.39(a) shows a transmitted THz waveform through a 2-cm-long sample. The input waveform is shown in the inset at the top. The effective and group indices of the photonic crystal fiber are obtained from the data. The measured values agree well with theoretical calculations, as shown in Fig. 5.39(b).

Fabrication of 3-D photonic crystals has been demonstrated in several different ways. A few notable techniques are summarized in Fig. 5.40. A relatively simple method is to stack up thin layers on which a 2-D structure is etched off. Overall symmetry of the resulting 3-D structure depends on the relative orientations and displacements between the layers as well as the 2-D structure. Figure 5.40(a) gives an example [159]. Conventional KOH etching is employed to construct a grating structure on 100-μm-thick (110) silicon wafers, taking off 185-μm-wide gaps separated by 50-μm-wide stripes. Four layers in the stacking direction make up one period: adjacent layers are 90 degrees off and alternate layers are displaced by a half-period of the grating. The transmission spectrum of a 16-layer sample shows a clear bandgap centered at 0.45 THz. The band edges agree well with the theoretical calculation indicated by the arrows. Figure 5.40(b) shows a construction scheme based on an epitaxial technique, which is capable of fabricating very fine structures. The scanning electron microscope image depicts a 3-D photonic crystal of

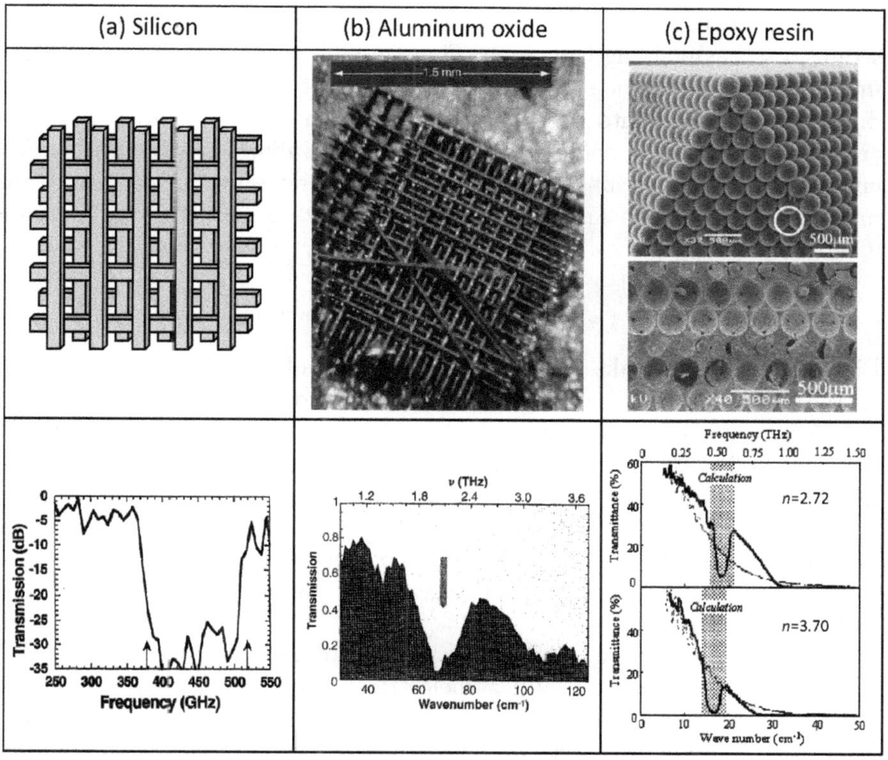

Fig. 5.40. Structures and transmission spectra of 3-D photonic crystals (a) Stack of micromachined (110) silicon wafers. Each layer (thickness of 100-μm) consists of 50-μm-wide rods separated by 185-μm-wide gaps. Four layers in the stacking direction make up one period. The transmission spectrum at normal incidence is for a 16-layer sample. The arrows indicate calculated band-edges. (Reprinted from [159].) (b) Photonic crystals of aluminium oxide fabricated by laser-assisted chemical vapor deposition. Each layer consists of 40-μm-diameter rods separated by 133 μm. The transmission spectrum at normal incidence shows a stopband at 2 THz. (From [160]. Reprinted with permission from AAAS.) (c) Photonic crystals of epoxy resin fabricated by inverting metal sphere templates. Scanning electron microscope images of the copper sphere (diameter of 267 μm) template and the cross section of an inverse crystal composed of air-spheres are shown in the top panel. Transmission spectra of two 13-layer samples for the waves propagating in the [111] direction are shown in the lower panel. The crystals have a dielectric constants of 2.72 (top) and 3.70 (bottom). Dotted lines represent bulk media without the crystal. (Reprinted with permission from [161]. ©2004, American Institute of Physics.)

aluminium oxide fabricated by laser-assisted chemical vapor deposition. The structure is composed of layers of parallel rods of 40-μm-diameter separated by 133-μm, forming a face-centered tetragonal lattice. Growing vertically, each rod is directly built from chemical vapors with the aid of a 2-mW, 488-nm laser beam focused on a spot size 5 μm in diameter. Figure 5.40(c) demonstrates a self-assembling technique. The scanning electron microscope (SEM) image at the top shows a metal template in which copper spheres of 267-μm diameter are packed into an fcc lattice (lattice constant of 378 μm). A reverse crystal of a dielectric medium is constructed by submerging the template in curable epoxy and etching away the copper spheres after curing. The transmission spectra of two samples (n=2.72 and 3.70) exhibit stopbands around 0.5 THz.

5.5.3 Plasmonics

Plasmonic devices make up an important part of THz applications. Some of the THz devices discussed in previous sections—bandpass filters (section 5.3.3), wire-grid polarizers (section 5.3.4), and metal-wire waveguides (section 5.4.5)–actually rely on phenomena associated with THz surface plasmons at metal-insulator interfaces. Phenomenological models were provided to account for how the devices function. In this section, we shall look into the underlying physical processes of surface plasmons at THz frequencies, and then discuss a few exquisite phenomena applicable to high-resolution imaging and nonlinear optics.

Surface plasmons are collective oscillations of free electrons at the boundary between a conducting and a dielectric media. The undulating surface charge density accompanies electromagnetic modes trapped at the interface. The strong coupling between a photon and a surface plasmon is called a surface plasmon polariton. Figure 5.41(a) illustrates the coupling between an electromagnetic wave and a charge density fluctuation. Due to the surface charges, the electric fields have components normal to the surface at the boundary, while the magnetic fields are transverse. The fields decay exponentially with distance from the surface as shown in Fig. 5.41(b).

As the surface plasmon propagates in the x direction, the electric and magnetic fields in the two media are expressed as

$$\mathbf{E}_d(\mathbf{r},t) = (\mathbf{e}_x E_{d,x} + \mathbf{e}_z E_{d,z})\, e^{-\kappa_d z} e^{i(k_x x - \omega t)}, \tag{5.66}$$

$$\mathbf{H}_d(\mathbf{r},t) = \mathbf{e}_y H_d\, e^{-\kappa_d z} e^{i(k_x x - \omega t)}, \tag{5.67}$$

$$\mathbf{E}_m(\mathbf{r},t) = (\mathbf{e}_x E_{m,x} + \mathbf{e}_z E_{m,z})\, e^{\kappa_m z} e^{i(k_x x - \omega t)}, \tag{5.68}$$

$$\mathbf{H}_m(\mathbf{r},t) = \mathbf{e}_y H_m\, e^{\kappa_m z} e^{i(k_x x - \omega t)}, \tag{5.69}$$

where $\mathbf{E}_d$ and $\mathbf{H}_d$ are in the dielectric medium ($z > 0$)), and $\mathbf{E}_m$ and $\mathbf{H}_m$ in the metal ($z < 0$)). The surface wave is characterized by the the wavenumbers k_x, κ_d, and κ_m for a given frequency ω. The generic wave equation, Eq. 2.15, yields the dispersion relations,

$$k_x^2 - \kappa_d^2 = \epsilon_d \frac{\omega^2}{c^2}, \tag{5.70}$$

$$k_x^2 - \kappa_m^2 = \epsilon_m \frac{\omega^2}{c^2}, \tag{5.71}$$

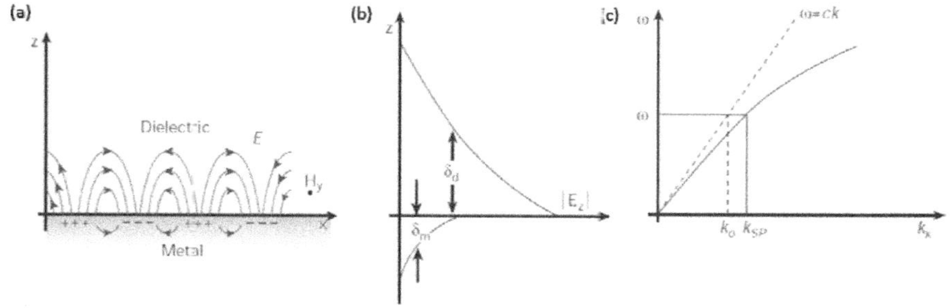

Fig. 5.41. Surface plasmons at the interface between a metal and a dielectric. (a) Electromagnetic wave coupled to surface charge density fluctuation. (b) Exponential decay of $|E_z|$ with decay lengths δ_d in the dielectric medium and δ_m in the metal. (c) Dispersion curve for surface plasmons (Reprinted by permission from Macmillan Publishers Ltd: Nature [162], ©2003.)

where ϵ_d and ϵ_m are the dielectric constants of the dielectric and the conducting media. Here we assume that the materials are nonmagnetic. Combining these two equations, we get the relation of the ratio κ_d/κ_m expressed in terms of k_x and ω:

$$\frac{\kappa_d^2}{\kappa_m^2} = \frac{k_x^2 - \epsilon_d\omega^2/c^2}{k_x^2 - \epsilon_m\omega^2/c^2}. \tag{5.72}$$

Applying Maxwell's equation, Eq. 2.3, we obtain the relations between E_x and H_y:

$$H_{d,y} = -i\epsilon_0\frac{\epsilon_d\omega}{\kappa_d}E_{d,x}, \tag{5.73}$$

$$H_{m,y} = i\epsilon_0\frac{\epsilon_m\omega}{\kappa_m}E_{m,x}. \tag{5.74}$$

H_y and E_x, which parallel to the surface, are continuous at the boundary. The boundary conditions lead to

$$\frac{\epsilon_d}{\epsilon_m} = -\frac{\kappa_d}{\kappa_m}. \tag{5.75}$$

Substituting Eq. 5.75 into Eq. 5.72, we obtain the dispersion relation of surface plasmons,

$$k_x = \frac{\omega}{c}\left(\frac{\epsilon_d\epsilon_m}{\epsilon_d+\epsilon_m}\right)^{1/2}. \tag{5.76}$$

As shown in Fig. 5.41(c), the surface plasmon wavenumber k_{SP} is always greater than that of the free-space wavenumber k_0 for a given frequency ω. This momentum mismatch means that surface plasmons are nonradiative, and, conversely, free-space radiation cannot directly excite them.

Figure 5.42 sketches a typical scheme used to couple light to a surface mode, which utilizes evanescent waves at an interface of dielectric media to

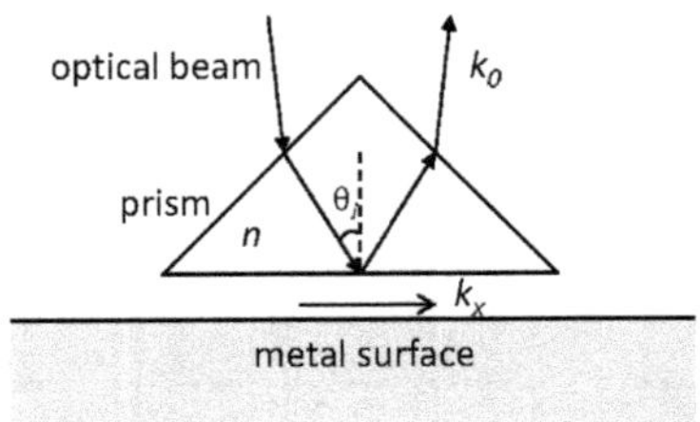

Fig. 5.42. Schematic for coupling light to a surface mode by use of evanescent waves at a prism-air interface

circumvent the momentum mismatch between surface plasmons and free-space radiation. With the angle of incidence θ_i beyond the critical angle of incidence ($\theta_c = \sin\frac{1}{n}$), the optical beam undergoes total internal reflection, which gives rise to an evanescent wave at the prism-air interface. Since $n \sin\theta_i > 1$, the evanescent wave can couple to a surface plasmon wave provided that $k_{SP} = k_x$, where $k_x = nk_0 \sin\theta_i (> k_0)$.

Substituting Eq. 5.76 into Eqs. 5.70 and 5.71, we obtain

$$\kappa_d^2 = -\frac{1}{\epsilon_m + 1}\frac{\omega^2}{c^2}, \tag{5.77}$$

$$\kappa_m^2 = -\frac{\epsilon_m^2}{\epsilon_m + 1}\frac{\omega^2}{c^2}, \tag{5.78}$$

We assume that the dielectric medium is air, i.e., $\epsilon_d = 1$. In the THz regime, ϵ_m is mostly imaginary and $|\epsilon_m| \gg 1$ (see section 5.2.3), thus the attenuation lengths δ_d and δ_m have the relations,

$$\delta_d = \frac{1}{\Re[\kappa_d]} \cong \frac{\sqrt{|\epsilon_m|}}{\sqrt{2}\pi}\lambda_0 \gg \lambda_0, \tag{5.79}$$

$$\delta_m = \frac{1}{\Re[\kappa_m]} \cong \frac{1}{\sqrt{2|\epsilon_m|}\pi}\lambda_0 \ll \lambda_0, \tag{5.80}$$

where $\lambda_0 = 2\pi/k_0$. δ_m is virtually identical to the skin depth of the metal. It is noteworthy that $k_x \approx k_0$, that is, surface plasmons on a flat metal surface propagate at velocities near the speed of light. Eq. 5.79 indicates that the surface modes at THz frequencies extend into the dielectric medium with distances of many wavelengths. Using $|\epsilon_m| \sim 10^6$ for common metals (see section 5.2.3), we estimate δ_d ~10 cm at THz frequencies. Propagation of THz surface waves on a metal surface have been observed in experimental studies [163, 164]. The measured attenuation length δ_d in air is ~1 cm, considerably shorter than the theoretical prediction. This indicates the difficulty of establishing and maintaining the large spatial extent of surface modes in realistic arrangements.

Plasmonics is of great interest in the optical regime, because surface modes at optical frequencies are confined to a small region of subwavelength scale,

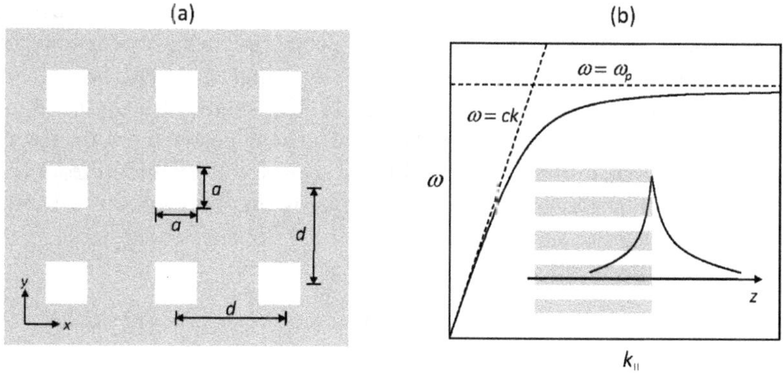

Fig. 5.43. Spoof surface plasmons on a structured surface [165]. (a) Perfect conductor with a square-hole lattice. (b) Dispersion curve for spoof surface plasmons. Inset shows the effective evanescent field of a surface mode.

which results in nanoscale spatial resolution and drastic field enhancement. The subwavelength confinement, however, is obtained only at frequencies near the plasma frequency, which is in the ultraviolet in most metals. In order to circumvent the limitation of low-frequency waves, THz plasmonic devices employ structured surfaces to localize and manipulate surface modes. A simple example is shown in Fig. 5.43 [165]. The structured surface is composed of a perfect conductor perforated with a square-hole array. The holes and their spacings are much smaller than the wavelength of radiation, $a < d \ll \lambda$, such that electromagnetic responses of the structure can be described by effective permittivity ($\epsilon_x = \epsilon_y = \epsilon_{\|}$ and ϵ_z) and permeability ($\mu_x = \mu_y = \mu_{\|}$ and μ_z) [165].

It is easy to evaluate the z-axis electromagnetic response of the structured material. Electrons can move freely inside a perfect conductor, and thus it is infinitely susceptible to external electric and magnetic fields, i.e., $\epsilon = \mu = \infty$. No restriction is yet imposed on electron motions in the z-axis, thus $\epsilon_z = \mu_z = \infty$. When it comes to the in-plane responses, calculations based on the effective medium model yield

$$\mu_{\|} = \mu_0 \left(\frac{8a^2}{\pi^2 d^2} \right) \tag{5.81}$$

and

$$\epsilon_{\|} = \epsilon_0 \frac{\pi^2 d^2}{8a^2} \left(1 - \frac{\omega_p^2}{\omega^2} \right), \tag{5.82}$$

where the effective plasma frequency ω_p is defined as

$$\omega_p = \frac{\pi c}{a}. \tag{5.83}$$

It is notable that ω_p is equivalent to the cutoff frequency of the square waveguide (see Eq. 5.35). While propagation of electromagnetic waves is forbidden below the cutoff frequency, exponentially decaying fields can exist near the opening of the waveguide (see section 5.4.1). Consequently, the holes permit effective homogeneous fields to penetrate into the effective medium for a finite distance, yet electric and magnetic fields are completely excluded from the perfect conductor. The penetration depth depends on the in-plane wavenumber $k_{\|}$ and frequency ω:

$$\delta_S = \left(k_{\|}^2 - \frac{\omega^2}{c^2}\right)^{-1/2} . \tag{5.84}$$

Applying boundary conditions together with the effective dielectric constant, we obtain the dispersion relation for the surface mode,

$$k_{\|}c^2 = \omega^2 + \left(\frac{64a^4}{\pi^4 d^4}\right)\frac{\omega^4}{\omega_p^2 - \omega^2}. \tag{5.85}$$

The dispersion curve is shown in Fig 5.43(b). It has the identical form of a typical dispersion relation for surface plasmons. The term *Spoof surface plasmon* was coined referring to surface plasmons on a structured surface. The frequencies near ω_p are associated with wavenumbers of large magnitude, and hence surface plasmons can be confined to a small area of the structured surface.

In this simple example, the effective plasma frequency is determined by the structural geometry and not by the intrinsic properties of the base material. This seminal property is generally applicable to any structured surface. An important consequence is that subwavelength confinement of surface plasmons is achievable in the low-frequency region by adjusting the effective plasma frequency to be near the radiation frequency of interest.

A fascinating phenomenon attributed to surface plasmons is that exceptionally high optical transmission is achievable through a periodic array of subwavelength-scale holes in a metal film in certain frequency bands [166]. The enhancement of transmission is associated with the photonic bands of surface plasmons formed by the periodic structure. At a resonant frequency, incident light matches its momentum to that of a surface plasmon together with the lattice momentum. The surface plasmon mode builds constructive interference, and radiates coherently into the other side of the conducting layer. The transmission, normalized by the ratio of hole area to the total area, can exceed unity at resonant frequencies. As we discussed in section 5.3.3, the resonantly enhanced transmission has been utilized for bandpass filters in the THz regime [131, 132]. A phenomenological model of resonant antennas was introduced to account for the extraordinary transmission. The concept of spoof surface plasmons gives insight into the underlying mechanisms of this extraordinary phenomenon.

It is instructive to examine the square hole array of Fig. 5.43(a) a little further. The periodic pattern on the metal surface gives rise to the formation of photonic bands of surface plasmons, supplying them with crystal momenta,

$$\mathbf{q} = l\, q_x \mathbf{e}_x + m\, q_y \mathbf{e}_y, \quad l, m = 0, \pm 1, \pm 2, \ldots, \tag{5.86}$$

where $q_x = q_y = 2\pi/d$. Incident light impinging on the surface can couple to surface plasmon modes with the momentum matching condition,

$$\mathbf{k}_{SP} = \mathbf{k}_{\parallel} + \mathbf{q}, \tag{5.87}$$

where $\mathbf{k}_{\parallel}$ is the component of the incident photon's wave vector parallel to the surface.

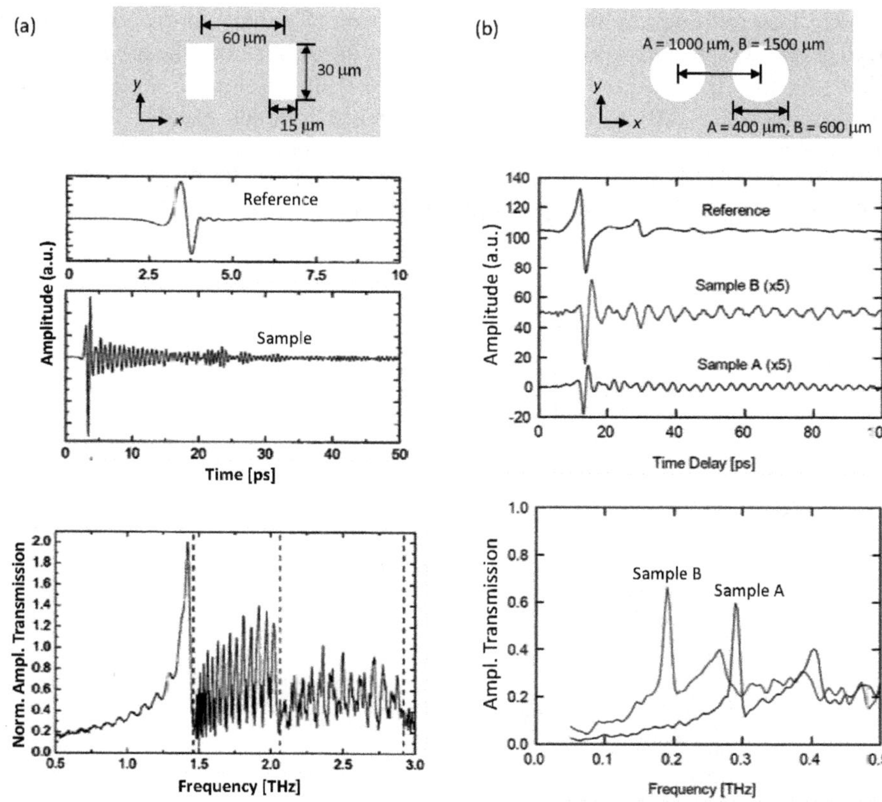

Fig. 5.44. Transmission resonances through a periodic array of subwavelength holes. (a) Rectangular hole array. Hole size is 15 μm×30 μm, and the grating period is 60 μm. (Reprinted with permission from [167]. ©2004, American Physical Society.) (b) Circular hole array. Hole diameter is 400 and 600 μm, and grating period is 1000 and 1500 μm for sample A and B, respectively. (Reprinted from [168].)

Figure 5.44 demonstrates experimental observations of transmission resonances at THz frequencies for periodic hole arrays on metal films [167, 168].

The transmission measurements were performed at normal incidence, and hence $k_{\|} = 0$. Consequently, the resonances occur at integral surface plasmon modes: $\mathbf{k}_{SP} = \mathbf{q}$. The rectangular hole array (hole dimension, 15 μm×30 μm, grating period, 60 μm) is constructed on a 0.25-μm-thick Al film deposited on a Si wafer (Fig. 5.44(a)), and the circular hole array (sample A: hole diameter of 400 μm, grating period of 1000 μm; sample B: hole diameter of 600 μm, grating period of 1500 μm) is perforated on a 75-μm-thick free-standing stainless steel film (Fig. 5.44(b)). The temporal waveforms of the incident and the transmitted THz pulses were measured by THz-TDS. The transmitted THz pulses have long tails of surface plasmon resonances, lasting a few hundred picoseconds. Subsequently, the transmission spectra contain strong resonance peaks. The rectangular hole array has a strongly enhanced transmission at 1.46 THz, which corresponds to the first integral surface plasmon mode of $(l, m) = (\pm 1, 0)$. The normalized amplitude transmission, $T \cong 2.0$, (power transmission is ∼4.0) at the resonant frequency is much greater than unity. The spectrum also shows the $(\pm 1, \pm 1)$ mode resonance at 2.06 THz. The resonances of the circular hole arrays (∼0.33 THz and ∼0.46 THz for sample A, ∼0.2 THz and ∼0.28 THz for sample B) correspond to $(\pm 1, 0)$ and $(\pm 1, \pm 1)$ modes.

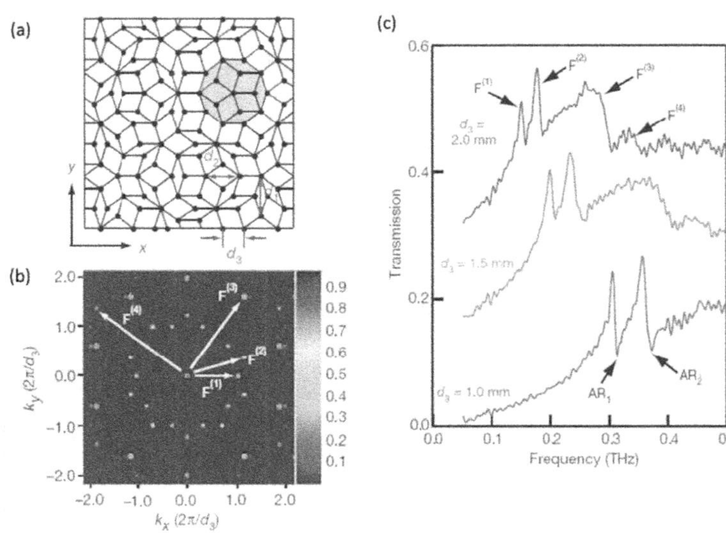

Fig. 5.45. Resonantly enhanced transmissions through circular holes patterned on a 2-D quasicrystal lattice structure. (a) A Penrose quasicrystal exhibiting local five-fold rotational symmetry (shaded area) with apertures at the vertices. The hole diameter is 400 μm, and the ratio of hole area to the total area is ∼0.12. (b) Geometrical structure factor of the quasicrystal. The reciprocal vectors, $\mathbf{F}(i)$, which exhibit ten-fold rotational symmetry, are assigned. (c) Transmission spectra of three Penrose type quasicrystal perforated films with different rhomb side lengths, d_3. (Reprinted by permission from Macmillan Publishers Ltd: Nature [169], ©2007.)

An interesting twist is that metal films with aperiodic hole arrays can also exhibit resonantly enhanced transmission. Figure 5.45(a) depicts a hole array patterned with a Penrose quasicrystal lattice having local five-fold rotational symmetry. The sample is a 75-μm-thick stainless steel film, the hole diameter is 400 μm, and the ratio of hole area to the total area is ~0.12. Figure 5.45(b) shows the geometrical structure factor of the quasicrystal calculated by the Fourier transform. It displays ten-fold rotational symmetry. The circular spots indicate the reciprocal vectors, $\mathbf{F}(i)$, satisfying the relation $e^{i\mathbf{F}\cdot\mathbf{R}} = 1$, where $\mathbf{R}$ is a lattice vector. Figure 5.45(c) shows the transmission spectra of three samples having different lattice constants, d_3=2000, 1500, and 1000 μm. The spectra exhibit transmission resonances corresponding to the reciprocal vectors, $\mathbf{F}(1)$, $\mathbf{F}(2)$, and $\mathbf{F}(3)$.

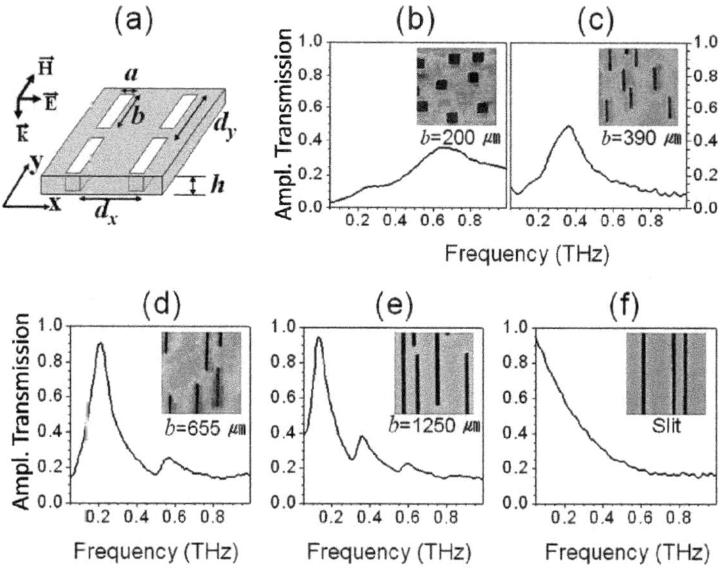

Fig. 5.46. (a) Schematic of a 2-D array of rectangular holes of width a and length b. The hole positions are randomized. (b)-(f) Amplitude transmission spectra, at normal incidence, of the random arrays of holes with a=70 μm and b=200, 390, 655, 1250, and 20000 μm, respectively. Insets are SEM images of the random arrays. (Reprinted with permission from [170]. ©2007, American Physical Society.)

Surprisingly, even a single hole can support resonant transmission, depending solely upon its shape. Figure 5.46 shows examples of shape-dependent transmission resonances [170]. Each sample is composed of a 2-D array of randomly located rectangular holes. The polarization of incident THz radiation is perpendicular to the long side of the holes (Fig. 5.46(a)). Transmitted THz pulses at normal incidence are measured by THz-TDS. While the transmission spectrum of the array of square holes features a weak and broad peak

(Fig. 5.46(b)), the arrays of rectangular holes exhibit sharp resonances determined by the long-side length, b, of rectangles. The fundamental shape resonance occurs at $\omega \approx \pi c/b$, the cutoff frequency of the rectangular waveguide, which is consistent with the effective medium model. The amplitude transmissions of Fig. 5.46(d)-(f) are near unity at the resonances. With the ratio of hole coverage $\sim$0.12 and b/a=9.4, the amplitude enhancement factor reaches up to 8, which confirms the theoretical prediction for a single rectangular hole, $3b/\pi a \cong 9$ [171].

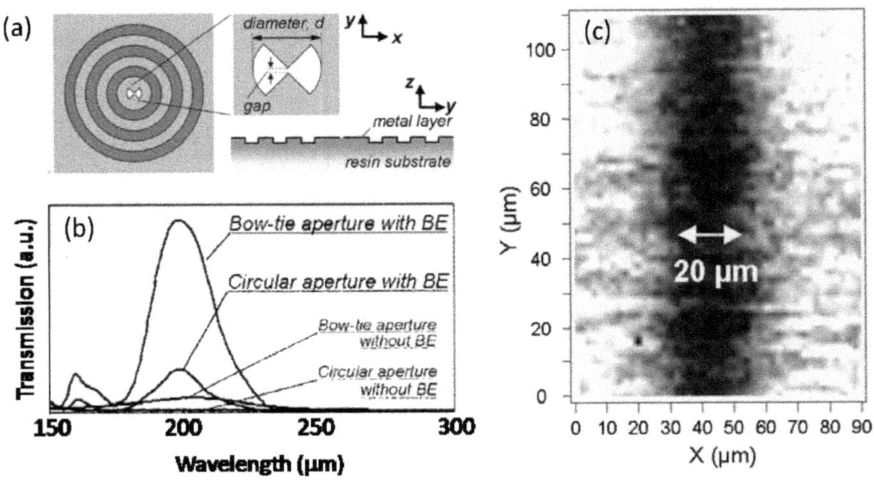

Fig. 5.47. THz imaging with subwavelength resolution by use of a metallic bull's eye (BE) structure. (a) Schematic diagram of a bow-tie aperture centered at the BE structure (E-field $\parallel$ y-axis). (b) Calculated transmission spectra for bow-tie and circular aperture with and without the BE. The aperture diameter and gap are 50 μm and 6 μm, respectively. The BE consists of six periodic concentric grooves: the depth is 13 μm and the period is 132 μm. The structure is patterned on a resin substrate with n=1.5 and is covered by a 2-μm-thick gold film. (c) Near-field image of 20-μm-wide Cr pattern at the wavelength of 207 μm (1.45 THz). (Reprinted with permission from [172]. ©2006, American Institute of Physics.)

Figures 5.47 and 5.48 show exemplary schemes to confine THz surface plasmons in a subwavelength scale [172, 173]. The periodic Bull's eye structure of Fig. 5.47(a) enhances the THz transmission through an aperture at the center. The transmission spectra shown in Fig. 5.47(b) indicate that the enhancement factor exceeds one order of magnitude and that the bow-tie aperture is much more efficient than the circular one. Figure 5.47(c) shows the near-field image of a 20-μm-wide Cr strip deposited on a resin substrate taken with a bow-tie aperture with a bull's eye at the wavelength of 207 μm. The spatial resolution of the image is 12 μm corresponding to $\lambda/17$. Figure 5.48 shows that subwavelength focusing is achievable by use of a periodically corrugated metallic cone. Surface plasmons are guided on the conical wire and superfocused at the tip of the cone.

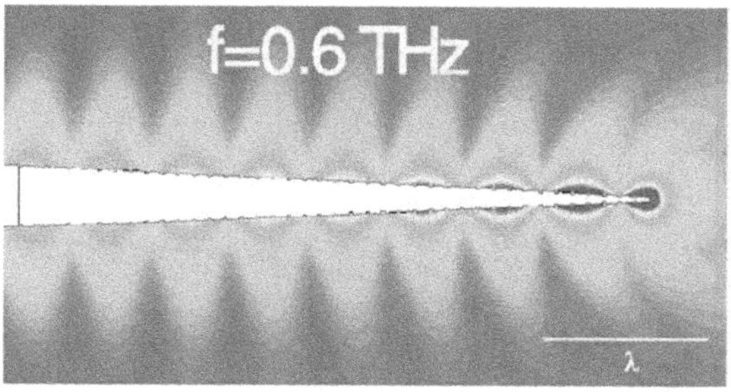

Fig. 5.48. Subwavelength focusing of THz surface plasmons on a corrugated cone. The length of the cone is 2 mm. The groove depth and period are 5 μm and 50 μm, respectively. The cone radius varies from 100 to 10 μm. The contour plot represents the E-field amplitude on a logarithmic scale of two orders of magnitude. (Reprinted with permission from [173]. ©2006, American Physical Society.)

5.6 Terahertz Phonon-Polaritons

In ionic crystals electromagnetic waves are strongly coupled to polar lattice vibrations near optical phonon resonances (see section 2.2.4). A phonon-polariton refers to a quasiparticle, resulting from strong coupling between a photon and an optical phonon. In noncentrosymmetric crystals, phonon-polaritons can be excited by femtosecond optical pulses via optical rectification, or, more precisely, impulsive stimulated Raman scattering. This is the same mechanism of THz generation we discussed in section 3.3. When a polariton wave impinges on a crystal/air interface, its electromagnetic component is coupled into air, and the transmitted part emerges as THz radiation in free space.

In a spectral region where an optical phonon mode is dominant, a dielectric constant $\epsilon_r(\omega)$ for an ionic crystal is given as

$$\epsilon_r(\omega) = \epsilon_r(\infty) + \frac{\epsilon_r(\infty) - \epsilon_r(0)}{\omega^2/\omega_T^2 - 1}, \tag{5.88}$$

where ω_T is the transverse-optical (TO) phonon frequency. Transverse electromagnetic waves can propagate only if the dispersion relation,

$$k = \frac{\omega}{c}\sqrt{\epsilon_r(\omega)}, \tag{5.89}$$

is satisfied. Figure 5.49 shows the phonon-polariton dispersion relation. The longitudinal-optical (LO) phonon frequency ω_L has a simple relation with ω_T:

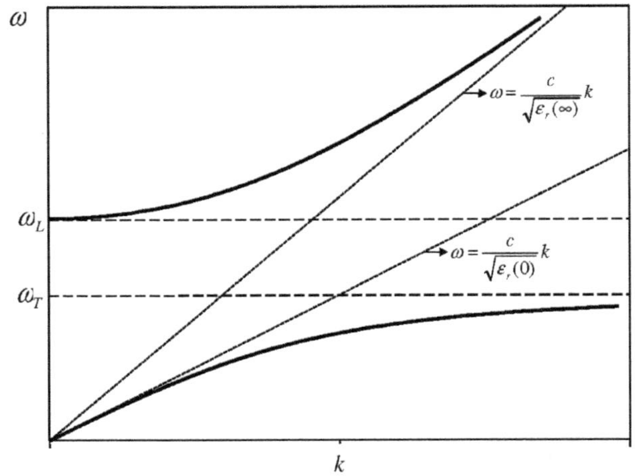

Fig. 5.49. Phonon-polariton dispersion relation

$$\omega_L = \sqrt{\frac{\epsilon(0)}{\epsilon(\infty)}}\,\omega_T. \tag{5.90}$$

The straight lines indicate the dispersion of uncoupled phonon and photon modes. On the high-frequency side, upper-branch polariton mode become photon-like and its propagation is characterized with dielectric constant $\sqrt{\epsilon(\infty)}$. As $\omega \to 0$, the lower-branch polariton mode behaves like an electromagnetic wave propagating at $c/\sqrt{\epsilon(0)}$. In the regions near ω_T and ω_L both polariton modes combine photon and phonon properties.

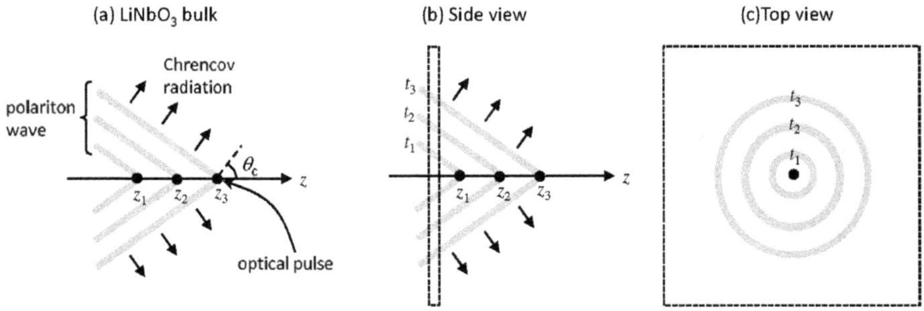

Fig. 5.50. (a) Polariton wave generated by a femtosecond pulse passing through a $LiNbO_3$ crystal. The black circles indicate a femtosecond optical pulse at different positions z_1, z_2, and z_3, and the gray lines depict wavefronts of Cherenkov radiation. The Cherenkov angle θ_c is 64°. (b) Side and (c) top view of the polariton propagation in the xy-plane.

Owing to their high optical transparency and strong nonlinearity, lithium niobate (LiNbO3) and lithium tantalate ($LiTaO_3$) crystals have been used to study optical excitation and detection of phonon-polaritons. Consider a femtosecond optical pulse propagating through a $LiNbO_3$ or $LiTaO_3$ crystal, which impulsively creates a THz nonlinear polarization via optical rectification (Fig. 5.50(a)). It is important to note that optical group velocity is significantly higher than THz phase velocity in these crystals. If the optical beam size is smaller than the wavelengths of THz radiation, the THz polarization behaves like a point source of a polariton wave which propagates through the crystal in the form of Cherenkov radiation. In a plane normal to the direction of optical pulse propagation, the Cherenkov cone forms an outgoing circular wave like one created by a pebble dropped into a pond, as shown in Fig. 5.50(b) and (c). Due to the huge velocity mismatch between optical and THz pulses, the Cherenkov angle is large ($\sim 64°$ in $LiNbO_3$ and $\sim 69°$ in $LiTaO_3$), and hence polariton waves propagate primarily in lateral directions.

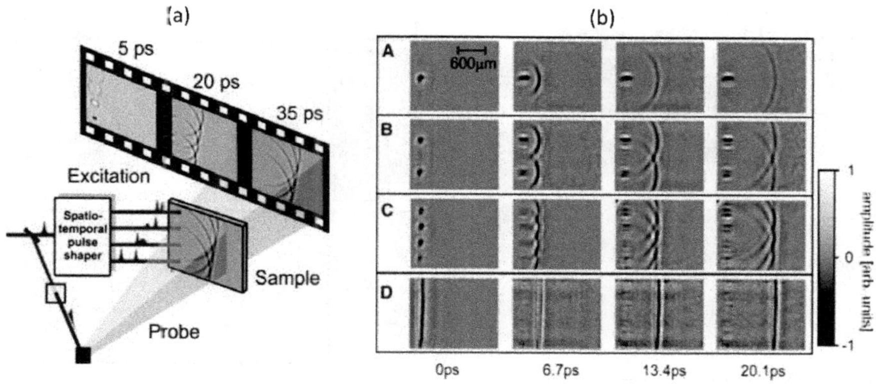

Fig. 5.51. (a) Schematic illustration of the spatiotemporal coherent control experiment. (b) Snap shots of polariton wave propagation in a $LiTaO_3$ after impulsive excitation with (A) one, (B) two, (C) four, and (D) nine excitation regions. (From [174]. Reprinted with permission from AAAS.)

Figure 5.51(a) illustrates an experimental scheme to directly visualize the evolution of polariton waves excited by femtosecond optical pulses. The experimental setup exploits a time-resolved pump-probe technique. Going through an optical pulse shaper, a pump pulse is transformed into a spatially and temporally controlled multiple pulses. The shaped optical pulses generate polariton waves in a thin $LiNbO_3$ or $LiTaO_3$ crystal, which replicates the spatial and temporal profile of the pump pulses. Usually, polariton wave frequencies ($\sim$1 THz) are far below the lowest phonon resonance (extraordinary $\omega_T/2\pi$ =7.4 THz for $LiNbO_3$ and 6.0 THz for $LiTaO_3$), thus the polariton waves propagate at a photon-like speed of $\sim c/\sqrt{\epsilon(0)}$. A spatially expanded probe beam transmits through the crystal and maps out the index change induced by the polariton waves. Temporal evolution of polariton waves can be recorded as controlling the relative time delay between pump and probe pulses.

Figure 5.51(b) shows sequences of snap shots depicting propagation of polariton waves in a $LiTaO_3$ crystal after optical pulses simultaneously illuminate one, two, four, and nine spots lined up with the crystal optic axis. The coherent polariton wavepackets emerging from the excitation spots undergo constructive and destructive interferences and form a superposed wavepacket in the far field.

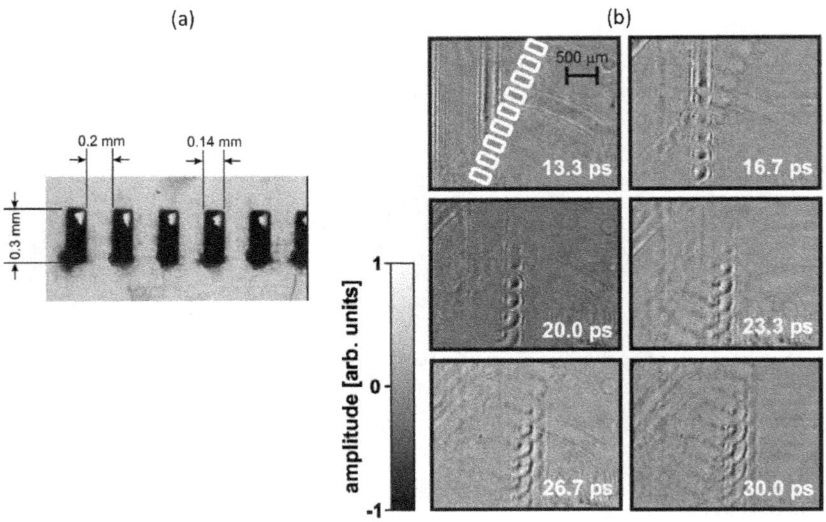

Fig. 5.52. (a) Optical micrograph of the laser-machined grating structure. (b) Evolution of a single-cycle polariton plane wave diffracted by the grating structure. (Reprinted with permission from [175]. ©2003, American Institute of Physics.)

In addition to the use of optical pulse shaping, permanent features patterned in a sample can manipulate polariton waves. A variety of polaritonic devices such as waveguides, resonators, and gratings have been demonstrated so far. Figure 5.52 shows an example that a polariton wave is diffracted by an integrated grating structure in a $LiNbO_3$ crystal. The grating pattern shown in Fig. 5.52(a) was imprinted in the sample by femtosecond laser machining. Figure 5.52(b) shows a sequence of snap shots illustrating that a single-cycle polariton plane wave is incident on the grating structure at an angle of 24°, passing through the structure while undergoing interferences and diffractions, and, eventually, building up the first- and second-order diffraction wavefronts.

6

Terahertz Spectroscopy of Atoms and Molecules

THz spectroscopy has been used to study a variety of physical phenomena from atomic transitions to dynamics of biological molecules, and hence involves a wide range of disciplines including physics, chemistry, engineering, astronomy, biology, and medicine. This chapter gives an overview of spectroscopic studies on atomic and molecular transitions at THz frequencies as well as technical schemes of spectroscopic instruments.

6.1 Manipulation of Rydberg Atoms

A Rydberg atom is an atom with electrons in states of high principle quantum number. Rydberg atoms provide tantalizing exhibitions of the transition from a microscopic quantum world into a macroscopic classical world. For example, a localized electron wavepacket (a superposition of multiple Rydberg states) was formed and manipulated to orbit around a nucleus by use of a series of laser pulses and a microwave field, mimicking Rutherford's classical atom [176]. Rydberg atoms supply an excellent test ground for the fundamental questions of light-matter interactions at the quantum-classical boundary. Figure 6.1 sketches a Rydberg atom having a single valence electron: the core electrons screen the nuclear charge, and hence the valence electron effectively sees a nucleus with one proton charge.

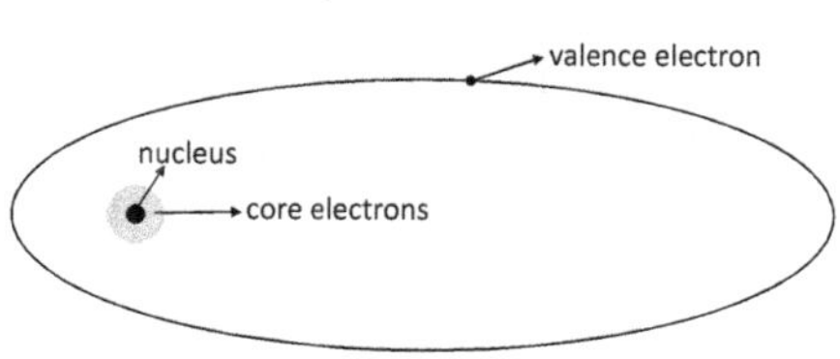

Fig. 6.1. Atom in a Rydberg state

The physical system of the Rydberg atom is well described by Bohr's semi-classical quantum theory of atomic electron orbits. The Bohr model depicts the hydrogen atom as an electron making a circular orbit around a nucleus with its angular momentum quantized. The consequent orbit is characterized by the quantized radius and energy level:

$$r_n = n^2 \frac{4\pi\epsilon_0\hbar^2}{m_e e^2} = n^2 a_0, \tag{6.1}$$

$$E_n = -\frac{m_e e^4}{32\pi^2\epsilon_0^2\hbar^2}\frac{1}{n^2} = -\frac{\mathcal{R}}{n^2}, \tag{6.2}$$

where the positive integer n is the principle quantum number,

$$a_0 = \frac{4\pi\epsilon_0\hbar^2}{m_e e^2} = 0.529\ \text{Å} \tag{6.3}$$

is the Bohr radius, and

$$\mathcal{R} = -\frac{\hbar^2}{2m_e a_0^2} = -13.6\ \text{eV} \tag{6.4}$$

is the Rydberg constant.

In Rydberg states of high n, binding energies decrease as $1/n^2$ and orbital radii increase as n^2, i.e., the valence electron has a large orbit and is loosely bound to the nucleus. For n=100, the geometrical size is estimated as $2n^2 a_0 \sim 1\ \mu$m. The consequent dipole moment, $d_R = n^2 e a_0 \sim 10^4$ D, is several orders of magnitude greater than those of small polar molecules such as CO and H_2O. The binding energy, $\mathcal{R}/n^2 \sim 1$ meV, is comparable to photon energies at THz frequencies. The large dipole moment and the small binding energy are favorable conditions for controlling Rydberg atoms with THz pulses. For example, the Rabi frequency (Eq. 2.138),

$$\omega_R = \frac{2d_R}{\hbar}E, \tag{6.5}$$

is in the THz region for a moderate electric field amplitude $E \sim 1-10$ kV/cm, i.e., a complete transition from one Rydberg state to another is accomplished on a picosecond time scale (see section 2.2.1 for more details about Rabi oscillations). From a classical point of view, slowly varying or static electric fields induce an ionization by deforming the Coulomb potential. The classical field ionization of Rydberg atoms has a threshold at field amplitudes in the range of 1-10 kV/cm, and occurs on a time scale of picoseconds. It is indeed tempting to consider how Rydberg atoms would respond to the short burst of electromagnetic fields in broadband THz pulses.

For experimental studies of THz interaction with Rydberg atoms, half-cycle THz pulses play an important role. Large-aperture photoconductive emitters produce high-power THz pulses with field amplitudes of 1-100 kV/cm (see section 3.2.4). As shown in Fig. 6.2, the pulse shape is characterized as a short and intense half-cycle (<1 ps) followed by a much longer and weaker half-cycle (>10 ps) of opposite polarity. Typical time scales of wave packet dynamics in Rydberg atoms are on the order of picoseconds, thus the effects of the long tail are often negligible and the THz pulses effectively function as half-cycle pulses.

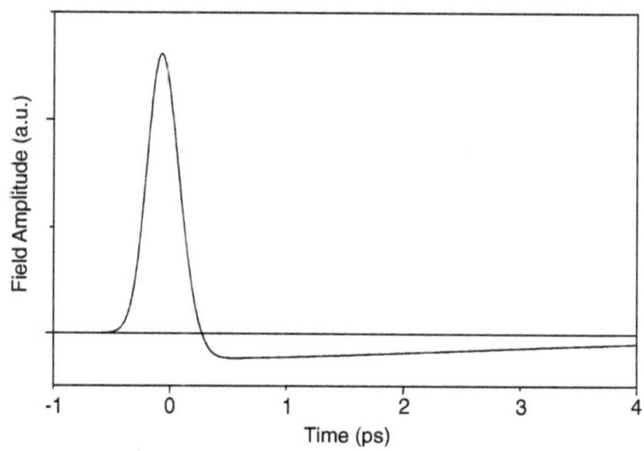

Fig. 6.2. Calculated waveform of a half-cycle THz pulse generated by a large-aperture photoconductive emitter

Ionization of Rydberg atoms by half-cycle THz pulses provide important insights into wave packet dynamics [177, 178]. The subpicosecond pulse duration of half-cycle THz pulses is considerably shorter than the orbital period of a classical electron in a Rydberg atom, $\tau_R = (2\pi m_e a_0^2/\hbar)n^3 \sim 10 - 100$ ps. Therefore, half-cycle THz pulses exert momentum kicks on the electrons during a small fraction of the orbital period. In classical mechanics, the energy gained by a momentum kick is expressed as the integral

$$U_K = -e \int \mathbf{E}_{THz}(t) \cdot \mathbf{v_e}(t) dt. \qquad (6.6)$$

If the energy gain is greater than the binding energy $U_B = \mathcal{R}/n^2$, the atom becomes ionized. The energy gain is maximized when the electron is at its maximum speed and is moving opposite to the field vector. The electron speed is highest at its closest distance to the nucleus and virtually independent of its total energy. Therefore, the ionization threshold, $U_K = U_B$, warrants that the threshold field amplitude is proportional to the binding energy, $E_{THz} \propto n^{-2}$. Only a small fraction of electrons meet the threshold condition. Figure 6.3 shows that the threshold fields for ionizing Na atoms in Rydberg states are indeed scaled as n^{-2} for 10 % ionization. Ionization of wavepackets at large distances from the nucleus requires higher field intensity than for those near the core. Figure 6.4 shows the fraction of atoms ionized by half-cycle THz pulses as a function of the field amplitude when the wavepacket is (A) near and (B) at a large distance from the nucleus. It is evident that ionization is more efficient for a wavepacket located at A than for one at B.

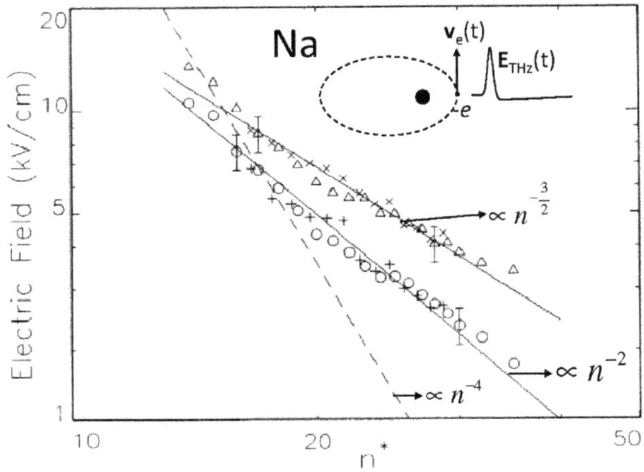

Fig. 6.3. Ionization thresholds of Na atoms in Rydberg states. The threshold electric fields are scaled as n^{-2} for 10% ionization of d (○) and s (+) states and as $n^{-3/2}$ for 50% ionization of d (△) and s (×) states. The dashed line represents the static field ionization limit, $E \sim n^{-4}$. (Reprinted with permission from [177]. ©1993, American Physical Society.)

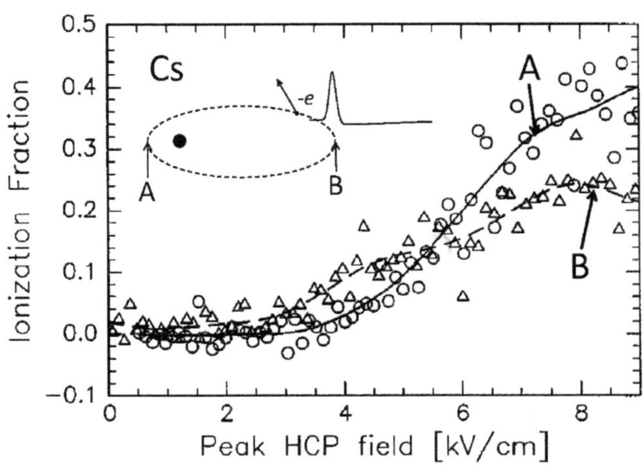

Fig. 6.4. Ionization of wavepackets near the core (A: ○) and near the outer turning point (B: △), as a function of peak THz field amplitude. (Reprinted with permission from [178]. ©1996, American Physical Society.)

A Rydberg wavepacket is a coherent superposition of multiple Rydberg states. Interacting with atoms, THz pulses coherently manipulate the wavepacket by redistributing the Rydberg states. Figure 6.5 shows an example. The Rydberg states of rubidium (Rb) atoms subject to a static bias field undergo Stark splitting. The Rb atoms are optically excited to the initial states of the n=40 manifold. Two half-cycle THz pulses with a time delay $\Delta\tau$ induce transitions from the initial states to the neighboring manifolds. The final state populations are measured by state selective field ionization, and the results are shown in Fig. 6.5(b). The fast oscillations correspond to the orbital period τ_R=9.7 ps. The slow modulation is accounted for by the phase retrieval of the wavepacket. Because the energy spacing ($\Delta E = E_n - E_{n-1}$) between the neighboring manifolds is nearly equal, the wavepacket evolves periodically with the phase retrieval period $\tau_{phase} = h/\Delta E \approx$130 ps. The coherent manipulation of quantum phase is applicable to quantum information processing. It has been demonstrated that information can be stored and retrieved through the manipulation of quantum phase in Rydberg wavepackets of Cs atoms [2].

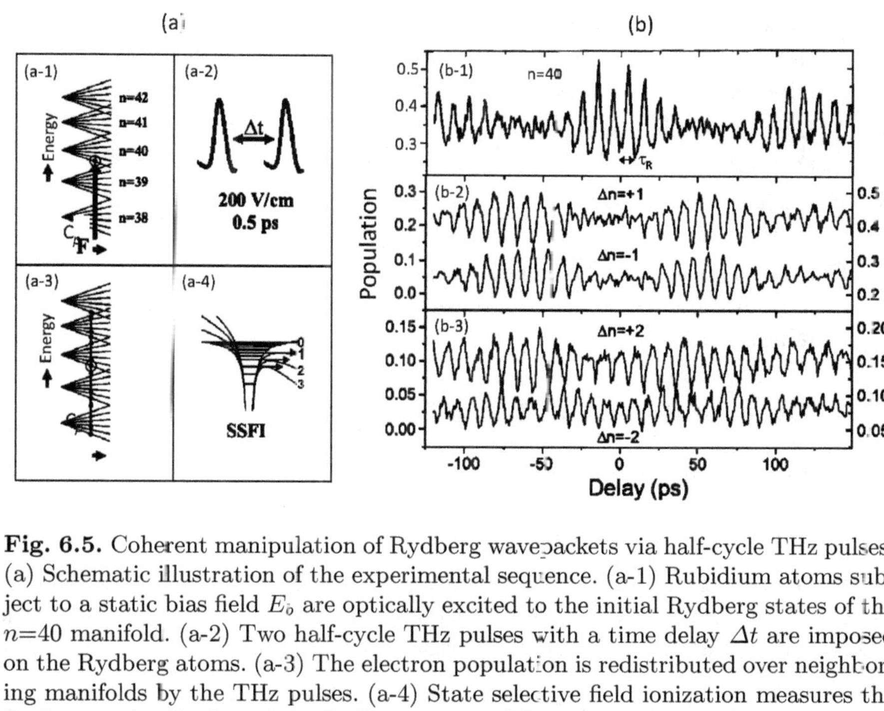

Fig. 6.5. Coherent manipulation of Rydberg wavepackets via half-cycle THz pulses. (a) Schematic illustration of the experimental sequence. (a-1) Rubidium atoms subject to a static bias field E_0 are optically excited to the initial Rydberg states of the n=40 manifold. (a-2) Two half-cycle THz pulses with a time delay Δt are imposed on the Rydberg atoms. (a-3) The electron population is redistributed over neighboring manifolds by the THz pulses. (a-4) State selective field ionization measures the final state distribution of the Rydberg atoms. (b) The population of the initial and the neighboring manifolds as a function of the delay between the two THz pulses for (b-1) n=40, (b-2) Δn=$\pm$1, and (b-3) Δn=$\pm$2. The curves are vertically offset for clarity. The left and right axes represent the populations for the higher- and lower-lying states, respectively. (Reprinted from [179].)

6.2 Rotational Spectroscopy

Rotational motion of molecules is a principal subject of molecular spectroscopy. In the microwave and THz regions, molecules in the gas phase exhibit narrow absorption lines corresponding to transitions between quantized rotational states (see section 2.2.3 for theoretical background). Each molecular species has its characteristic rotational energy levels that uniquely determine the precise frequencies of the absorption lines. The unambiguous signatures of rotational spectra are utilized to identify chemicals. This process is applied to environmental monitoring, atmospheric remote sensing, and analysis of the interstellar medium. One of the most notable works associated with the remote sensing of molecular species is the detection and monitoring of the depletion of ozone in Antarctica [180, 181, 182].

For the last several decades, many researchers have made innumerable contributions to collecting comprehensive data sets of molecular rotational transition lines. The experimental and theoretical data are catalogued in several databases:

- HITRAN (http://cfa-www.harvard.edu/hitran//)
  - **Hi**gh-resolution **Tran**smission Molecular Absorption Database
  - maintained by the Harvard-Smithsonian Center for Astrophysics
- JPL Spectral Line Catalog (http://spec.jpl.nasa.gov/)
  - maintained by the Jet Propulsion Laboratory
- Physical Reference Data (http://physics.nist.gov/PhysRefData/)
  - maintained by the NIST Physics Laboratory
- CCCBBD (http://cccbdb.nist.gov/),
  - **C**omputational **C**hemistry **C**omparison and **B**enchmark **D**ata**B**ase
  - maintained by NIST

6.2.1 Basics of Rotational Transitions

The rotational energy levels of a symmetric-top molecule (see Eq. 2.177) are expressed as

$$E_{rot}(J,K) = BJ(J+1) + CK^2, \tag{6.7}$$

where B and C are rotational constants, and $J(=0,1,2,\ldots)$ and $K(=0,1,2,\ldots,J)$ are the angular momentum quantum numbers. For $K=0$, the thermal distribution of the rotational states at temperature T has the relation,

$$N_J \propto (2J+1)e^{-BJ(J+1)/k_BT}. \tag{6.8}$$

The rotational transitions of the molecule are subject to the selection rules, $\Delta J = \pm 1$ and $\Delta K = 0$, and the transition energy has the simple expression, $\Delta E_J = 2B(J+1)$. As an example, Fig. 6.6 shows the rotational absorption spectrum of carbon monoxide (CO) at room temperature (T=298 K). The absorption lines are equally spaced by $2B = 0.1194$ THz.

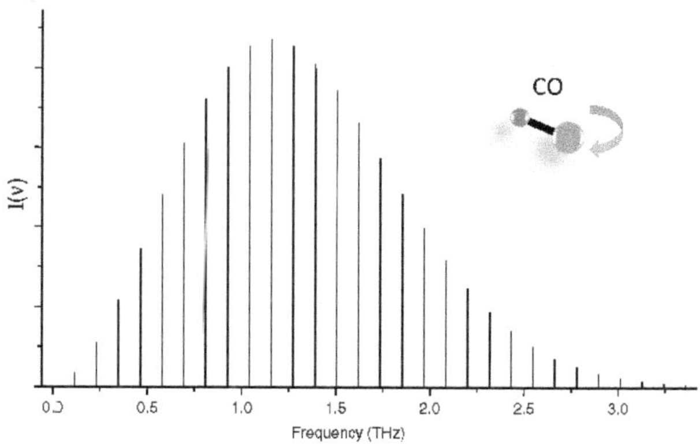

Fig. 6.6. Calculated rotational absorption lines of CO (B=1.93 cm^{-1}=0.0597 THz) at T=298 K.

There is no simple rule to describe the rotational energy levels of an asymmetric-top molecule. Conventionally, the rotational states are characterized by the three quantum numbers $J_{K_{-1},K_{+1}}$ in the King-Hainer-Cross notation. These molecules generally exhibit complicated absorption lines, even for the very simple ones. As an example, Fig. 6.7 shows a typical absorption spectrum of water vapor in the spectral range from 0.3 to 6 THz.

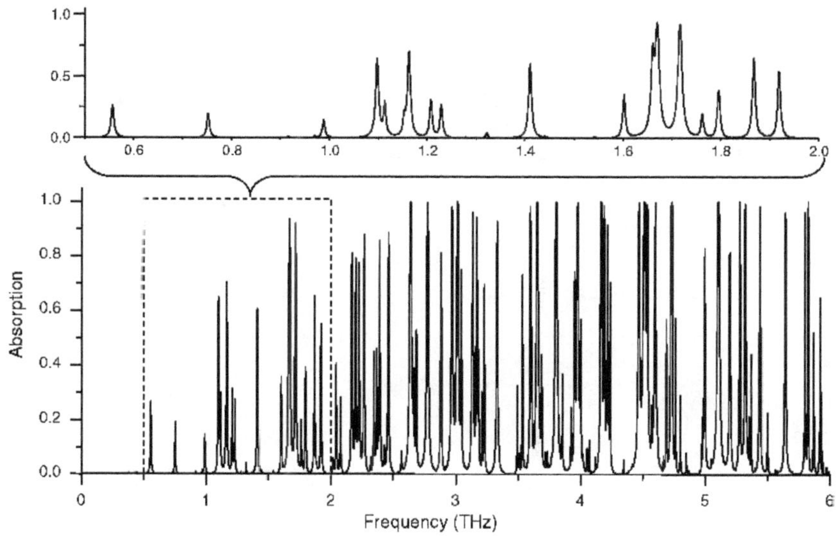

Fig. 6.7. Water absorption spectrum for 10-cm path at atmospheric pressure in the spectral range 0.3-6 THz. The top panel shows an extension between 0.5 and 2 THz.

In the low density limit, the linewidth of a rotational transition is largely governed by Doppler broadening,

$$\Delta\nu = \nu\sqrt{\frac{k_B T}{Mc^2}}, \tag{6.9}$$

where M is the molecular mass. At room temperature, Doppler broadening is $\Delta\nu/\nu \sim 10^{-6}$, and hence the linewidth in the THz region is ~1 MHz. As pressure increases, the linewidth broadens due to collisions between the molecules. The collisional broadening is approximately proportional to the pressure,

$$\Delta\nu = \gamma_c P, \tag{6.10}$$

where the broadening coefficient is γ_c ~10 MHz/Torr. The rotational linewidth near atmospheric pressure is typically 1-10 GHz, dominated by pressure broadening.

6.2.2 High-Resolution Spectroscopy

Rotational spectroscopy is of great importance for studying basic physical concepts. Rotational spectra provide crucial information to determine molecular structures. Furthermore, resolving absorption line-shapes in detail, high-resolution spectroscopy sheds light on the microscopic mechanisms of molecular collisions. We discuss a couple of representative examples to elucidate the power of high-resolution rotational spectroscopy.

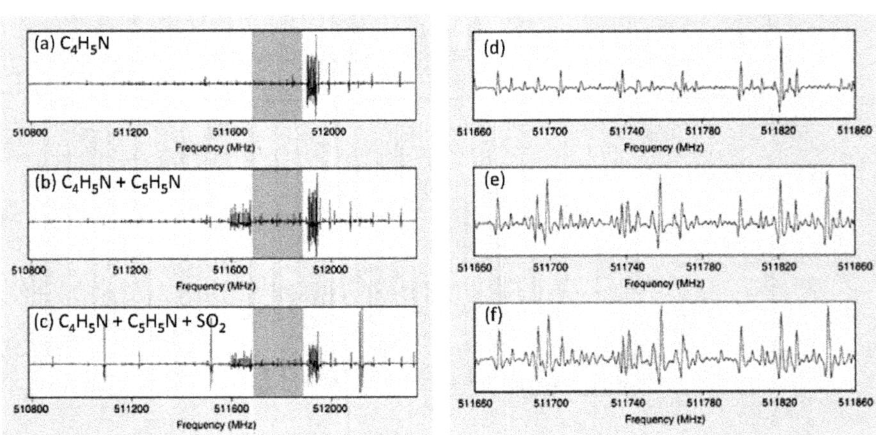

Fig. 6.8. Absorption spectra (1-GHz window near 0.512 THz) of (a) 10 mTorr of pyrrole, (b) 10 mTorr of pyrrole + 20 mTorr of pyridine, and (c) 10 mTorr of pyrrole + 20 mTorr of pyridine + 20 mTorr of sulfur dioxide. (d), (e) and (f) The graphs show the shaded regions of 0.2-GHz window. (Reprinted with permission from [183]). ©1998, American Chemical Society.

Figure 6.8 shows an example of the detection and identification of chemical agents by use of a high-resolution spectroscopy technique [183]. The source of the THz radiation is a backward wave oscillator (BWO), and the signal transmitted through a gas cell is detected by a L-He InSb hot-electron bolometer. A small portion of the radiation is coupled into a Febry-Perot cavity for use as a reference. The spectral resolution is ~10 KHz, limited by the linewidth of the BWO. The samples are mixtures of pyrrole (C_4H_5N), pyridine (C_5H_5N), and sulfide dioxide (SO_2). It is evident that the absorption lines of each molecular species exhibit clearly discernible spectral signatures. Adding pyridine to pyrrole results in the strong absorption band around 0.5116 THz as shown in Fig. 6.8(b). Having a relatively simple structure, SO_2 has a small number of strong lines in the 1-GHz spectral band (Fig. 6.8(c)) and none in the narrower region of the 0.2-GHz window (Fig. 6.8(f)).

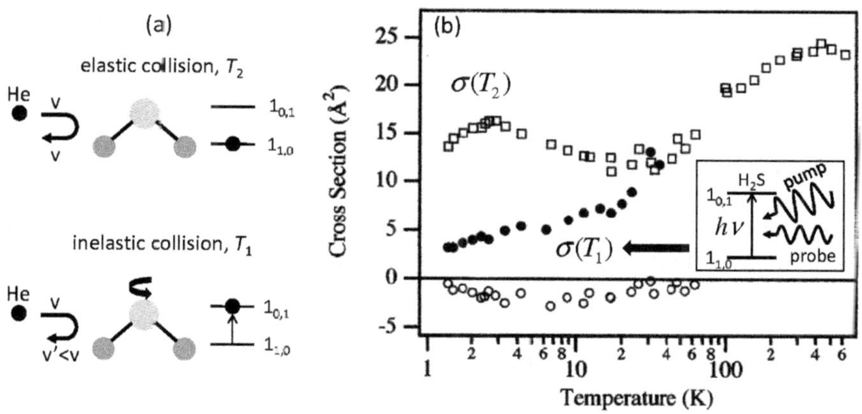

Fig. 6.9. Quantum collisional signature of rotational absorptions (a) Schematics of resonant pump-probe experiment for the measurement of population decay time T_1 of the $1_{1,0} - 1_{0,1}$ transition of H_2S. (b) Cross sections for collisions of H_2S with He: (solid circles) $\sigma(T_1)$, rotationally inelastic, (open squares) $\sigma(T_2)$ pressure broadening, and (open circles) pressure shift. (Reprinted with permission from [184]. ©1998, American Physical Society.)

The absorption characteristics of rotational transitions are sensitive to collisional interactions between molecules. Because of this, many efforts have been dedicated to investigating pressure-dependent rotational line-shapes in search of the underlying microscopic mechanisms of molecular collisions [185, 186]. Under the extreme condition of low temperature and low density, molecular collisions are drastically different from the classical stochastic process. Figure 6.9 demonstrates that rotational absorptions exhibit characteristics of quantum collision in an astrophysical environment. The sample gas molecules are hydrogen sulfide (H_2S) injected into cold helium (He) gas at a pressure of 1-10 mTorr. At low temperature, H_2S provides a virtual two-level system composed of $1_{1,0}$ and $1_{0,1}$ states[1]. The energy gap between the two levels is 19 cm^{-1} which is equivalent to the thermal energy at 27 K. As shown in Fig. 6.9(a), collisions of H_2S with He are either elastic or inelastic depending on whether they cause the $1_{1,0} - 1_{0,1}$ transition. The elastic collision is a stochastic process that gives rise to pressure broadening $\Delta\nu = 1/2\pi T_2$, where T_2 is the dephasing time (see section 2.2.1). The inelastic collision involves the quantum transition between the two levels, and hence is suppressed at low temperature, below which the thermal energy is lower than the transition energy. The inelastic collisional rate is expressed as $1/2\pi T_1$, where T_1 is the population decay time (see section 2.2.1). A time-resolved resonant pump-probe experiment was employed to measure T_1 of the $1_{1,0} - 1_{0,1}$ transition as shown in the inset of Fig. 6.9(b). Figure 6.9(b) shows the transition between classical and quantum mechanical collision processes, where the rotationally inelastic and pressure broadening cross sections, $\sigma(T_1)$ and $\sigma(T_2)$, are determined by T_1 and T_2. The inelastic cross section gradually falls off as temperature decreases because of the quantum collision suppression.

6.2.3 Atmospheric and Astronomical Spectroscopy

Monitoring the Earth's atmosphere and observing molecules in the interstellar medium are the cardinal applications of remote sensing in the THz region. Several outstanding space programs have developed and implemented state-of-the-art THz instruments as their key observation apparatus. We survey two eminent THz instruments currently in operation.

Earth Observing System Microwave Limb Sounder

The Earth Observing System (EOS) Microwave Limb Sounder (MLS) on NASA's Aura satellite observes thermal emission from the edge of the Earth's atmosphere (http://mls.jpl.nasa.gov/). The data on chemical composition, temperature, and humidity of the atmosphere are used for monitoring the stratospheric ozone layer, climate change, and global air quality.

Table 6.1. EOS MLS Radiometers and Primary Measurement Targets

Radiometer	Measurement Target
R1 (118 GHz)	temperature, geopotential height
R2 (190 GHz)	H_2O, HNO_3, HCN
R3 (240 GHz)	O_3, CO
R4 (640 GHz)	HCl, ClO, HOCl, HO_2, BrO, N_2O, volcanic SO_2
R5 (2.5 THz)	OH

[1] Like H_2O, H_2S is a asymmetric top molecule and the rotational energy levels are usually expressed by the King-Hainer-Cross notation, J_{K_-,K_+} (section 2.2.3).

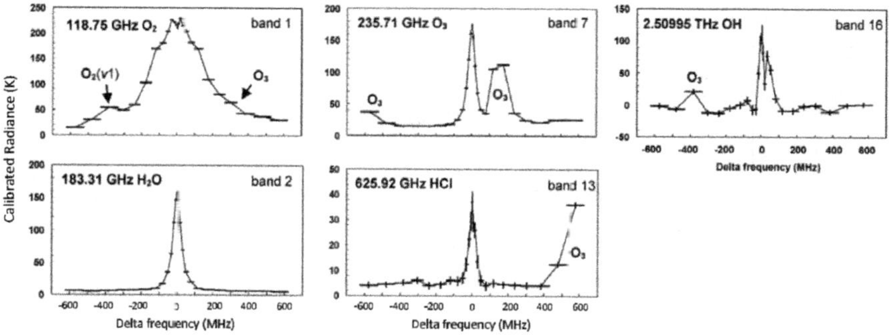

Fig. 6.10. Target spectral lines measured by the R1-R5 radiometers and 25-channel spectrometers. (Reprinted from [187]. ©2006 IEEE)

The EOS MLS is equipped with seven THz receivers that measure radiation in five spectral bands centered near 0.118, 0.19, 0.24, 0.64 and 2.5 THz [187]. The THz receivers are heterodyne radiometers operating at ambient temperature (see section 4.10). The 118-GHz radiometer (R1) covers the strong O_2 line at 118 GHz, and measures temperature and pressure. The 190-GHz radiometer (R2) measures the H_2O line at 183 GHz as well as HNO_3 lines in the spectral region. The main targets of the 240-GHz radiometer (R3) is the strong O_3 and CO lines. The 640-GHz radiometer (R4) covers the lowest HCl line at 626 GHz, the strong ClO line at 634 GHz, and lines of BrO, N_2O and HO_2. The primary target molecule of the 2.5-THz radiometer (R5) is OH which exhibits a strong doublet at 2.510 and 2.514 THz. The R1-R4 radiometers consist of solid-state local oscillators and Schottky diode mixers. The R5 THz radiometer utilizes a methanol (CH_3OH) gas laser as a local oscillator. Table 6.1 lists the EOS MLS Radiometers and their primary measurement targets. Figure 6.10 shows example spectra in each of the five spectral regions measured at tangent heights in the middle to upper stratosphere [187].

Figure 6.11 features the signal flow in the EOS MLS and provides a comprehensive picture of how THz remote-sensing spectrometers operate [187]. A three-reflector antenna system is employed to collect atmospheric signals for the R1-R4 radiometers, while a THz scanning-mirror and telescope assembly collects signals for the R5 radiometer. The signals are calibrated against radiation from cold space and blackbody calibration targets. To accomplish this, switching mirrors coupled to the collecting optics direct the reference radiation to the radiometers at times. An optical multiplexer arranges optical paths to feed the signals into the radiometers. The spectrometer module consists of four types of spectrometers with different resolutions and bandwidths, which cover different altitude ranges.

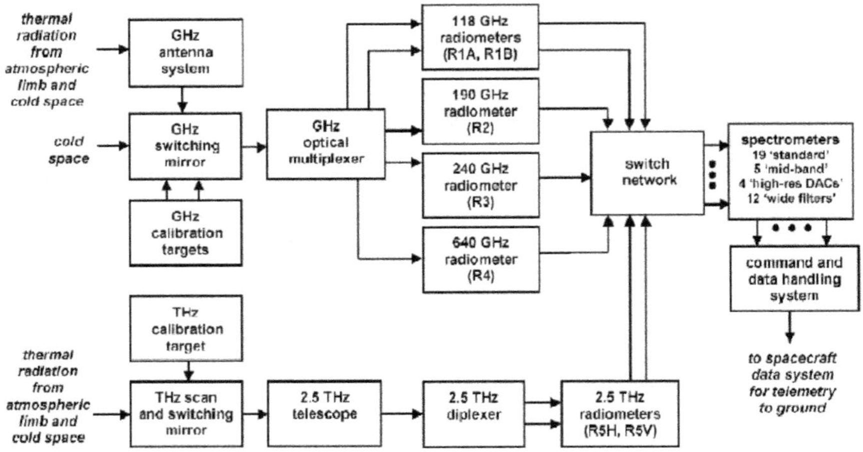

Fig. 6.11. Signal flow block diagram of the EOS MLS (Reprinted from [187]). ©2006 IEEE

Herschel Space Observatory

The Herschel Space Observatory (http://herschel.esac.esa.int/), the fourth 'cornerstone' mission of the European Space Agency's Horizon 2000 program, is a space based THz telescope that will be launched in early 2009. The primary scientific objectives of the Herschel telescope are (i) to search for the earliest stage of proto-galaxies, (ii) to trace the evolution of the symbiosis of active galactic nuclei (AGN) and starburst, (iii) to uncover the mechanism behind how stars and planetary systems formed, (iv) to probe the chemical composition of the atmospheres and surfaces of astronomical objects such as planets, satellites, comets, and asteroids, and (v) to investigate the molecular chemistry of the interstellar medium. Its huge 3.5-m-diameter mirror collects signals from deep space, and on-board THz instruments detect and analyze the THz radiation emitted by the cool and dusty interstellar medium.

The Herschel telescope is equipped with, arguably, the most advanced and sensitive THz instruments presently available:

- Heterodyne Instrument for the Far Infrared (HIFI)
  - very high resolution heterodyne spectrometer
  - 0.48-1.25 THz, 1.41-1.91 THz
- Photodetector Array Camera and Spectrometer (PACS)
  - short wavelength camera and spectrometer
  - 1.4-5.5 THz
- Spectral and Photometric Imaging Receiver (SPIRE)
  - long wavelength camera and spectrometer
  - 0.45-0.95 THz, 0.92-1.5 THz.

These three THz instruments detect many molecular rotational lines and atomic fine-structure lines with very high spectral and spatial resolutions. The data are analyzed to determine the precise gas dynamics of the interstellar medium such as chemical composition, velocity structures, and collisional conditions, which provide crucial information used to accomplish the scientific objectives mentioned above. Figure 6.12 illustrates the Herschel service module which carries the THz instruments. The optical bench the THz instruments are mounted on is contained within the Herschel cryostat of superfluid helium that cools down the instruments as low as 1.7 K. The SPIRE and PACS bolometers are further cooled to 0.3 K by ^{3}He sorption coolers.

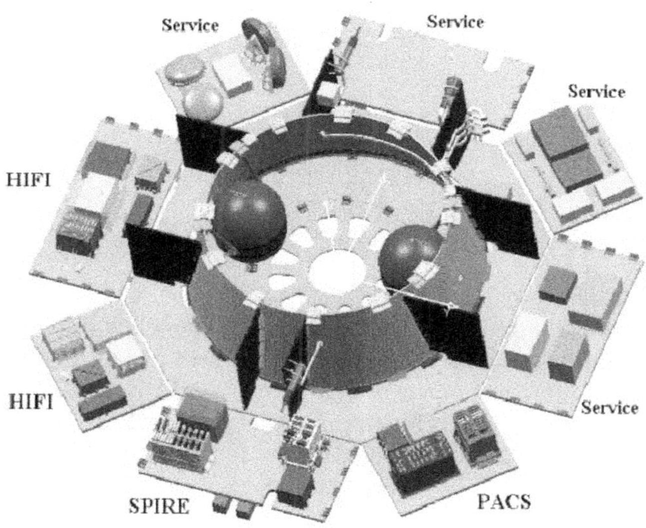

Fig. 6.12. The Herschel service module. (Reprinted from [188].)

The **He**terodyne **I**nstrument for the **F**ar **I**nfrared (**HIFI**) is a very high resolution heterodyne spectrometer: velocity resolution is in the range 0.3-300 km/s. The technical principles of HIFI are similar to those of the EOS MLS. Its seven mixer bands cover the frequency ranges of 0.48-1.25 THz and 1.41-1.91 THz, utilizing two low-noise, orthogonally-polarized mixers for each spectral band. Semiconductor-insulator-semiconductor (SIS) mixers are used for the five bands between 0.48 and 1.25 THz, and hot-electron-bolometer mixers for the two bands between 1.41 and 1.91 THz.

Figure 6.13 illustrates the arrangement of the HIFI components [189]. The five subsystems include the Local Oscillator and Focal Plane Units (LOU and FPU), the Wide-Band and High-Resolution Spectrometers (WBS and HRS), and the Instrument Control Unit (ICU). The LOU utilizes solid-state THz sources based on frequency multiplication of microwaves. The reference beams from the LOU are coupled to the FPU in the Herschel cryostat. The FPU splits

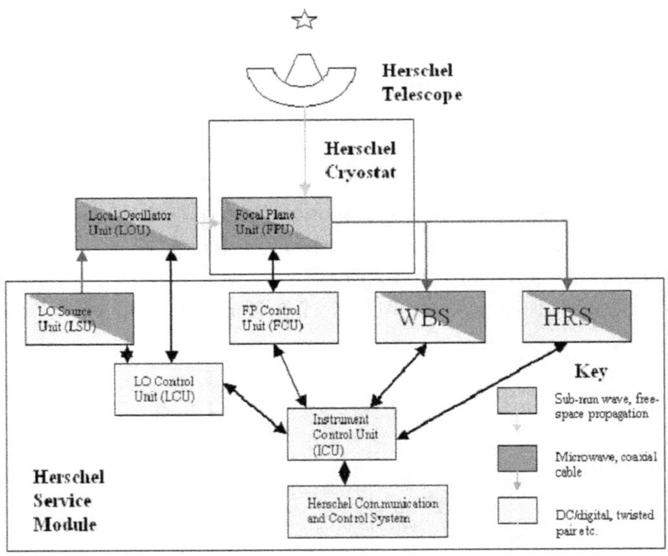

Fig. 6.13. General HIFI component diagram. (Reprinted from [189].)

the astronomical signal into seven beams, combines them with the reference beams, splits each beam into two linearly polarized beams, and focuses them into two mixer units. The output signals from the FPU are measured by two spectrometers. The WBS is a four-channel acousto-optical spectrometer. The sample bandwidth is 4-8 GHz, and the resolution is ~1 MHz. The HRS is a highspeed digital autocorrelator. It covers narrower bands with resolution up to 140 kHz. The ICU orchestrates the operation of the measurement units and the signal flow between the subsystems.

The **P**hotodetector **A**rray **C**amera and **S**pectrometer (**PACS**) is an imaging photometer and medium resolution spectrometer for the high-frequency spectral band from 1.4 to 5.5 THz. The camera and spectrometer modes are mutually exclusive. Figure 6.14 illustrates the schematics of the PACS functional units and the optical beam paths [190]. Passing through the entrance optics unit, the signal beam is split into the spectrometer and the photometer trains.

In the spectrometer train, the image slicer rearranges the two-dimensional signal image into a linear 1×25 pixel image and delivers it to the Littrow-mounted grating. The beam diffracted off the grating is split into the spectral bands, 1.4-2.9 THz ('red' band) and 2.9-5.5 THz ('blue' band), by a dichroic beamsplitter. The two beams are fed into the Ge:Ga photoconductive detector arrays. The stressed/unstressed Ge:Ga detectors have photoconductive thresholds at 1.4 and 2.3 THz, respectively. Figure 6.15(a) illustrates the concept of the integral-field spectrometer. The 5×5-pixels field-of-view image is

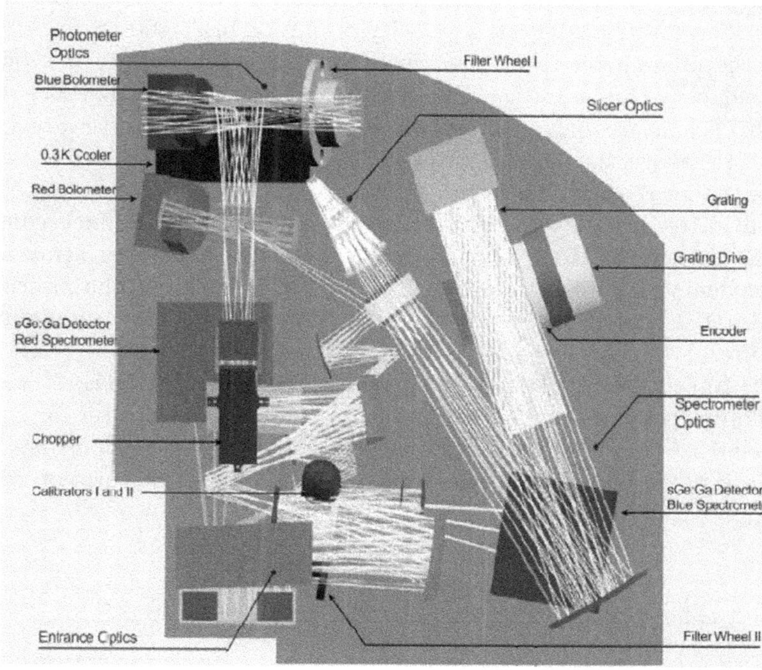

Fig. 6.14. PACS components and optical layout. (Reprinted from [190].)

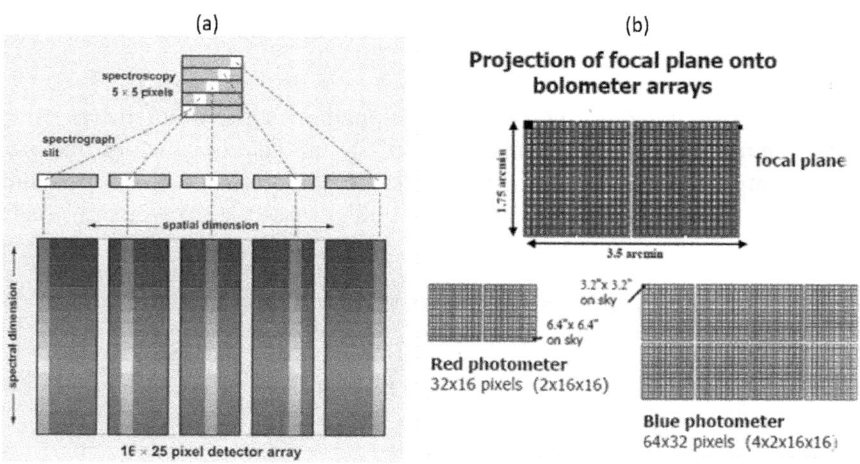

Fig. 6.15. (a) Projection of the focal plane onto the Ge:Ga detector arrays in the spectrometer. (b) Dimensions of the bolometer arrays and the projection of the focal plane. (Reprinted from [190].)

rearranged into a linear array and projected onto the 16×25 Ge:Ga detector arrays. The spectrometer has a velocity resolution of 75-300 km/s.

In the photometer train, the signal beam is also split into the low- and high-frequency bands, 1.4-2.3 THz ('red' band) and 2.3-5.5 THz ('blue' band). The two beams are magnified and directed onto the Si bolometer arrays. Figure 6.15(b) shows the dimensions of the two bolometer arrays and the projection of the focal plane. They sample the same 1.75×3.5-arcmin field of view. The bolometer array for the high-frequency band has 64×32-pixels composed of 4×2 matrices of 16×16 pixels. The 32×16-pixel bolometer array for the low-frequency band combines two 16×16-pixel assemblies. The matrices attached to 0.3-K multiplexers are thermally isolated from the surrounding 2-K structure.

The **S**pectral and **P**hotometric **I**maging **RE**ceiver (**SPIRE**) is an imaging photometer and an imaging Fourier transform spectrometer for frequencies below 1.5 THz. Both the camera and the spectrometer employ hexagonal 'spider-web' Ge bolometer arrays cooled to 0.3 K by a ^{3}He sorption refrigerator.

Table 6.2. SPIRE imaging photometer

Bolometer Array	PLW	PMW	PSW
Center Frequency (THz)	0.6	0.9	1.2
Bandwidth (THz)	0.24	0.26	0.36
Number of Pixels	43	88	139

The SPIRE imaging photometer, equipped with three bolometer arrays, makes simultaneous observations on a 4×8-arcmin field of view in three broad bands centered near 0.6, 0.9, and 1.2 THz. Characteristics of the bolometer arrays are shown in Table 6.2. Figure 6.16 illustrates the array arrangements.

Table 6.3. SPIRE Fourier transform spectrometer

Bolometer Array	SLW	SSW
Spectral Band	0.446-0.949 THz	0.926-1.55 THz
Resolution	1.2, 7.5, or 30 GHz at 1.2 THz	
Number of Pixels	19	37

The SPIRE Fourier-transform Spectrometer utilizes a Mach-Zender interferometer. Two bolometer arrays, SLW (19 pixels) and SSW (37 pixels), simultaneously measure the interferogram of a 2.6-arcmin-diameter field of view in the spectral bands of 0.446-0.949 THz and 0.926-1.55 THz. Table 6.3 shows the

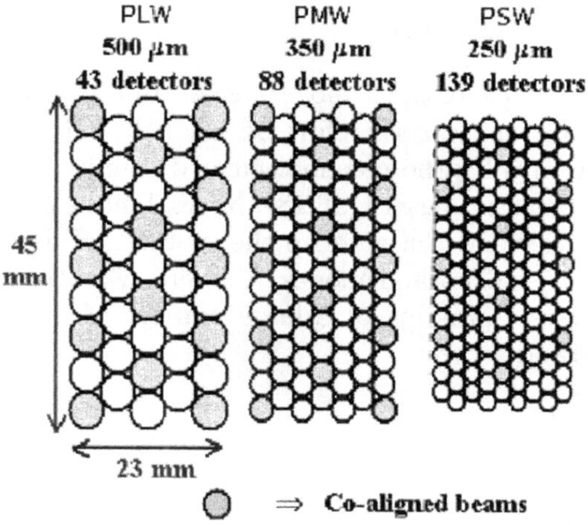

Fig. 6.16. Schematic diagram of the photometer bolometer arrays. The three detector arrays sample the 4×8-arcmin field of view on same sky positions. (Reprinted from [191].)

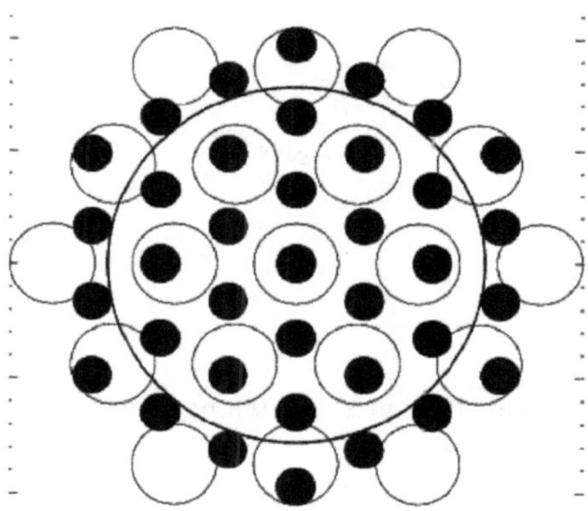

Fig. 6.17. Schematic diagram of the spectrometer bolometer arrays. SSW: filled circles (37 pixels, 16-arcsec diameter). SLW: open circles (19 pixels, 35-arcsec diameter). The large circle indicates 2.6-arcmin-diameter field of view on the sky. (Reprinted from [191].)

spectrometer characteristics. The detectors form hexagonal arrays as shown in Fig. 6.17. The detector diameters of SSW and SLW are 16 and 34 acrsec, respectively. The spacing between pixels is ~2 beam widths: 50.5 acrsec for SLW and 32.5 arcsec for SSW.

6.3 Biological Molecules

Vibrational motion in biological molecules is often characterized on the picosecond time scale, and can therefore be studied by THz spectroscopy. The vibrational modes involve the intramolecular dynamics of stretch, bend, and torsion among bonded atoms (Fig. 6.18). Normal mode analysis is a standard technique to identify and characterize the vibrational dynamics. In general, molecules have $3N-6$ normal modes, [2] where N is the number of atoms. Biological molecules consisting of a large number of atoms, therefore, have complicated mode structures.

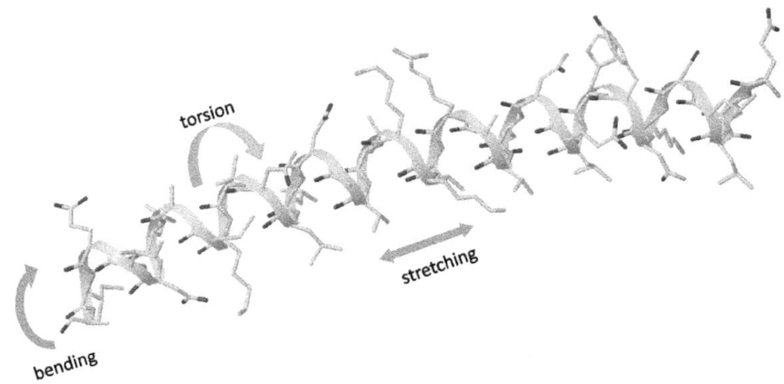

Fig. 6.18. Macromolecular dynamics: stretching, bending, and torsional vibrations

Normal mode analysis based on the harmonic approximation, however, fails to describe some slow and large amplitude motions of macromolecules, where the potential is largely governed by anharmonicity. These global fluctuations are particularly interesting in the THz regime. They are an important piece of the puzzle for understanding biochemical processes, yet details of the mechanisms remain to be studied. No other technique is as sensitive as THz time-domain spectroscopy (THz-TDS) to directly resolve the slowest motions in macromolecular systems.

In condensed-phase biosamples, intermolecular interactions including van der Waals forces and hydrogen bonding modify the mode structure of intramolecular vibrations and also give rise to additional vibrational modes involving collective dynamics of several molecules. The intermolecular interactions are usually weaker than intramolecular interactions, and the characteristic signatures of the intermolecular modes often emerge in the THz region.

[2] Linear molecules have $3N-5$ normal modes.

The spectral features of the low-frequency modes strongly depend on temperature: the absorption lines undergo severe broadening and shift as temperature varies.

All natural biosystems contain water. The exceptional properties of water as a solvent are critical for understanding the dynamics of biomolecules. The behavior of biomolecules in water is drastically different from that in the solid phase because water molecules have strong a influence on the molecular structure and interaction via hydrogen bonding. Since the conformational changes and the large amplitude motions evolve with strong damping, the THz modes form a continuous and smooth spectrum. It is also of great importance that water molecules interacting with biomolecules respond differently to THz radiation depending on the local environment. In fact, the interaction of THz radiation with biological systems is usually dominated by and extremely sensitive to water, hence the THz response to water provides indirect yet crucial information about biomolecular dynamics in aqueous solutions.

6.3.1 Liquid Water

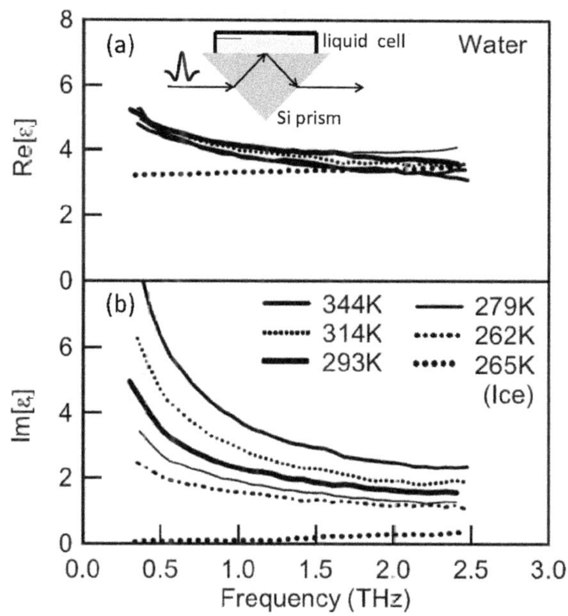

Fig. 6.19. (a) Real and (b) imaginary part of the dielectric constant of water and ice at various temperatures. The measurement scheme of total internal reflection is shown in the inset. (Reprinted from [192].)

Concerning the cardinal importance of water in biological systems, we first discuss the interaction of THz radiation with water in its liquid phase. Figure 6.19 shows the real and imaginary part of the dielectric constant of water at several temperatures [192]. Measurements on ice are also presented. The data were obtained by THz-TDS in a total reflection geometry as shown in the inset. The imaginary part of dielectric constant undergoes a substantial reduction as temperature is decreased, while the real part is largely independent of temperature.

The dielectric response of liquid water is governed by several physical processes. Major contributions are attributed to two types of relaxation dynamics characterized by fast (∼10 fs) and slow (∼10 ps) relaxation times. The slow relaxation is associated with rotational dynamics, yet the origin of fast relaxation is not yet clearly understood. Strong hydrogen-bond interactions between water molecules also make a sizable contribution. The intermolecular stretch mode is resonant at 5.6 THz and has a broad line width. Putting all the contributions together, we can write the dielectric constant of water as

$$\epsilon_r(\omega) = \epsilon_\infty + \frac{\epsilon_s - \epsilon_1}{1 - i\omega\tau_D} + \frac{\epsilon_1 - \epsilon_\infty}{1 - i\omega\tau_2} + \frac{A_T}{\omega_T^2 - \omega^2 - i\omega\gamma_T}. \tag{6.11}$$

The second and third terms represent rotational relaxation with the relaxation time τ_D and strength $\epsilon_s - \epsilon_1$ and fast relaxation with the relaxation time τ_2 and strength $\epsilon_1 - \epsilon_\infty$, respectively. The last term describes the dispersion of the intermolecular stretch mode with the resonance frequency ω_T and the damping constant γ_T.

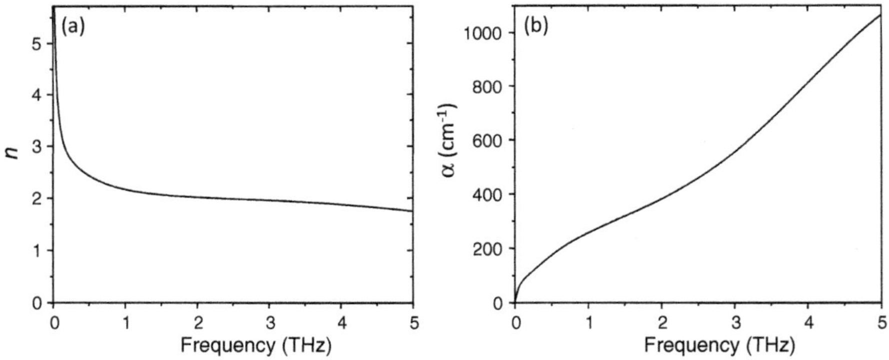

Fig. 6.20. Frequency-dependent (a) refractive index n and (b) absorption coefficient α of water at room temperature

At room temperature the dielectric response of water in the frequency range 0-5 THz is well described by this theoretical model with the following parameters [192]:

- τ_D=9.36 ps, τ_2=0.3 ps
- ϵ_∞=2.5, ϵ_s=80.2, ϵ_1=5.3

- $\omega_T/2\pi$=5.6 THz, $\gamma_T/2\pi$=5.9 THz, $A_T/4\pi^2$=38 THz^2.

Using this model, we obtain the optical constants of water in the THz region. Figure 6.20 shows the refractive index n and the absorption coefficient α, where

$$n(\omega) = \Re[\epsilon_r(\omega)], \tag{6.12}$$

$$\alpha(\omega) = 2\frac{\omega}{c}\Im[\epsilon_r(\omega)]. \tag{6.13}$$

Figure 6.21 presents several sets of experimental data of THz absorption at room temperature measured by different THz spectroscopic techniques [193]. The theoretical curve (solid gray) matches very well with the experimental results.

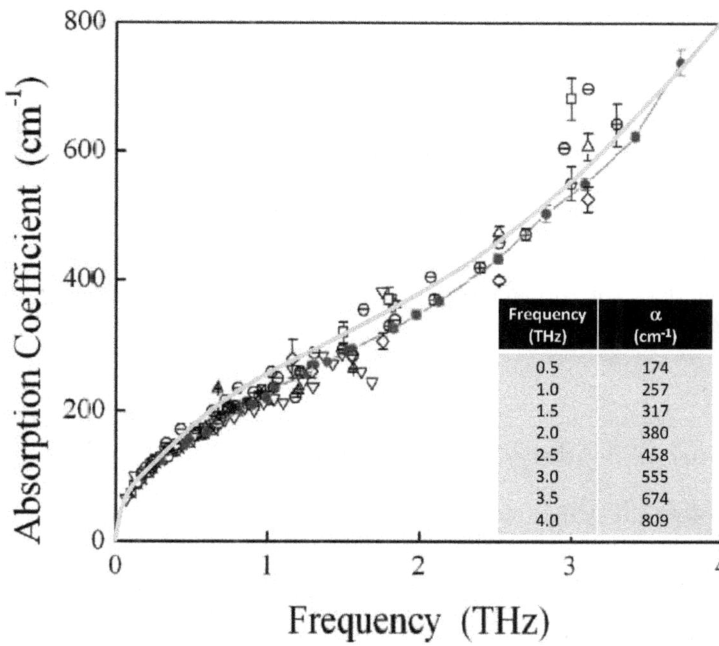

Fig. 6.21. Absorption spectra of liquid water between 0 and 4 THz at room temperature. The thick gray curve indicates calculations based on the model of Eq. 6.11. The table includes the calculated values at several frequencies. (Reprinted with permission from [193]. ©2006, American Institute of Physics.)

Finally, we look into the temperature dependence of THz absorption. Figure 6.22 shows the integrated absorption from 0.06 to 1 THz as temperature varies between 270 and 315 K [194]. The data are normalized at 314 K. The solid curve is a guide for the eye. Concerning practicality, it should be noted that the absorption varies widely, ~1.5% per degree around room temperature, therefore precise control of temperature may be necessary for experimental studies, in particular for those to quantify small changes in absorption spectra.

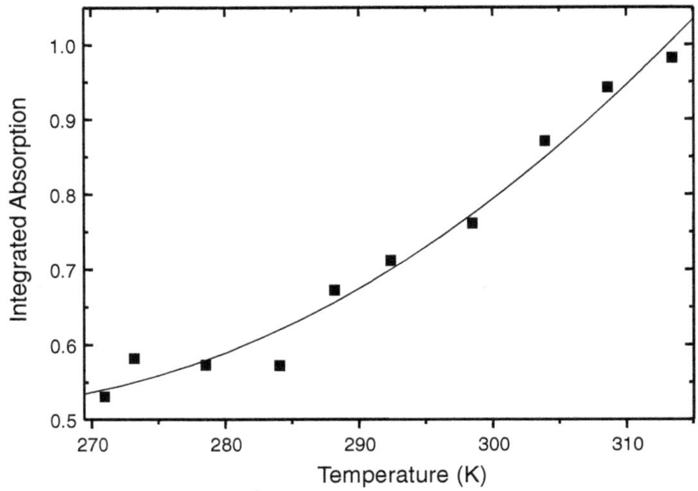

Fig. 6.22. Integrated absorption between 0.06 and 1 THz as a function of temperature. The data are normalized at 314 K. The solid curve is a guide for the eye. (Data from Ref. [194])

6.3.2 Normal Modes of Small Biomolecules

Intramolecular Vibration

We begin with a relatively simple biological system. For a small biomolecule in its stable conformation, a harmonic approximation fits well with the potential energy surface near its minima. Normal mode analysis, therefore, provides a good framework to describe its vibrational dynamics. In the limit of weak intermolecular interactions, theoretical calculations based on density-functional theory (DFT) provide a reasonable interpretation identifying the intramolecular vibrational modes.

Figure 6.23 shows the experimental and computational absorption spectra of three isomers of dicyanobenzene (DCB) immersed in a solid matrix (polyethylene) and a liquid solvent (chloroform) at room temperature [195]. Polyethylene and chloroform are commonly used materials for THz spectroscopy of biological systems because they have a constant refractive index and low absorption in the THz region. The molecular formula of DCB is $C_8H_4N_2$. As shown in the insets of Fig. 6.23, the isomers have simple planar structures having two cynides (CN) attached to a benzene ring. The experimental data are obtained by THz-TDS and Fourier transform infrared spectroscopy (FTIR). The DFT calculations exhibit qualitative agreements with the experimental data except for the broad peaks in the solid-phase spectra below 100 cm^{-1} (3 THz). The absence of such modes in the liquid-phase and computational spectra indicates that the low-frequency modes must arise from intermolecular interactions. The three isomers have markedly different spectral features in both experiment and theory, especially in the low-frequency region below 250 cm^{-1} (7.5 THz), which demonstrates the potential of THz spectroscopy for identification of small biomolecules.

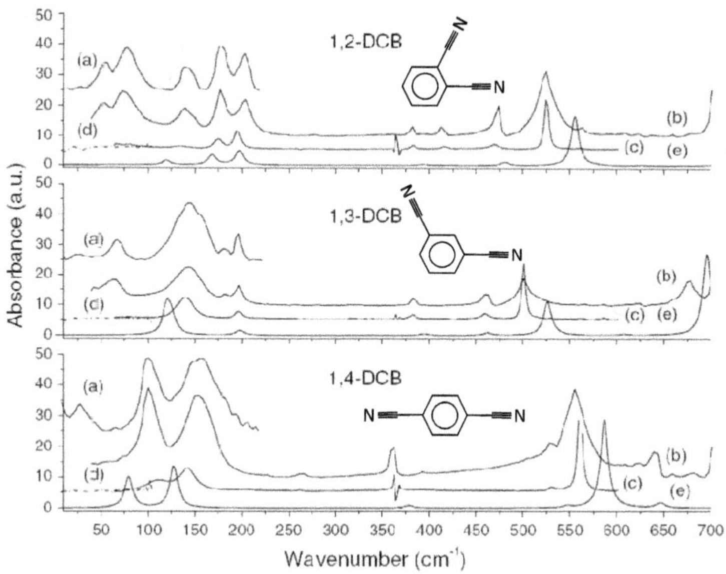

Fig. 6.23. Experimental and computational absorption spectra of 1,2-, 1,3-, and 1,4-dicyanobenzene isomers at room temperature. (a) THz-TDS and (b) FTIR spectra of solid phase in polyethylene matrix. (c) THz-TDS and (d) FTIR spectra of solution phase in chloroform. (e) DFT simulations of gas phase. Spectral intensities of solid and solution phase are multiplied by 2 and 10, respectively. Spectra are vertically offset for clarity. (Reprinted with permission from [195]. ©2007, Elsevier.)

Hydrogen Bonds

A hydrogen bond forms via an attractive intermolecular interaction between an electronegative atom and a hydrogen atom bonded to another electronegative atom. Usually the electronegative atom is either oxygen or nitrogen with a partial negative charge. Positively charged hydrogen is nothing more than a bare proton with little screening, hence the hydrogen bond is much

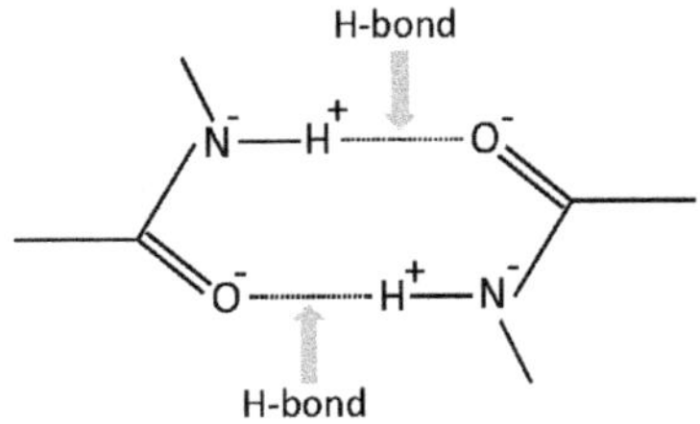

Fig. 6.24. Hydrogen bond

stronger than nominal dipole-dipole intermolecular interactions or van der Waals forces. Hydrogen bonding energy is about a tenth of that for covalent or ionic bonds. Because of the relatively strong interaction, the hydrogen bond plays a crucial role in biophysical and biochemical processes in naturally occurred biosystems. A salient example is protein folding with highly periodic structures of α-helix and β-sheet.

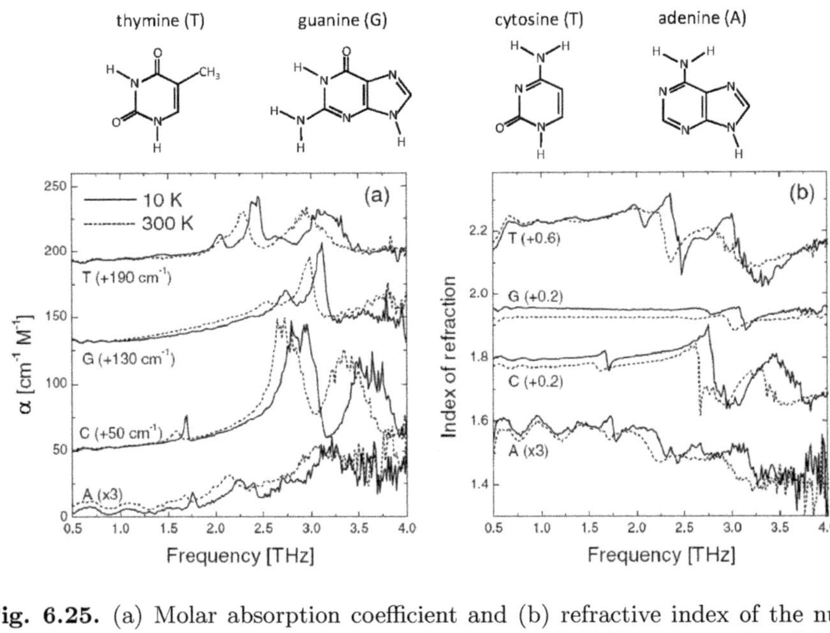

Fig. 6.25. (a) Molar absorption coefficient and (b) refractive index of the nucleobases guanine, adenine, cytosine, and thymine, at 10 K (solid curves) and 300 K (dashed curves) measured by THz-TDS. The spectra are vertically offset for clarity. The absorption curve for adenine is multiplied by a factor of 3, and the index is expanded by a factor of 3 around a value of 1.58. Molecular structures are shown on the top. (Reprinted from [196].)

As with DCB, intramolecular vibrational resonances of small biomolecules usually lie in the high-frequency region above 10 THz. When these molecules are closely packed and interact each other, they often form hydrogen bonds. Since the hydrogen bond is weaker than covalent and ionic bonds, vibrational modes mediated by hydrogen bonds have resonant frequencies lower than the typical intramolecular resonant frequencies and fall into the low-frequency region below 10 THz.

The collective nature of intermolecular hydrogen-bond vibrations is clearly seen in the THz responses from crystalline nucleobases [196]. The building blocks of DNA include guanine (G), adenine (A), cytosine (C) and thymine (T). Figure 6.25 shows the THz-TDS spectra of the absorption coefficient and the refractive index of the nucleobases in the spectral region 0.5-4.0 THz. The samples are polycrystalline powders embedded in a polyethylene matrix. Each molecular system forms a unique conformational state maintained by hydrogen bonds, and its spectrum exhibits distinctive features of vibrational resonances. The resonant peaks undergo a blue shift of ~5% as temperature decreases from 295 K to 10 K. The line shift arises from two mechanisms: temperature-dependent lattice dilatation and phonon-phonon scattering associated with lattice vibration anharmonicity. A detailed study of the temperature dependence will be discussed later in this section.

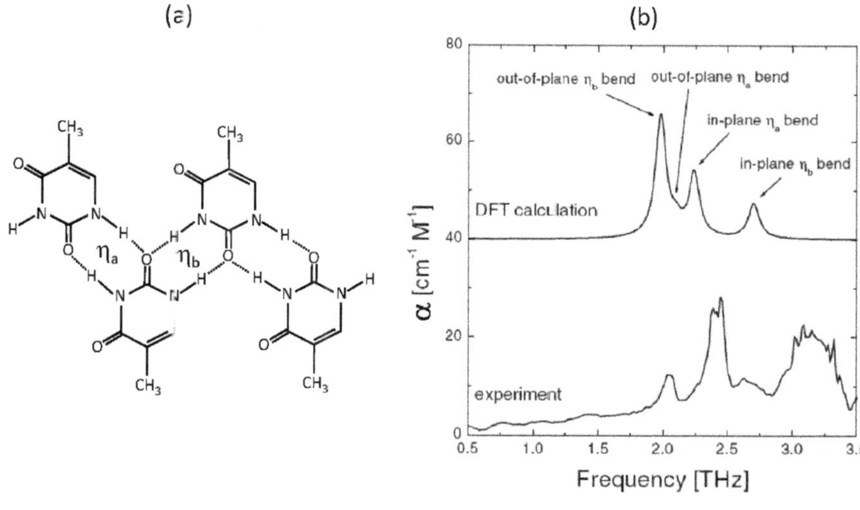

Fig. 6.26. (a) Hydrogen-bond systems, η_a and η_b, in crystalline thymine. Dotted lines indicate the hydrogen bonds. (b) Computational and experimental absorption spectra of crystalline thymine. Vibrational modes of the hydrogen-bond systems are assigned to the absorption peaks of the calculated spectrum. (Reprinted from [196].)

A theoretical analysis of vibrational modes has been carried out for thymine. Figure 6.26(a) depicts the two hydrogen-bond systems in crystalline thymine. The experimental absorption spectrum is compared with the DFT calculation in Fig. 6.26(b). The calculated resonance lines are convoluted with a phenomenological line width of 0.1 THz, agreeable with the experimental line-shape. The calculated spectrum presents qualitative agreement with the experimental observation. Although a definitive assignment of the resonance peaks with normal modes is not possible, it is certain that the intermolecular hydrogen-bond interactions are the origin of the low-frequency vibrational modes and, further, that normal mode analysis is useful to identify and characterize them.

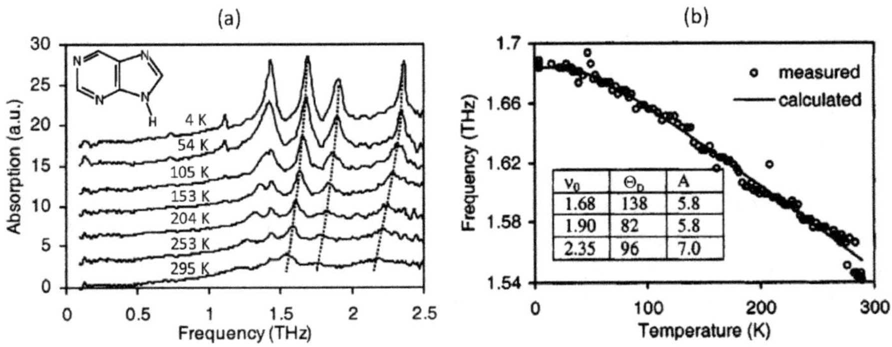

Fig. 6.27. (a) Absorption spectra of purine at 4, 54, 105, 153, 204, 253, and 295 K. Spectra are vertically offset for clarity. (b) Experimental (open circle) and calculated (solid line) resonance frequency of the peak near 1.6 THz as a function of temperature. The best-fitting parameters of ν_0 (THz), Θ_D (K), and A (10^{-4} THz/K) are tabulated in the inset. (Reprinted with permission from [197]. ©2003, American Institute of Physics.)

Temperature dependence of hydrogen-bond vibrations are examined with purine [197]. Purine is the skeleton of adenine and guanine: its molecular structure is depicted in the inset of Fig. 6.27(a). Figure 6.27(a) shows the absorption spectra measured at 4, 54, 105, 153, 204, 253, and 295 K. While resonance peaks are barely discernible at room temperature, they emerge sharp and strong and shift to the high-frequency side as temperature decreases. The major contribution to the line shift comes from phonon-phonon scattering due to the anharmonicity of the vibrational potentials. The temperature dependence of resonance frequency is expressed by the empirical formula

$$\nu(T) = \nu_0 - \frac{A\Theta_D}{e^{\Theta_D/T} - 1}, \tag{6.14}$$

where ν_0 is the resonance frequency at 0 K, A is a constant, and Θ_D is the Debye temperature. Figure 6.27(b) shows the experimental line shift of the peak near 1.6 THz and the fit to Eq. 6.14. The fitting result agrees well with experimental observation.

Solvation Dynamics

The dynamics of water molecules interacting with biomolecules is of fundamental importance for understanding the physical and chemical processes in biological systems. The mutual interaction influences the rotational relaxation and hydrogen-bond stretching of water molecules in an aqueous environment. Since the key mechanisms of water molecule dynamics are on the picosecond time scale (see section 6.3.1), THz spectroscopy can make a direct observation of how water molecules in the vicinity of biomolecules behave differently from those in bulk water.

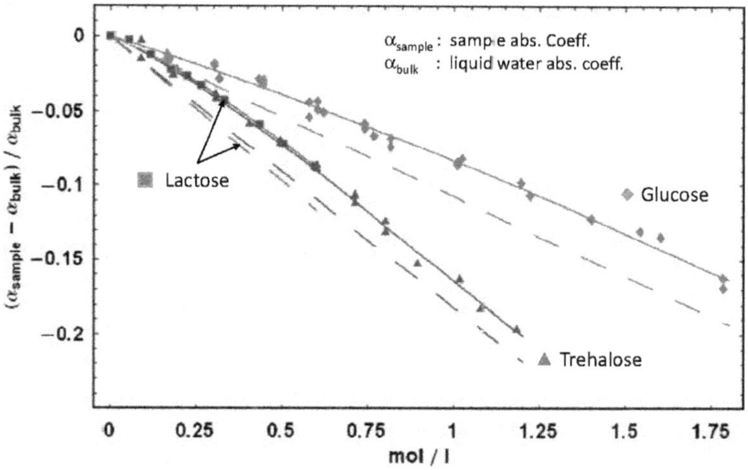

Fig. 6.28. Normalized differential absorption coefficient of water solutions of trehalose, lactose, and glucose at 273 K as a function of solute concentration. The absorption coefficients are obtained from the integrated absorption between 2.1 and 2.8 THz. The dashed lines denote the predicted concentration dependence of the THz absorption of a two-component model with non-interacting ingredients. (Reprinted with permission from [198]. ©2008, American Chemical Society.)

Figure 6.28 shows the concentration-dependent THz absorption of water solutions of three different sugars, trehalose, lactose, and glucose [198]. The normalized differential absorption coefficient,

$$\Delta\alpha_N = \frac{\alpha_{\mathrm{sample}} - \alpha_{\mathrm{bulk}}}{\alpha_{\mathrm{bulk}}}, \tag{6.15}$$

where α_{sample} and $\alpha_{\mathrm{bulk}}(=420\ \mathrm{cm}^{-1})$ are the integrated absorption coefficients of the sample and liquid water from 2.1 to 2.8 THz, measures the solute-induced absorption change relative to the absorption by bulk water. As a reference, the dashed line for each solution indicates the prediction of the two-component model which assumes that the two ingredients are separated by sharp interfaces and do not interact with each other. Under this condition, the overall THz absorption linearly decreases with increasing concentration of the solute because THz absorption by sugar molecules is largely negligible compared with that by water. Yet the measured THz absorbance exhibits an onset of nonlinearity at a certain concentration, which indicates the existence of dynamical hydration shells. The shell of water molecules in the vicinity of a protein surface forms a region of biological water. Both being disaccharides, trehalose and lactose show a similar concentration dependence, meaning that they exhibit a similar long-range order of surrounding hydration shells. The relatively small $\Delta\alpha_N$ of glucose arises because the simple sugar of monosaccharide is smaller and has fewer O-H bonds than the disaccharides. As a consequence, its influence on the surrounding water network reaches a shorter distance. A quantitative analysis shows that the average number of carbohydrate-water hydrogen bonds is an excellent scaling factor to determine the long-range influence on solvation dynamics.

Including the effect of the dynamical hydration shell, a three-component model provides the overall absorption expressed as

$$\alpha_{\text{total}}(c_s, \delta R) = \frac{V_{\text{solute}}(c_s)}{V_{\text{total}}}\alpha_{\text{solute}} + \frac{V_{\text{shell}}(c_s, \delta R)}{V_{\text{total}}}\alpha_{\text{shell}} + \frac{V_{\text{water}}(c_s, \delta R)}{V_{\text{total}}}\alpha_{\text{bulk}}, \tag{6.16}$$

where α_{total}, α_{solute}, α_{shell} and α_{bulk} are the absorption coefficients of the solution, the solute, the solvation water, and the bulk water, respectively. V_{total}, V_{solute}, and V_{shell} are the volume of the solution, the solute, and the dynamical hydration shell. The two parameters, c_s and δR, are the solute concentration and the shell thickness. The best fit to the data produces α_{shell} and δR listed in Table 6.4. An important observation is that α_{shell} is larger than α_{bulk} for all of the samples.

Table 6.4. α_{shell} and δR of the Dynamical Hydration Shell

Solute	α_{shell} (cm^{-1})	δR (Å)
glucose	437	3.7
lactose	435	5.7
trehalose	429	6.5

This is somewhat counterintuitive. Biological water is expected to be less responsive to THz radiation because its motions are hindered by protein surfaces. One plausible explanation is that sugar molecules may become more flexible in an aqueous environment than in a solid state. No microscopic mechanism, however, has yet been identified to explain the increase of absorption. It surely is an interesting subject for future studies. For trehalose and lactose, the dynamical hydration shell extends to a range of 6-7 Å from the surface of the sugar molecules, which corresponds to roughly two layers of water molecules. Glucose has a thinner monolayer hydration shell extending 3-4 Å.

Explosives Detection

THz spectroscopy is a promising new technique for the detection of concealed explosive materials. Common energetic explosives are characterized by their unique spectral signatures in the THz region. Their strong absorption bands differentiate the explosives from typical barrier materials such as clothing, plastic, and paper which are largely transparent in the spectral range. Detection schemes using THz spectroscopy are free of typical safety concerns because THz radiation is nonionizing and noninvasive.

Figure 6.29 shows the absorption spectra of the common explosives, 1,3,5-trinitro-1,3,5-triazacyclohexane (RDX)[3], tetranitro-tetracyclooctane (HMX), pentaerythritol (PETN), trinitrotoluene (TNT) obtained by THz-TDS [199]. The samples are polycrystalline explosives pressed into polyethylene pallets. Sample preparation conditions may alter the spectral features to an extent. Nevertheless, some absorption bands consistently appear in several independent studies. A few pronounced absorption peaks of the explosives are listed in Table 6.5.

Table 6.5. Absorption peaks of explosives

Explosive	Absorption Peak Position (THz)
RDX	0.8, 1.4, 2.0, 3.0
HMX	1.8, 2.8
PETN	2.1, 2.8
TNT	1.7, 2.2

Due to the strong absorption in the THz region, a transmission geometry is only applicable to detecting a small amount of explosives. For bulky targets, a reflection mode should be employed. Furthermore, under realistic circumstances, it is safe to assume that an explosive target has a rough and non-flat surface. Figure 6.30 demonstrates that THz spectroscopy is capable of detecting and identifying explosives from the measurement of diffuse reflection from a granular surface. Because the phase information is lost in the complicated reflections, the absorption spectra obtained from the reflection measurements are recovered by use of the Kramers-Kronig dispersion relations. Figure 6.30(a) shows a comparison between the absorption spectra of RDX obtained from the separate THz-TDS measurements of transmission and reflection modes. The spectral signature of RDX is still pronounced in

[3] RDX makes up around 91% of C-4.

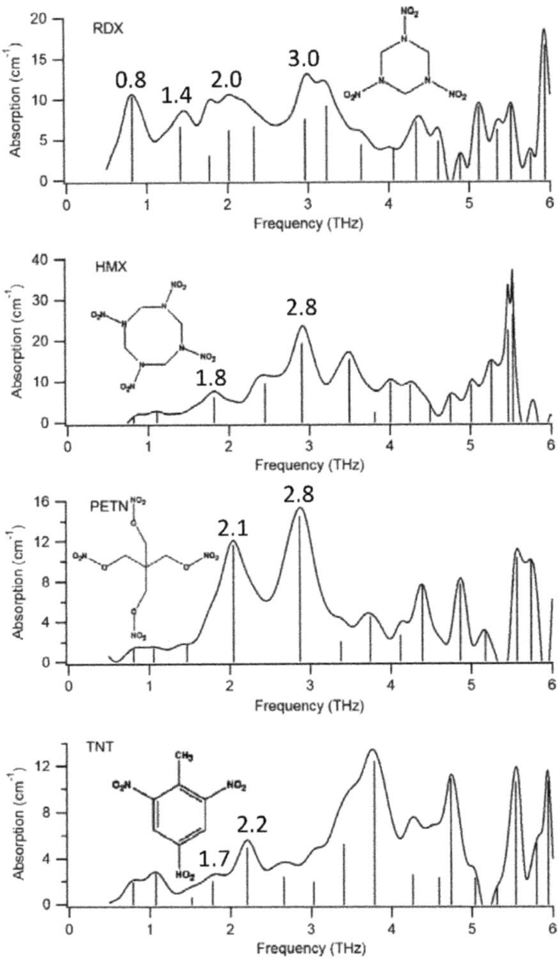

Fig. 6.29. Absorption spectra of the explosives, RDX, HMX, PETN, and TNT in the spectral region 0-6 THz at room temperature. Molecular structures are shown in the insets. The vertical lines indicate the result of a Lorentzian fit to the spectra. (Reprinted with permission from [199]. ©2007, Elsevier.)

the spectrum of the reflection mode. The RDX spectrum is compared against those of two common materials, polyethylene and flour in Fig. 6.30(b). Transmission spectra of polyethylene and flour are mostly featureless in this spectral range, yet the absorption spectra of the reflection mode contain some small spectral variations. The surface roughness of the samples may contribute to the discrepancy. Nevertheless, the RDX spectrum is clearly discernible from the others.

It is sensible to assume that a real explosive target will be concealed under barrier materials. Figure 6.31 shows the absorption spectra of RDX covered

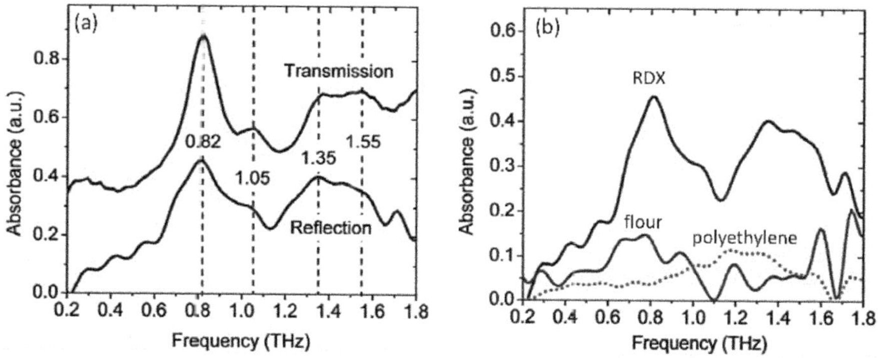

Fig. 6.30. (a) Absorption spectra of RDX from the transmission and diffuse reflection measurement. The transmission spectrum is vertically offset for clarity. (b) Absorption spectra of RDX, polyethylene and flour from the reflection spectra. (Reprinted from [200]. ©2007 IEEE)

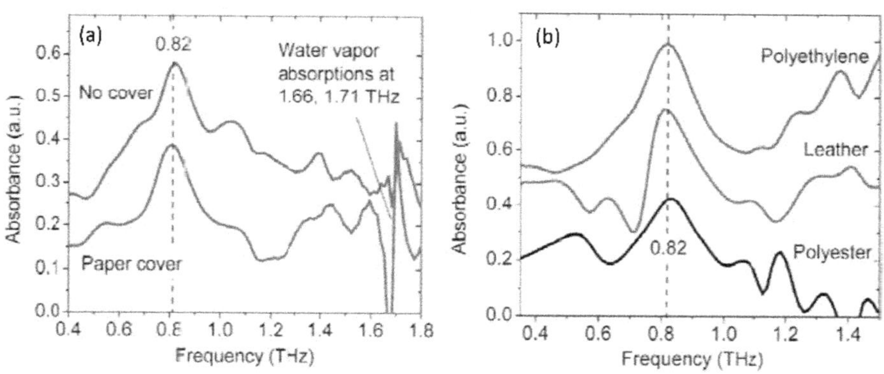

Fig. 6.31. (a) Absorption spectra of RDX obtained from the diffuse reflection measurements with and without a paper cover. (b) Absorption spectra of RDX under different covers, polyethylene, leather, and cloth. The spectra are vertically offset for clarity. (Reprinted from [200]. ©2007 IEEE)

with four different barrier materials: ~0.05-mm-thick white paper, ~0.1-mm-thick black polyethylene, ~0.3-mm-thick yellow leather, and ~0.4-mm-thick green polyester cloth. These covers prohibit visual inspections, yet are all partially transparent to THz radiation. Although some features are missing in the spectra of the covered RDX samples, the strong peak at 0.82 retains its characteristics throughout the measurements.

Drug Test

As with identifying and characterizing spectral fingerprints of biomolecules, THz spectroscopy has enormous potential as a probing tool for biomedical and bioengineering applications. One of the promising applications is the characterization of pharmaceutical materials. In particular, because of its sensitivity to hydrogen-bonding interactions and lattice vibrations of molecular crystals, it is useful for differentiating pharmaceutical polymorphs [4] and measuring compound crystallinity.

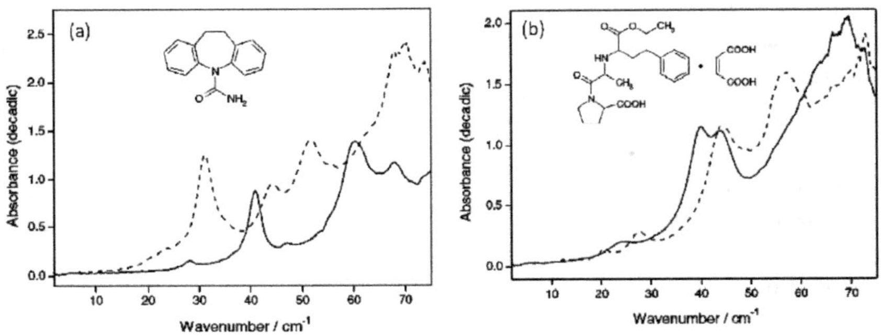

Fig. 6.32. Absorbance spectra of (a) carbamazepine form III (solid line) and form I (dashed line) and (b) enalapril maleate form I (solid line) and form II (dashed line). (Reprinted with permission from [201]. ©2004, Elsevier.)

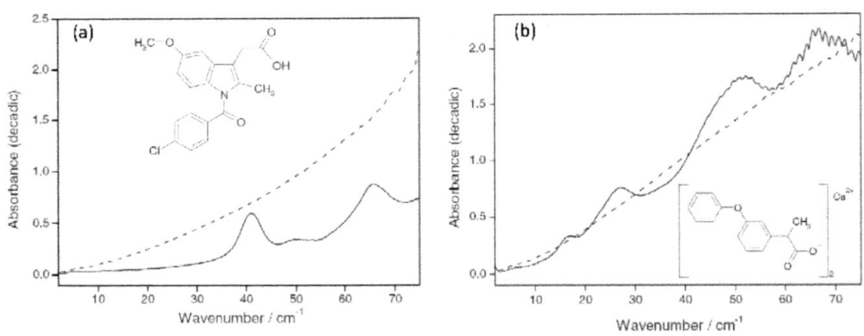

Fig. 6.33. Absorbance spectra of (a) indomethacin crystalline (solid line) and amorphous (dashed line) and (b) fenoprofen calcium crystalline hydrate (solid line) and liquid crystalline anhydrate (dashed). (Reprinted with permission from [201]. ©2004, Elsevier.)

Carbamazepine (CBZ), enalapril maleate (EM), indomethacin (IM), and fenoprofen calcium (FC) are pharmaceutical materials in which several poly-

[4] Polymorphism here refers to materials of the same chemical composition that take different physical structures. It has very important therapeutic implications for pharmaceutical drugs.

morphic and crystalline states coexist. As shown in Fig 6.32, polymorphism is pronounced in the absorption spectra of CBZ form III (P-monoclinic) and form I (triclinic) and of EM form I and form II as it displays markedly different spectral features. Figure 6.33 demonstrates the influence of crystalinity on intermolecular vibrational modes. The absorption spectra of IM and FC in a crystalline state exhibit distinctive mode structures, while the spectra of IM in a amorphous form and FC in a liquid crystalline state are largely featureless.

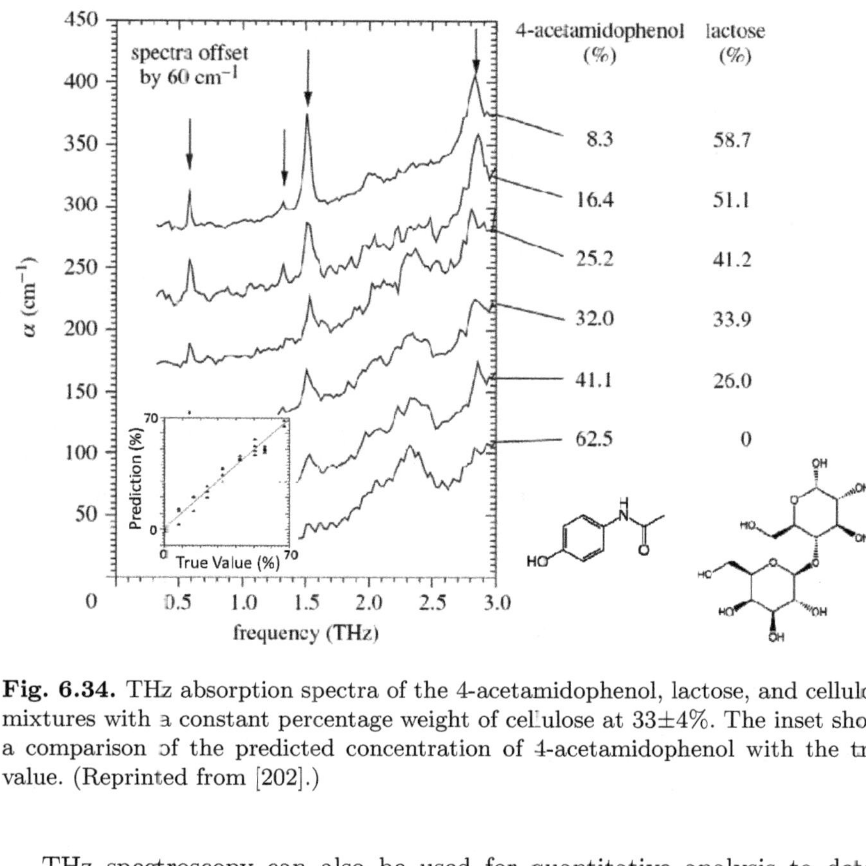

Fig. 6.34. THz absorption spectra of the 4-acetamidophenol, lactose, and cellulose mixtures with a constant percentage weight of cellulose at 33±4%. The inset shows a comparison of the predicted concentration of 4-acetamidophenol with the true value. (Reprinted from [202].)

THz spectroscopy can also be used for quantitative analysis to determine the concentration of active ingredients in a pharmaceutical product. Figure 6.34 shows the spectra of THz absorption coefficients of several mixtures containing 4-acetamidophenol, lactose, and cellulose [202]. Each mixture has a different concentration of 4-acetamidophenol and lactose, while the concentration of cellulose was kept constant at 33±4%. The peaks marked with arrows correspond to vibrational modes in lactose. Quantifying the spectral variation, a chemometric model predicts the concentration of the ingredients.

The inset shows a very good agreement between the predicted and true values of the 4-acetamidophenol concentration.

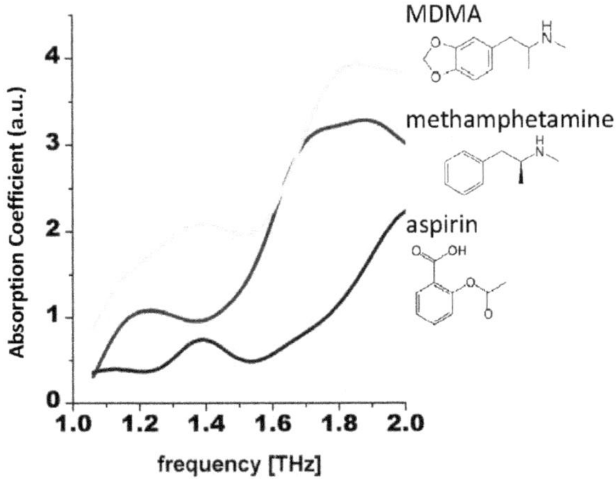

Fig. 6.35. Absorption spectra of MDMA, methamphetamine, and aspirin. (Reprinted from [203].)

Another practical application with a huge potential is noninvasive detection and identification of concealed illegal drugs. Figure 6.35 shows the results of a proof-of-principle experiment, where the absorption spectra of MDMA a.k.a. Ecstasy and methamphetamine are compared with the spectrum of aspirin [203].

6.3.3 Dynamics of Large Molecules

Protein Structure

Since the first three-dimensional image of a protein structure was produced by x-ray crystallography in 1958, nearly 40,000 crystal structures of biomolecules have been determined. From the intensive studies of several decades, we now have a very good understanding of the complex atomic arrangements of large biological molecules. Protein structure is organized into four levels. Primary structure refers to the linear chain of amino acids or the peptide sequence. Secondary structure contains a periodic formation of polypeptides linked by hydrogen bonds. The most common types of secondary structures are α-helices and β-sheets. The tertiary structure of protein molecules is formed as secondary structures fold into a unique three-dimensional globular structure. Quaternary structure refers to a protein complex formed by interactions among protein molecules. The three-dimensional structure of proteins is crucial information to understand how they function in biological systems. Yet it is not sufficient to gain a complete understanding of the processes, because the proteins are not entirely rigid, and the relative motions between the functional groups play a very important role. The functional rearrangements of structure are called *conformational changes*, and the resulting structures are called *conformations*. The conformational changes often occur on the picosecond timescale, and hence the large-amplitude vibrational modes lie in the THz region.

An example of protein structure is shown in Fig. 6.36 depicting a myoglobin (Mb) tertiary structure. Mb, a relatively small protein, has been intensely investigated, and hence its chemical and physical properties are very well known. Mb consists of eight α-helices which fold into a compact globular structure. It is water soluble because hydrophillic (polar) amino acids form outside of the structure. The hydrophobic (non-polar) groups face toward the inside of the structure and form a cleft where an oxygen-carrying heme group is attached. The hydrophobic effect makes a crucial contribution in maintaining the stability of the folded protein. The eight α-helices can be treated as a set of rigid rods, and the relative motion of each chain within the globular structure can be modelled as highly anharmonic vibrations. This motion is believed to be in the THz region.

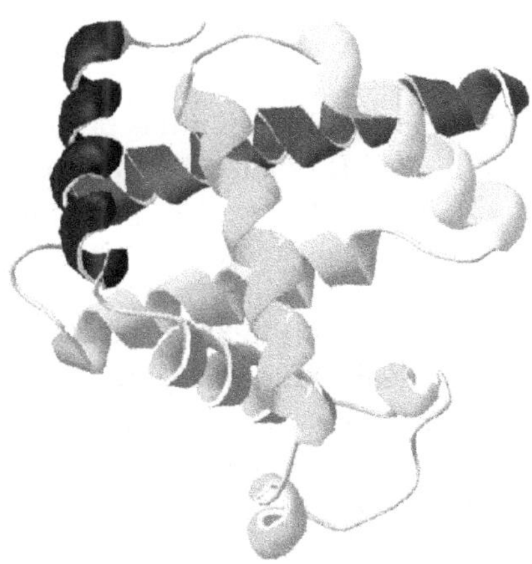

Fig. 6.36. Structure of myoglobin

Figure 6.37 shows the comparison between the calculation of the normal mode distributions and the experimental observation of THz absorbance of Mb. The calculation was performed using CHARMm (Chemistry at HARvard Molecular Mechanics), a molecular simulation program, and the experimental data was taken by THz-TDS. The largely continuous absorbance spectrum exhibits no sharp spectral features. This is expected, because the modes are highly sensitive to the local environment, and the consequent spectral shifts and broadening easily wash out any sharp features in the mode structure. The general spectral features are common for large biomolecules. In the following sections, we will see that conformational dynamics of large molecules is affected by different environmental conditions, and THz spectroscopy is highly sensitive in detecting these changes.

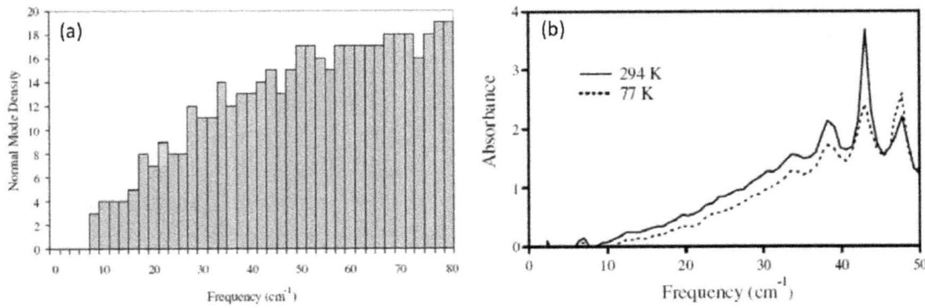

Fig. 6.37. Spectra of (a) normal mode density and (b) THz field absorbance for horse heart myoglobin. (Reprinted from [204].)

Conformational Changes

Bacteriorhodopsin (bR) is a small membrane protein that changes its conformation by absorbing light. bR, composed of seven α-helices and a cofactor retinal (Fig. 6.38(a)), has a strong absorption band peaked at 570 nm. As it absorbs a photon, the retinal molecule changes its conformation. This triggers a photochemical cycle in which the protein undergoes a sequence of structural transformations. Figure 6.38(b) illustrates the schematic of the photocycle. The conformational changes also induce spectral shifts of the absorption band, and hence the intermediate states denoted as J, K, L, M, N, and O, have different absorption maxima. Because of the huge line shift and index change, bR has been under vigorous investigation for applications to optical memory and optical switching.

The M state has a long lifetime of milliseconds at room temperature and is frozen into a metastable state by cooling it to 233 K. Figure 6.39 demonstrates that the conformational change can be detected by observing the vibrational modes in the THz region. The data show a cycle of the experimental procedure. (i) As temperature falls from 295 K to 233 K, the absorption by bR in the ground state decreases all over the spectral region. (ii) Illuminated by light of λ <640 nm, the bR undergoes a conformational change to the M state, which induces the increase in absorption. The temperature is maintained at 233 K. (iii) Warmed up to 295 K, the bR returns to its initial state. The significance of this measurement is that THz spectroscopy clearly distinguishes

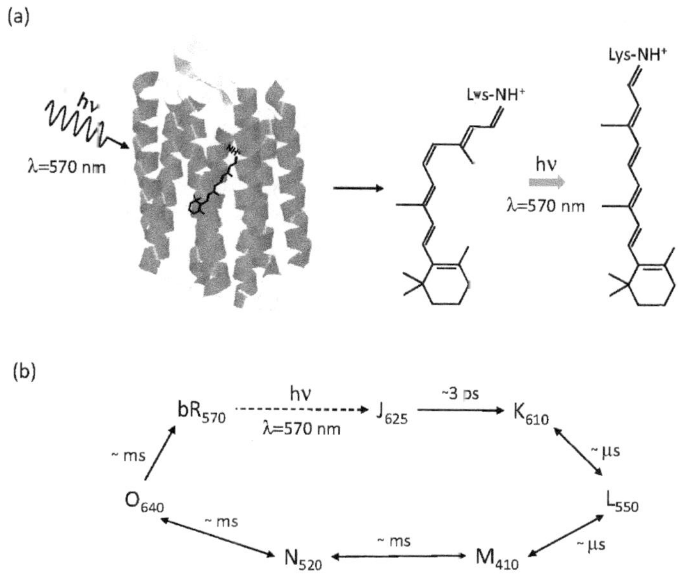

Fig. 6.38. Photo-induced conformational changes in bacteriorhodopsin. (a) Retinal photoisomerization. The retinal molecule changes its conformation by absorbing a photon at 570 nm. (b) Schematic of the photochemical cycle. Subscripts denote the peak wavelengths of the absorption band in nm.

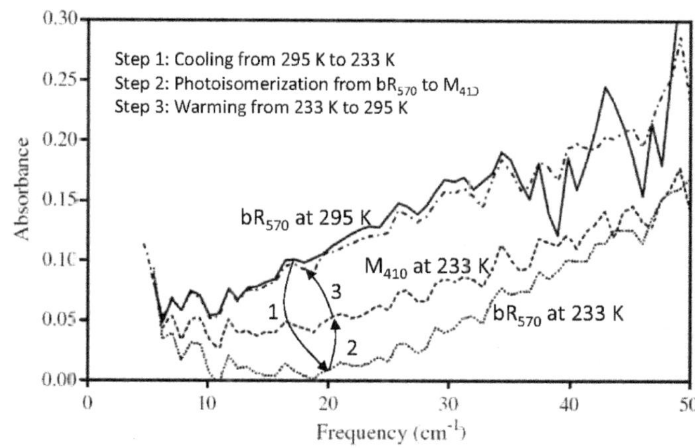

Fig. 6.39. THz field absorbance for bR in a cycle: ground state room at 295 K → ground state at 233 K → M state at 233 K → ground state at 295 K. (Reprinted from [204].)

the M state from the ground state, and it provides solid evidence of high THz sensitivity to conformational changes.

The Asp^{96} to Asn (D96N) mutation is generated by a single residue mutation in a wild-type bR molecule. There is virtually no difference between the two structures of wild-type bR and D96N except for the residue change. Nevertheless, the photocycling time of D96N is prolonged over 1000 times by the mutation, because the M state of D96N is much more stable than that of wild-type bR. The augmented stability of the M state implies that the conformational changes of D96N are less flexible than those of wild-type bR. The experimental observation that THz absorption of the mutant is smaller than that of the wild type (Fig. 6.40) is consistent with the hypothesis that the mutant has reduced conformational flexibility.

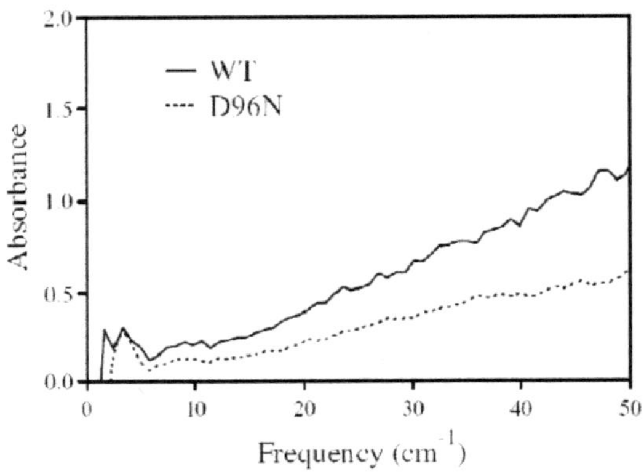

Fig. 6.40. THz field absorbance for wild-type bR and the D96N mutant. (Reprinted from [204].)

Proteins and Water

Protein dynamics in aqueous environments are of enormous interest, and THz spectroscopy provides a sensitive probe of the interaction between proteins and water molecules. Water induces changes in the physical and chemical properties of proteins, and proteins simultaneously affect the nearby hydrogen-bond network of water molecules. The one or two layers of water molecules in the vicinity of a protein surface form a region of biological water which has distinctive physical properties from bulk water (Fig. 6.41).

Figure 6.42 shows the absorption spectra of Mb aqueous solutions and powders at various water concentrations. The spectra are nearly continuous and the absorption coefficients gradually rise with an increase of frequency. The nonpolar Mb molecule has much lower absorption in the THz region when compared with water. Consequently, absorption by water dominates the absorption spectra of the Mb-water mixtures, which explains the overall trend in the experimental data that the absorption coefficient escalates as the water

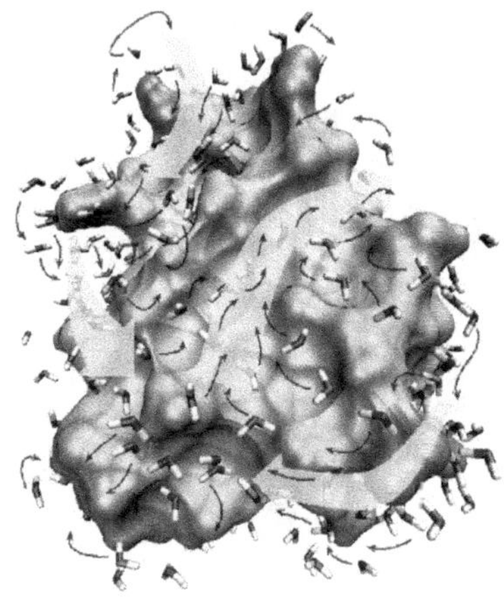

Fig. 6.41. Water molecules dancing around a protein. (Reprinted from [205].)

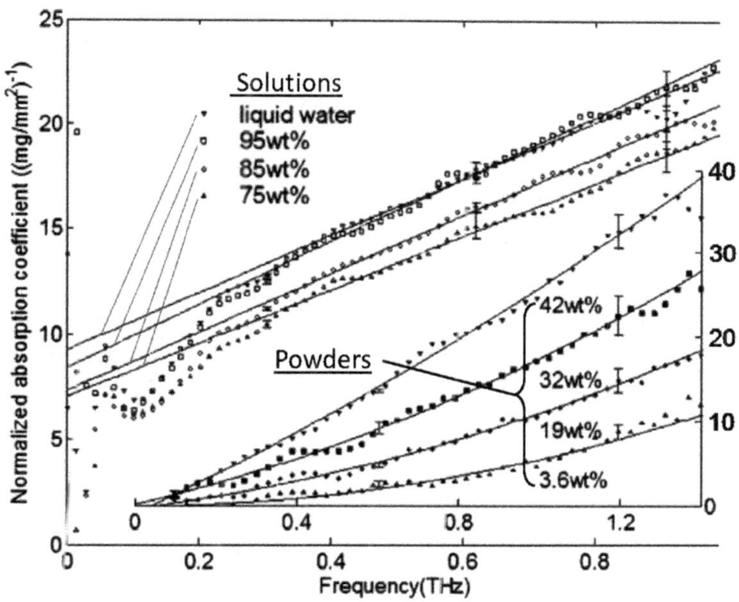

Fig. 6.42. (a) Absorption spectra of Mb aqueous solutions and powders at different water concentrations. The solid line is a polynomial fit for each sample in the range 0.3-0.8 THz for solutions and 0.1-1.2 THz for powders. (Reprinted with permission from [206]. ©2006, American Chemical Society.)

concentration increases (Fig. 6.43(a)). The three-component model of Eq. 6.16 is applied to analyze the data. Figure 6.43(b) shows the molar absorptivity of Mb as a function of water concentration at 0.35, 0.5, and 0.8 THz. The dashed line denotes the absorptivity of dry Mb. The huge enhancement of the Mb absorptivity in aqueous environments indicates that biological water in the vicinity of Mb molecules has higher absorption than bulk water. As in the case of sugar waters, this result is somewhat puzzling if the hindered motions of biological water are considered. A plausible explanation is that the tight globular structure of Mb loosens up when it interacts with water molecules and, subsequently, large amplitude motions become more readily available.

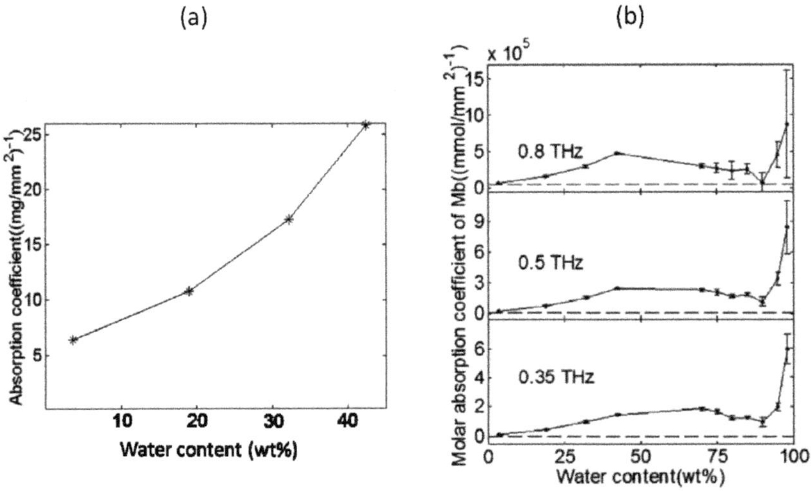

Fig. 6.43. (a) Absorption coefficient at 1.0 THz for powder samples. (Reprinted with permission from [206]. ©2006, American Chemical Society.) (b) Molar absorptivity of Mb as a function of water content at frequencies of 0.35, 0.5, and 0.8 THz. The dashed line indicates the absorptivity of dry Mb as a base level. (Reprinted with permission from [207]. ©2004, American Chemical Society.)

Another approach to account for mysterious protein hydration dynamics is to employ a microscopic model of water molecules in the hydration shell. In a study of solvation dynamics around the five helix bundle protein λ^*_{6-85}, experimental measurements are compared with molecular dynamics simulations [208]. Figure 6.44 shows the difference in the THz absorption coefficient at 2.25 THz relative to bulk water, $\Delta\alpha = \alpha_{\text{sample}} - \alpha_{\text{bulk}}$, as a function of protein concentration measured at 15°C, 20°C, and 22°C. The temperature-dependent offset of $\Delta\alpha$ mainly comes from α_{bulk} being sensitive to temperature. The seminal feature of these experimental data is the nonmonotonic behavior of $\Delta\alpha$ at relatively low concentrations (1 mM corresponds to 0.65 wt%). The simple model of noninteracting components, therefore, cannot explain the nontrivial concentration dependence on THz absorption. This is a strong indication that the absorbance of hydration water may be sensitive to the the distance between protein molecules.

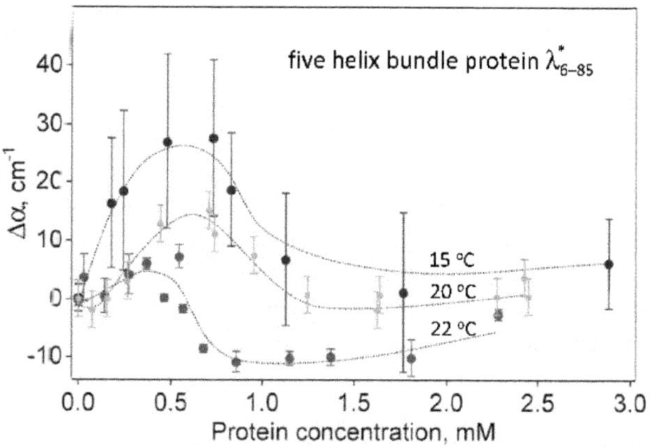

Fig. 6.44. Differential absorption coefficient ($\Delta\alpha = \alpha_{sample} - \alpha_{bulk}$) of λ^*_{6-85} at 2.25 THz as a function of protein concentration at 15 °C, 20 °C, and 22 °C. (Reprinted from [208].)

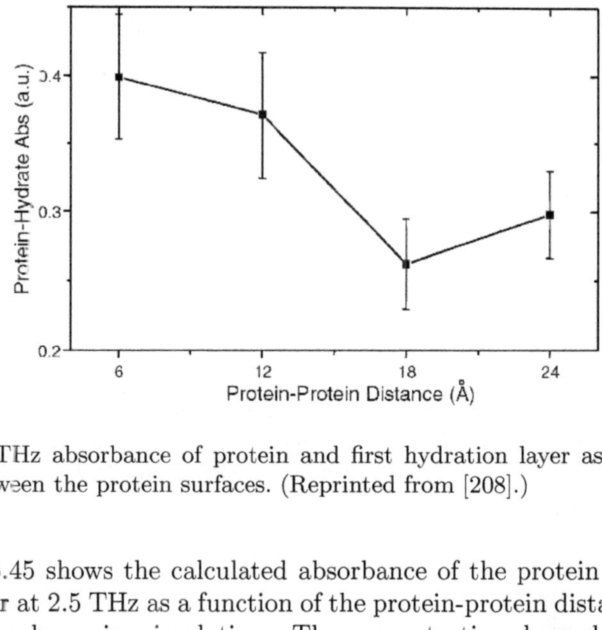

Fig. 6.45. THz absorbance of protein and first hydration layer as a function of distance between the protein surfaces. (Reprinted from [208].)

Figure 6.45 shows the calculated absorbance of the protein and first hydration layer at 2.5 THz as a function of the protein-protein distance obtained by molecular dynamics simulations. The concentration dependence that the absorbance decreases as molecules are brought closer together from 24 to 18 Å and increases at shorter distances is consistent with the overall trend of experimental observation, namely, the decrease of $\Delta\alpha$ in the region 0.5-1.0 mM and the gradual increase at higher concentrations. This trend is relevant to increasing overlap between solvation layers as the proteins get closer. The molecular dynamics calculations and the experimental data also suggest that the influence of a protein surface on the surrounding water network motion exceeds ~10 Å.

Hydration dynamics exhibit drastically different traits as the protein goes through an unfolding transition at low pH. The denaturing process exposes the hydrophobic residues of the protein to surrounding water, retarding the motions of water molecules in the vicinity of the protein surface. Figure 6.46 shows that the THz absorption of λ^*_{6-85} strongly depends on pH, while the absorption of the buffer alone hardly changes over the same pH range. The long-range solvation effect of λ^*_{6-85} is substantially reduced by the exposure of the hydrophobic residues. As a consequence, the proteins at pH 2 and 5 are nearly transparent in the THz region and the concentration dependence of the THz absorption follows the similar trend with the two-component model.

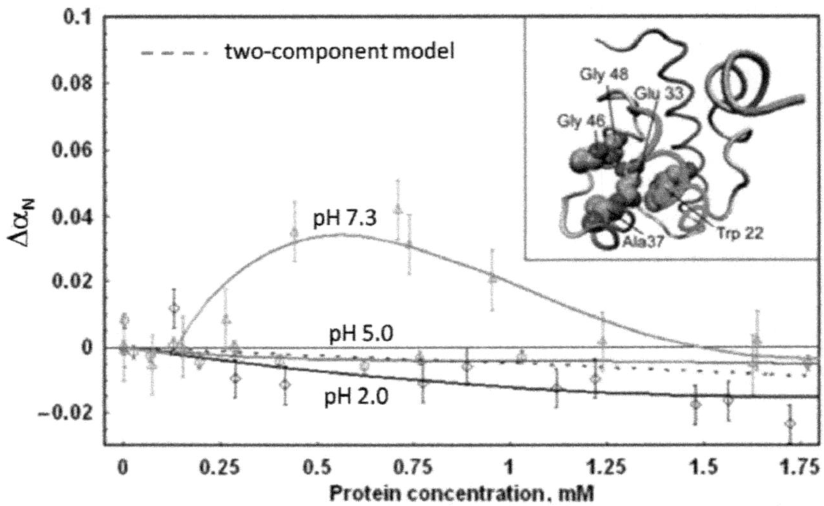

Fig. 6.46. Normalized differential absorption coefficient $\Delta\alpha_N = \Delta\alpha/\alpha_{bulk}$ of λ^*_{6-85} at pH 2.0, pH 5.0, and pH 7.3. The temperature is kept at 20 °C. The inset shows the protein structure. (Reprinted with permission from [209]. ©2008, American Chemical Society.)

Label-Free DNA Sensor

An intriguing application of THz spectroscopy of biomolecules is to directly probe the binding states of genetic material without using fluorescent labels. In particular, prototypical biochips capable of identifying DNA states at the femtomole level have been demonstrated [210].

Figure 6.47 shows the complex refractive index, $\tilde{n}(\nu) = n(\nu) + i\kappa(\nu)$, of hybridized (double stranded) and denatured (single stranded) DNA films measured by conventional THz-TDS with a free-space optical arrangement. Both

$n(\nu)$ and $\kappa(\nu)$ of denatured DNA are considerably smaller that those of hybridized DNA, meaning that the interaction of THz radiation with denatured DNA is much weaker than that with hybridized DNA. It can be accounted for by the fact that double-stranded DNA structure surely has a higher vibrational mode density than single-stranded in the THz region. Regardless of the microscopic nature causing the difference, the ultimate outcome of this experimental observation is that THz spectroscopy is applicable to label-free identification of DNA binding states.

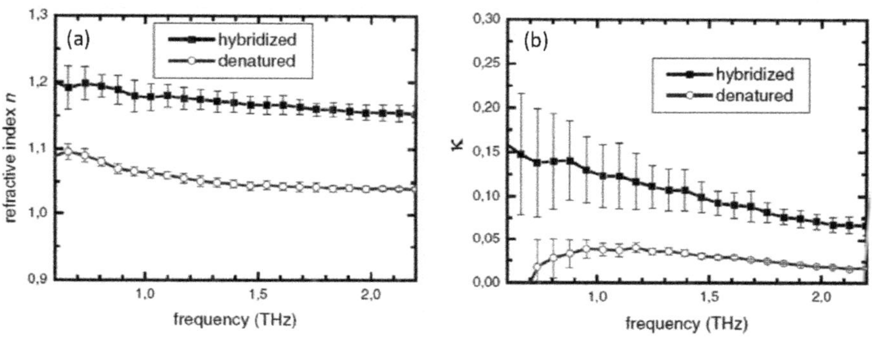

Fig. 6.47. (a) Real and (b) imaginary part of the refractive index $\tilde{n}(\nu) = n(\nu) + i\kappa(\nu)$ of hybridized and denatured DNA films (Reprinted from [210].)

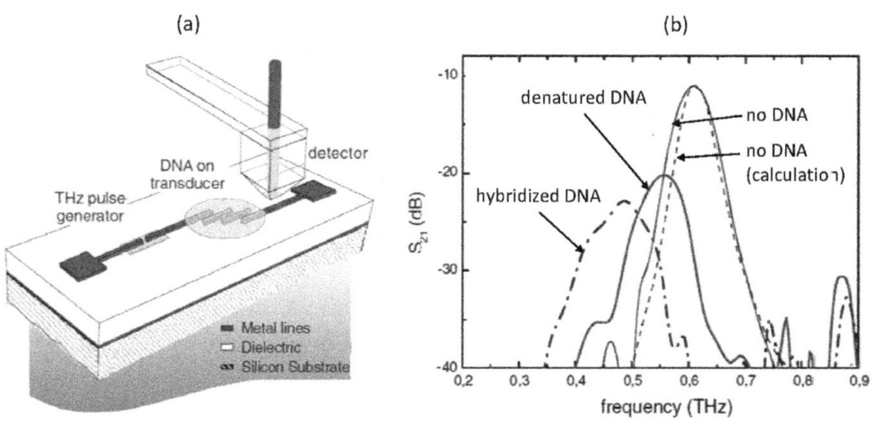

Fig. 6.48. (a) Scheme of an integrated DNA sensor consisting of a photoconductive THz source, a sensing THz resonator a the freely positionable electro-optic detector. (b) Spectra of the THz transmission (S_{21}) through the DNA sensor without DNA, with hybridized and with denatured DNA. (Reprinted from [210].)

The detection sensitivity is dramatically improved by use of integrated THz sensors exploiting surface plasmon waves localized in a microresonator. Figure 6.48(a) illustrates the scheme of using an integrated THz sensor consisting of a photoconductive THz emitter, a coupled metal-line THz resonator, and a movable electro-optic probe. The THz radiation generated by the photoconductive emitter is directly coupled to the resonator in which THz surface plasmon waves interact with DNA samples. The contrast between the transmission spectra of hybridized and denatured samples is remarkable. Because of the larger refractive index, the transmission spectrum of hybridized DNA undergoes a stronger redshift than that of denatured DNA. The amount of DNA deposited in the resonator is estimated as $\sim$1 femtomole.

7

T-Ray Imaging

Modern imaging technologies are interwoven into the fabric of our everyday lives. Television and computer screens continuously spit out all sorts of images; radar imaging is used for weather forecast; x-ray imaging, magnetic resonance imaging, and ultrasound imaging are indispensable tools for medical diagnosis, etc. T-ray imaging, or terahertz imaging, is the latest entry into the crowded field of imaging technologies. Many applications are emerging for the relatively new technology, yet the technological development is still in its infancy, and it is hard to predict the scale and impact of the technology when it matures. Nevertheless, T-ray imaging is unique enough in its characteristics that it has already found some niche applications inaccessible by other imaging technologies.

7.1 Introduction

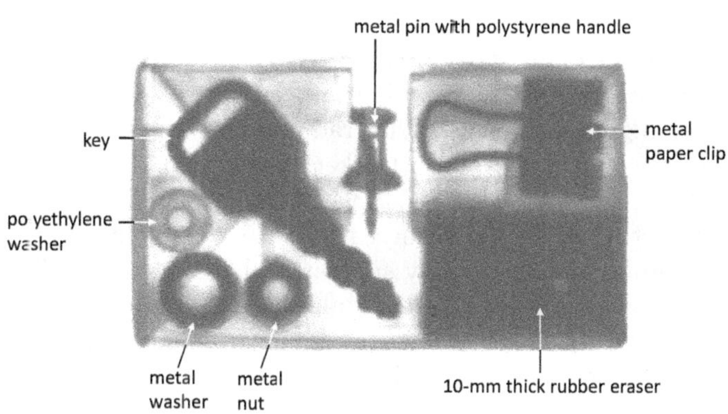

Fig. 7.1. (a) THz image of a closed cardboard box containing several metal and plastic objects.(Reprinted from [211].)

THz radiation penetrates deep into nonpolar and nonmetallic materials such as paper, plastic, clothes, wood, and ceramics that are usually opaque at optical wavelengths. These are also common packaging materials, and hence T-ray imaging devices have been used for nondestructive testing to inspect sealed packages. Figure 7.1 shows a THz image of several metal and plastic objects concealed in a closed cardboard box. Optical probes cannot see through the cardboard box, but it is useless to try to hide the contents from the T-ray imaging. Metal has a very short penetration depth and is highly reflective in the THz region, and hence the metallic objects completely block the THz radiation, while the plastic ones are partially transparent. The clear contrast between metal and plastic facilitates the inspection of circuits within a plastic package. Figure 7.2 shows a contactless smart card in which a radio-controlled integrated-circuit is embedded. The plastic packaging material shows little absorption, whereas the metal circuit is completely opaque.

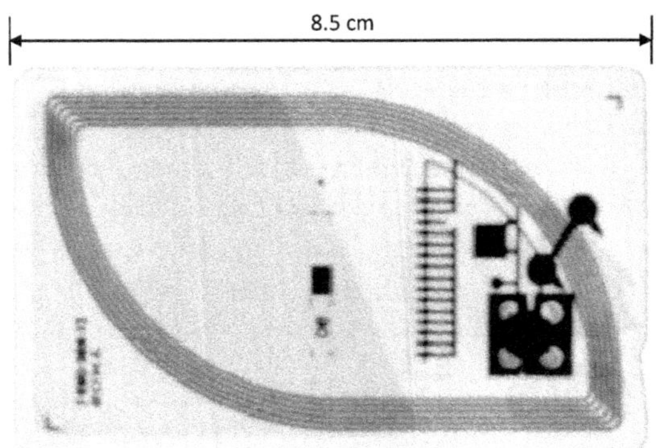

Fig. 7.2. THz transmission image of a contactless smart card embedded with a metallic circuit. (Reprinted from [211].)

Because water is highly absorptive in the THz region, hydrated substances exhibit strong contrast to surrounding materials in a THz image. Figure 7.3 illustrates the high sensitivity of THz radiation to water, showing the water content of a leaf on a living plant. Darker shade in the image indicates more water. The line scans shown in Fig. 7.3(b) demonstrate that the water intake in the leaf structure is gradually increasing after the plant was watered.

Besides metal and water, many substances having spectral fingerprints in the THz frequency region are also detectable by T-ray imaging. In this context, T-ray imaging and sensing technologies for defense and security applications have been under intense investigation. It has been demonstrated that T-ray imaging devices can effectively detect and identify weapons, explosives, and chemical and biological agents concealed underneath various covering materials.

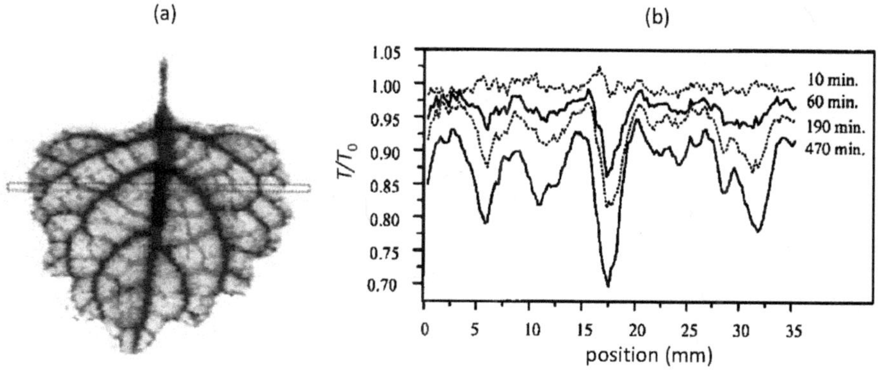

Fig. 7.3. (a) THz image of a Coleus leaf. The gray scale is correlated with water content, with darker shade indicating more water. (b) Line scans along the dashed box in the THz image. The scans were performed at 10, 60, 190, and 470 min after the plant had been watered. As more water is taken into the leaf structure, transmission falls off where the leaf stems are located. (Reprinted from [212]. ©1996 IEEE)

Some speculate that the biggest opportunity for T-ray imaging technologies to make a huge impact will be found in medical applications. A big advantage of THz applications in medicine is that a T-ray imaging device utilizes a noninvasive, nonionizing, and noncontact probe. A medical imaging device based on THz radiation is, therefore, inherently gentle to the human body. So far, T-ray imaging has been applied to spotting the onset of cancer, characterizing burn injuries, detecting tooth decay, etc.

An imaging system is usually composed of various subsystems. Most of the THz devices discussed in the preceding chapters are potentially a part of a T-ray imaging system. Accordingly, T-ray imaging technologies are largely categorized as pulsed THz time-domain imaging and continuous-wave THz imaging. This chapter will discuss how the T-ray imaging systems perform their functions as well as their applications in various fields, in particular, security and medicine.

7.2 Imaging with Broadband THz Pulses

7.2.1 Amplitude and Phase Imaging

A straightforward implementation of pulsed THz imaging is accomplished by raster-scanning of a target at the focal plane of a THz time-domain spectroscopy (THz-TDS) system. Figure 7.4 sketches the scheme of pulsed THz imaging in transmission geometry. The target image is obtained by analyzing the transmitted THz waveforms. Changes in the amplitude $A(x, y)$ and the phase $\phi(x, y)$ of the THz pulses map out the spatial inhomogeneity of the target, characterized by material properties such as refractive index $n(x, y)$, absorption coefficient $\alpha(x, y)$, and thickness $d(x, y)$.

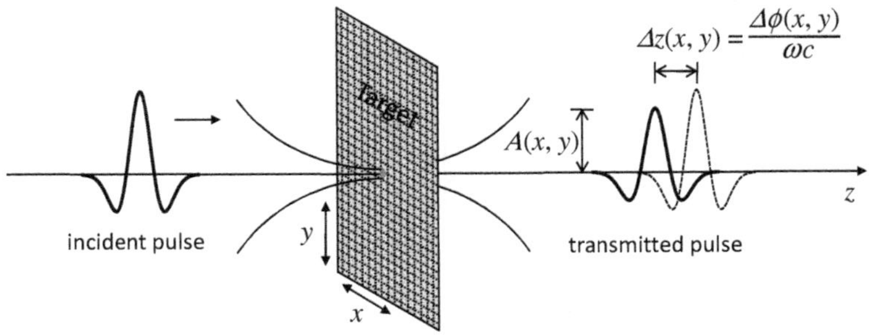

Fig. 7.4. Raster-scan imaging with THz pulses in transmission geometry

Pulsed THz imaging is inherently associated with versatile visualization schemes because a THz-TDS image contains much more information than a typical 2-D image containing the same number of pixels. Each pixel of a THz-TDS image contains a whole waveform in the time domain. The excessive information provides many different display options for a THz-TDS image. In particular, it contains spectroscopic information. Figure 7.5 illustrates some commonly used display schemes. In the time domain, the maximum amplitude $A_a(x, y)$ or the arrival time $t_a(x, y)$ of the waveform is used to form a 2-D image. If the target material has a complicated dispersion relation, a wavelet analysis can reveal important properties of the sample. Frequency-domain imaging is also possible with the spectrum of the waveform obtained by the Fourier transform. If the sample contains a material having spectral fingerprints, the material's characteristics can be brought out in the image by use of the amplitude or the phase at the specific frequencies of the spectral fingerprints.

Figure 7.6 demonstrates the versatility of pulsed THz imaging. The two images of a chocolate bar are formed by use of (a) peak-to-peak amplitude and (b) arrival time, i.e., phase of the transmitted THz pulse. Darker shade indicates lower transmission in the amplitude image and later arrival time in the phase image. Overall, the chocolate is fairly transparent in the THz region. The two images, however, bring out different and nuanced aspects of the sample. First, the four almonds in the chocolate bar are clearly discernible in the amplitude image. Because of the higher water content in almonds than in chocolate, THz absorption is stronger in almonds. Consequently, the absorption profile of the almonds is displayed as the dark spots in the amplitude image. The region of the embossed letters is relatively thinner than the surrounding area, but the difference in absorption is negligible. The letters are visible in the amplitude image largely because of scattering at the stepped edges. Second, in the phase image, the sample thickness is a crucial parame-

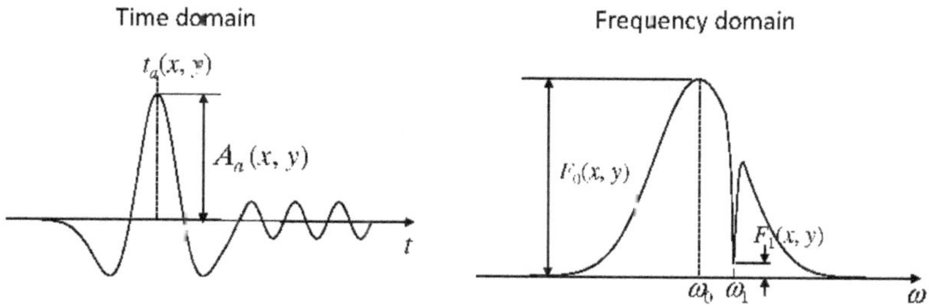

Fig. 7.5. Display options for THz-TDS imaging: maximum amplitude $A_a(x, y)$ and arrival time $t_a(x, y)$ in the time domain, maximum amplitude $F_0(x, y)$ and amplitude at resonance $F_1(x, y)$ in the frequency domain, and total energy $I(x, y) = \int |A(x, y, t)|^2 dt$.

ter to determine the arrival time, hence the letters are clearly distinguished from their surroundings. On the other hand, almonds and chocolate have similar refractive indices, and hence there is little contrast between them in the phase image.

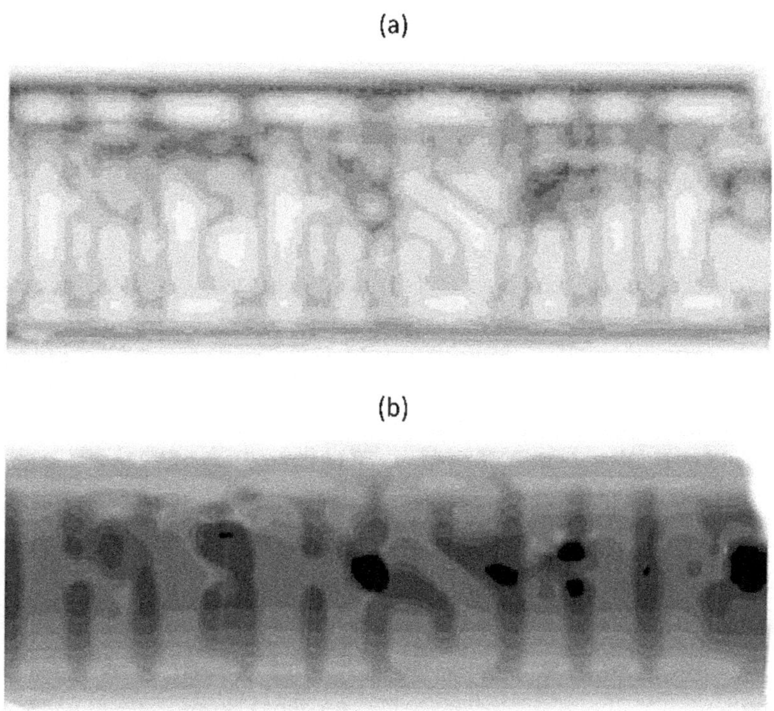

Fig. 7.6. THz images of a chocolate bar constructed by (a) peak-to-peak amplitude and (b) arrival time of the transmitted waveform. (Reprinted from [213].)

7.2.2 Real-Time 2-D Imaging

One drawback of raster-scan imaging is that the image taking process is too slow—typical acquisition time for a 2-D image is at least a few minutes—to capture the image of a moving object. The speed of image taking is greatly improved by use of 2-D electro-optic (EO) sampling [214]. The free-space EO sampling technique is described in detail in section 3.4.

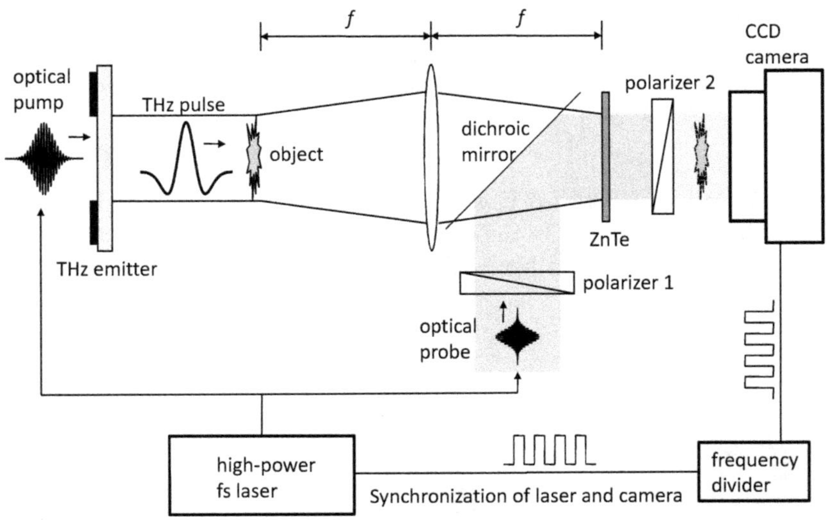

Fig. 7.7. Schematic diagram of real-time 2-D THz imaging system

Figure 7.7 depicts the scheme of a real-time THz imaging system using 2-D EO sampling. Real-time imaging requires strong THz radiation, and hence the optical source is high-power femtosecond pulses from a laser amplifier. For practical applications, the pulse energy should be at least $\sim$1 mJ. At present, the most efficient table-top high-power THz source is the large-aperture photoconductive emitter (see section 3.2.4). An alternative way to generate high-power THz pulses is optical rectification in a large-area EO crystal. The THz beam size is adjusted to cover the entire object. The object is placed in the focal plane of a lens or an imaging system, and the THz beam delivers its image onto a large-area EO crystal. The linearly-polarized probe beam is combined with the THz beam and copropagates in the detection crystal. The THz field induces a transient birefringence in the EO crystal, and subsequently modulates the probe polarization. Only the modulated part of the probe beam, perpendicular to the initial polarization, passes through the second polarizer and is detected by the CCD camera. Timing electronics synchronize the laser and CCD camera.

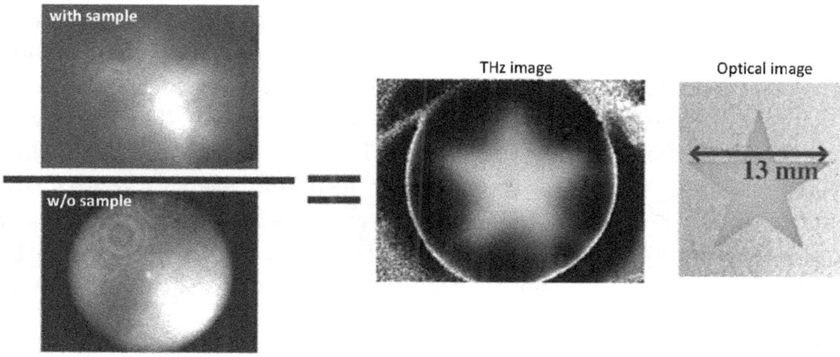

Fig. 7.8. CCD images with and without the sample of a metal film with a star-shaped aperture. Normalizing the image data produces the THz image shown in the middle. An optical image of the sample is also shown for comparison. (Reprinted from [215].)

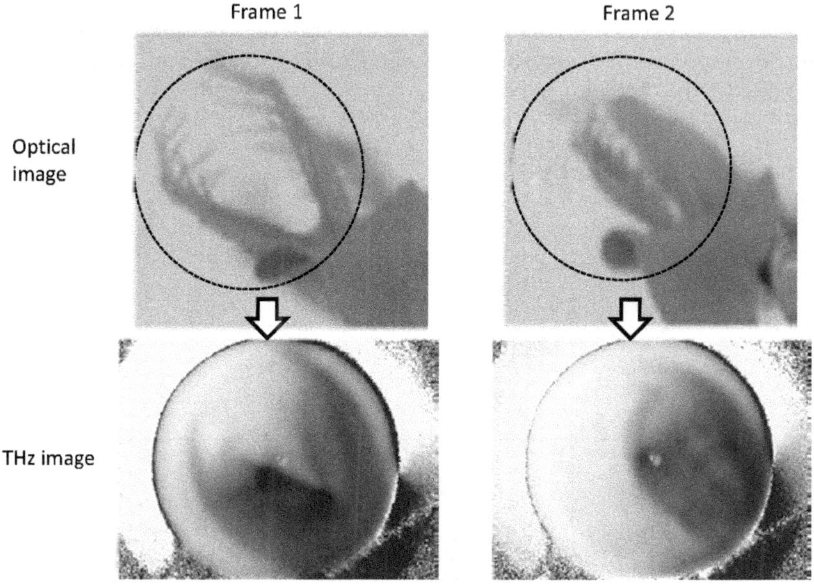

Fig. 7.9. Snapshots from a THz movie showing a moving object. The images are the initial and final steps of the folding leaves of a Venus Flytrap. Optical images are shown for comparison. (Reprinted from [215])

Figure 7.8 shows a 2-D image taken by a real-time THz imaging system [215]. The sample is a metal film with a star-shaped aperture. The optical source is a 1-kHz Ti:sapphire regenerative amplifier, and the THz emitter is a large-aperture photoconductive emitter with a 15-mm gap. The EO crystal is a 2-mm-thick ⟨110⟩ ZnTe crystal. The CCD camera operates at or below 30 frames per second (fps). Because the probe beam and the crystal are spatially inhomogeneous, the raw CCD images are corrected by background subtraction. Further, in order to clean up the THz image, the image data with the sample are normalized by those without. The processed THz image is compared with the optical image of the sample.

Video-rate THz imaging is demonstrated in Fig. 7.9 when the real-time imaging system operates at a frame rate of 10 fps. The moving object is a Venus Flytrap, a carnivorous plant. The imaging system captured the motion of the two leaves, initially open, closing after being touched with tweezers. The THz images of Fig. 7.9 show the first and last frames of the movie. Optical images are also shown for comparison.

7.2.3 T-Ray Tomography

Tomography refers to thin-slice cross-sectional imaging of the internal structure of a 3-D object. The complete 3-D object structure can be reconstructed by stacking up the 2-D image slices. Several T-ray tomography techniques have been developed exploiting the high transparency of non-polar, non-metallic materials in the THz range.

Reflection Tomography

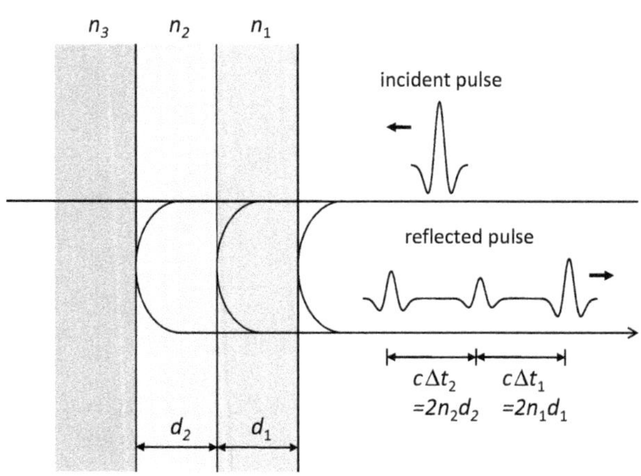

Fig. 7.10. Reflected THz pulse from dielectric interfaces

The time-resolved detection scheme of THz-TDS is directly applicable to measure the depth profile of an object with a multilayer structure. When a short THz pulse is incident on the object, the reflected waveform consists of a series of pulses reflected from the interfaces. A time-of-flight analysis of the waveform can map out the internal layer structure. Because the temporal resolution of the THz pulse arrival time is a few femtoseconds, the time-of-flight analysis can determine the positions of the interfaces at a resolution of about one micrometer. The refractive-index profile of the medium can also be obtained by analyzing the reflectivities at the interfaces.

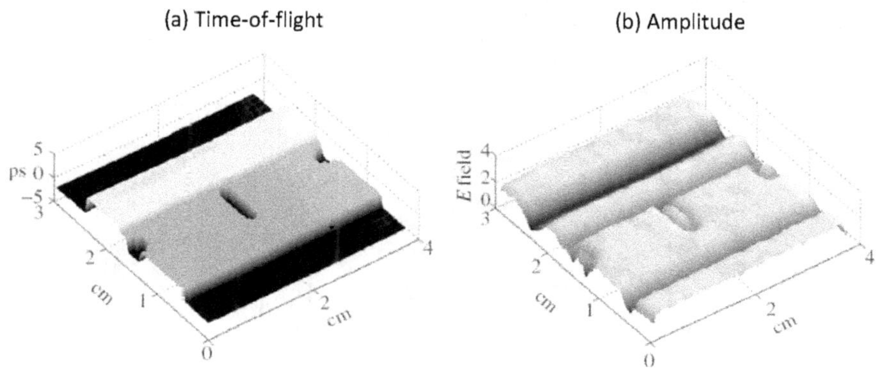

Fig. 7.11. THz reflective images of a razor formed by (a) the arrival times and (b) the peak amplitudes of the reflected pulses. Darker shade indicates longer delay time for the time-of-flight image and weaker intensity for the amplitude image. (Reprinted from [216].)

An example of reflective imaging is demonstrated in Fig. 7.11 showing two THz reflective images of a razor: (a) tomographic image formed by the arrival times and (b) amplitude image formed by the peak amplitudes. The razor attached on a metal mirror contains three reflection surfaces of different heights: the metal holder, the razor blade, and the metal mirror. The time-of-flight image contrasts the surfaces by distinguishing the surface heights. In the amplitude image, the three surfaces look almost identical, because the reflectivities of the metal surfaces are nearly unity. The step edges, however, are recognizable because of scattering losses.

Figure 7.12 demonstrates the 3-D tomographic imaging of a 3.5-in. floppy disk by analyzing the time-of-flight delay of THz waveforms returned from the multilayer target. The THz image formed by the total reflected power is shown in Fig. 7.12(a). Figure 7.12(b) shows the cross-sectional depth profile near the center of the floppy disk, marked as the dashed line in (a). At each horizontal position x, a THz waveform is measured, and processed by frequency filtering to get arrival times with better accuracy. The tomographic image is reconstructed by putting together the processed THz waveforms. The surfaces of the main parts, the front and back covers, the magnetic diskette, and the metallic hub, are clearly seen.

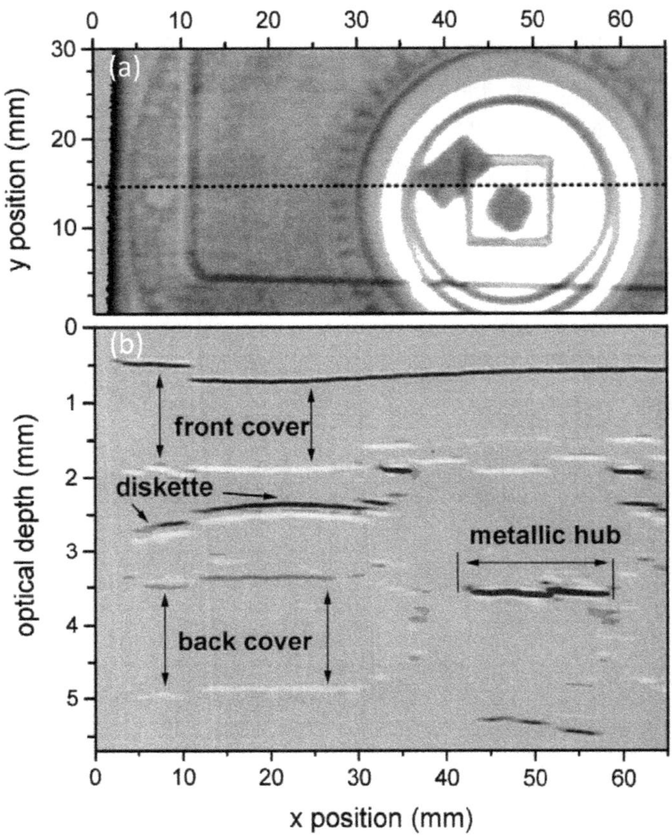

Fig. 7.12. (a) THz image of a 3.5-in. floppy disk formed by the total reflected power and (b) tomographic image at $y = 15$ mm in (a), marked by the dashed line. Darker and lighter stripes indicate positive and negative refractive-index steps, $\Delta n > 0$ and $\Delta n < 0$, respectively. (Reprinted from [217].)

THz time-of-flight tomography is sensitive in detecting small changes in optical properties of media such as the refractive index and absorption coefficient. Figure 7.13 demonstrates its application to nondestructive identification of defects in space shuttle foam insulation [218]. The NASA investigators believe that the Columbia space shuttle crash may have been caused by foam insulation breaking away and striking the left wing. It is therefore critical to detect defects in insulating foam prior to launch. It is difficult to locate defects in the low-density, low-absorption medium using other nondestructive inspection techniques such as X-rays or ultrasound. In general, two kinds of defects are formed in form insulation: voids and delaminations. Delamination refers to a structural failure where a part of the structure is separated by a thin layer and loosely attached to surroundings. These defects cause significant loss of mechanical toughness. The sample shown in Fig. 7.13(a) is made of lightweight polyurethane sprayed-on foam insulation (SOFI). The refractive index of SOFI is close to unity (n=1.02), and the absorption coefficient is less than 1 cm^{-1} below 1.5 THz. Figure 7.13(b) and (c) show the tomographic images of a part of the sample obtained by changes in amplitude and pulse energy, respectively. Three voids and one delaminated area are recognized in both images.

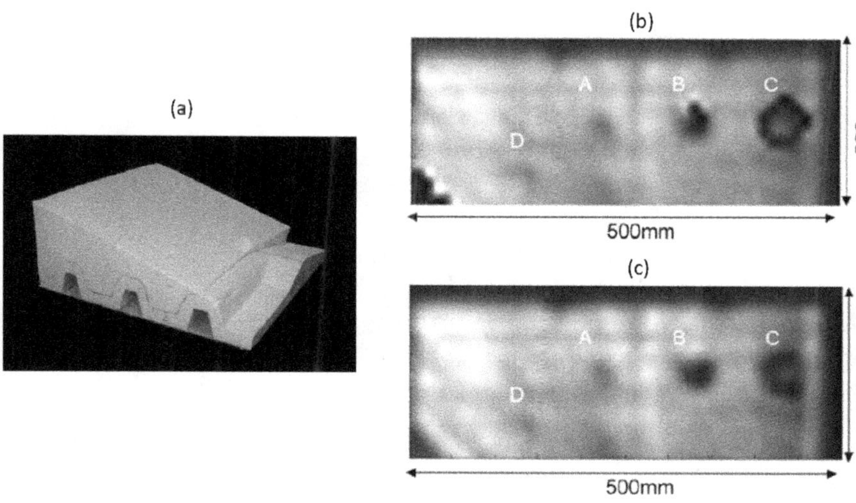

Fig. 7.13. (a) Sample of sprayed-on foam insulation. The size of the sample is 600×600 mm. (b) Amplitude and (c) pulse-energy images of a cross-sectional depth profile. A part of the sample with a 220×500-mm area is displayed. Defects are clearly seen in the tomographic images: A, B, and C are voids of 12.5, 25, and 37.5 mm, respectively, and D is a 50×50-mm delamination. (Reprinted from [218]. ©2005 IEEE)

Computed Tomography

Computed tomography (CT) usually refers to an imaging technique for producing 3-D images of an object from cross-sectional X-ray images. A CT system takes 2-D shadowgraphs of an object while rotating the object on a turntable. A single shadowgraph, with the 3-D structure compressed onto a 2-D plane, is not sufficient to figure out the entire structure of the object, but the series of the shadowgraphs taken around the axis of rotation contain enough information to generate cross-sectional images via specialized mathematical analysis.

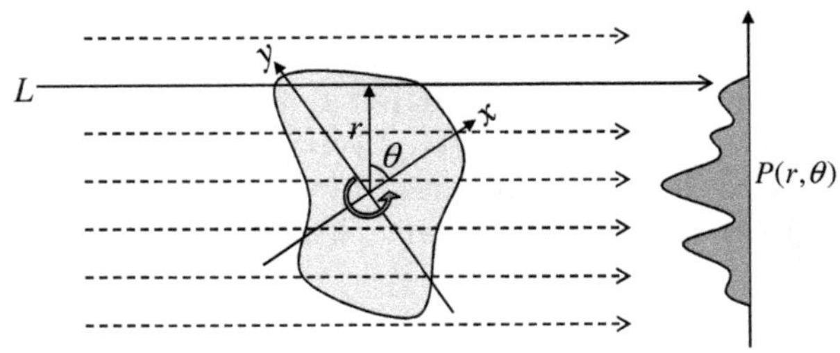

Fig. 7.14. Schematic of computed tomography

Figure 7.14 illustrates the basic CT scheme, where $P(r, \theta)$ is the shadow image or projection of the object at the projection angle θ. r is the distance from the axis of rotation. The key mathematical tool of CT is the Radon transform:

$$P(r, \theta) = \int_{L(r,\theta)} f(x, y)\, ds, \tag{7.1}$$

where $f(x, y)\, ds$ is the probability that a ray is absorbed or deflected in the segment ds along a straight line L. $f(x, y)$, therefore, depicts the local optical property of the object at (x, y). A CT image is reconstructed by an inverse transformation of the integral equation.

The basic principles of X-ray CT can be applied to T-ray imaging. Because T-ray CT measures the entire waveform of the transmitted radiation, its images are inherently multifaceted, e.g., a 3-D map of the refractive index or a spectral signature of the sample. Figure 7.15 shows an example of T-ray CT imaging. The tomographic image of a hollow dielectric sphere is reconstructed by using the amplitudes of the transmitted THz pulses as the input of the filtered backprojection algorithm, the most common algorithm used in the tomographic reconstruction.

Diffraction Tomography

The integral along a straight line in Eq. 7.1 is valid only in the regime of geometrical optics. If an object is characterized by spatial features comparable to the wavelength of incident radiation, the projection of a transmitted beam through the object is strongly affected by diffraction along the propagation.

Figure 7.16 illustrates that an incident plane wave is diffracted by an object, and the diffracted wave is projected onto a measurement plane. Neglecting polarization, the radiation field amplitude $u(\mathbf{r})$ satisfies the scalar wave equation,

$$\left(\nabla^2 + k^2\right) u(\mathbf{r}) = o(\mathbf{r}) u(\mathbf{r}) \tag{7.2}$$

Fig. 7.15. T-ray CT image of a hollow dielectric sphere. The sphere attached to a rotating plastic rod is scanned with a 1-mm step size and 18 different projection angles. The amplitudes of the transmitted THz pulses are used as the input to the inverse Radon transform. (Reprinted from [216].)

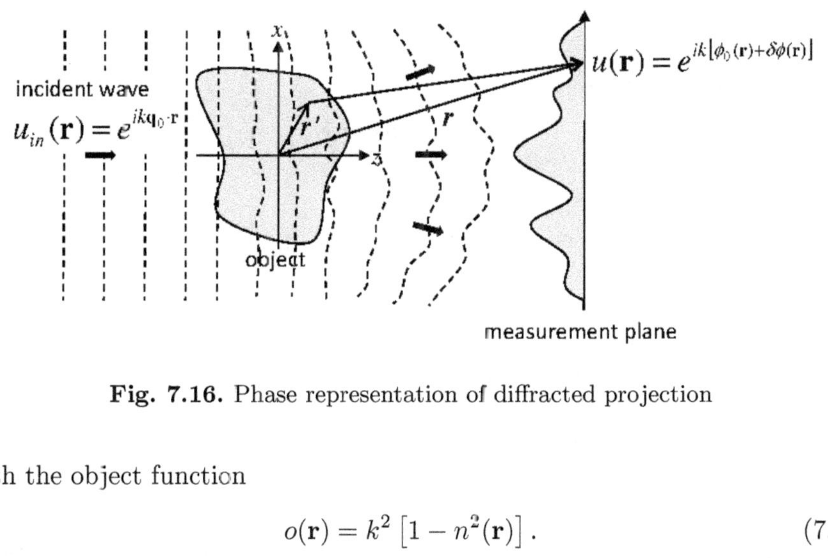

Fig. 7.16. Phase representation of diffracted projection

with the object function

$$o(\mathbf{r}) = k^2\left[1 - n^2(\mathbf{r})\right]. \tag{7.3}$$

Diffraction tomography is used to reconstruct the 3-D image of the refractive index $n(\mathbf{r})$ by use of a backpropagation algorithm taking the diffracted projection $u(\mathbf{r})$ as an input.

Here we introduce the Rytov approximation to solve the wave equation. Conceptually, the Rytov approximation is similar to the Born approximation. In fact, the two approximations become identical in the far field. It has been shown, however, that the Rytov model is far superior to the Born approximation for reconstructing the image of a large object. Assuming a small phase perturbation, the first-order Rytov approximation produces a linear relation between the object function and the phase perturbation.

Complex phase representation of the incident and diffracted waves has the forms,

$$u_{in}(\mathbf{r}) = e^{ik\mathbf{q}_0\cdot\mathbf{r}} = e^{i\phi_0(\mathbf{r})}, \tag{7.4}$$

$$u(\mathbf{r}) = e^{i\phi(\mathbf{r})}, \tag{7.5}$$

where $\mathbf{q}_0$ is a unit vector in the direction of the incident wave propagation. We assume that the object is a weak scatterer, i.e. $n(\mathbf{r}) \approx 1$, then the object function can be approximated as

$$o(\mathbf{r}) \approx -2k^2\delta n(\mathbf{r}), \tag{7.6}$$

where $\delta n(\mathbf{r}) = n(\mathbf{r}) - 1$. In the weak scattering limit, we can rewrite $\phi(\mathbf{r})$ as

$$\phi(\mathbf{r}) = \phi_0(\mathbf{r}) + \delta\phi(\mathbf{r}), \tag{7.7}$$

where $\delta\phi(\mathbf{r})$ is the perturbation introduced by the diffraction. Inserting Eq. 7.5, Eq. 7.6, and Eq. 7.7 into the wave equation, Eq. 7.2, we obtain the Rytov equation,

$$\nabla^2\delta\phi(\mathbf{r}) + 2ik\mathbf{q}_0 \cdot \nabla\delta\phi(\mathbf{r}) = k^2\delta n(\mathbf{r}). \tag{7.8}$$

Introducing $f(\mathbf{r})$, given as

$$\delta\phi(\mathbf{r}) = e^{-ik\mathbf{q}_0\cdot\mathbf{r}}f(\mathbf{r}), \tag{7.9}$$

we can modify the Rytov equation as

$$\left(\nabla^2 + k^2\right) f(\mathbf{r}) = 2ik^2\delta n(\mathbf{r})e^{ik\mathbf{q}_0\cdot\mathbf{r}}. \tag{7.10}$$

The solution of Eq. 7.10 is

$$f(\mathbf{r}) = 2ik^2 \int G(\mathbf{r} - \mathbf{r}')\delta n(\mathbf{r}')e^{ik\mathbf{q}_0\cdot\mathbf{r}'}d\mathbf{r}'^3, \tag{7.11}$$

where the Green's function $G(\mathbf{r})$ is the solution to the linear differential equation

$$\left(\nabla^2 + k^2\right) G(\mathbf{r} - \mathbf{r}') = \delta(\mathbf{r} - \mathbf{r}'). \tag{7.12}$$

Consequently, we get the linear relation between the phase perturbation $\delta\phi$ and the differential refractive index δn,

$$\delta\phi_{\mathbf{q}_0}(\mathbf{r}) = 2ik^2e^{-ik\mathbf{q}_0\cdot\mathbf{r}} \int G(\mathbf{r} - \mathbf{r}')\delta n(\mathbf{r}')e^{ik\mathbf{q}_0\cdot\mathbf{r}'}d\mathbf{r}'^3. \tag{7.13}$$

While the single projection $\delta\phi_{\mathbf{q}_0}(\mathbf{r})$ at a given $\mathbf{q}_0$ is insufficient to form a tomographic image, multiple projections at different angles provide enough information for reconstruction algorithms.

Tomographic imaging was demonstrated with a test object of complicated structure, using a diffraction tomography system based on the 2-D electro-optic imaging system (Fig. 7.7) [219]. The test object shown in Fig. 7.17(a) is composed of three polyethylene bars whose thickness, 1.5 mm, is comparable to the wavelengths of the incident THz pulses. Figure 7.17(b) shows the tomographic 3-D image of the test object, formed by stacking up 2-D horizontal slices reconstructed from diffraction tomography.

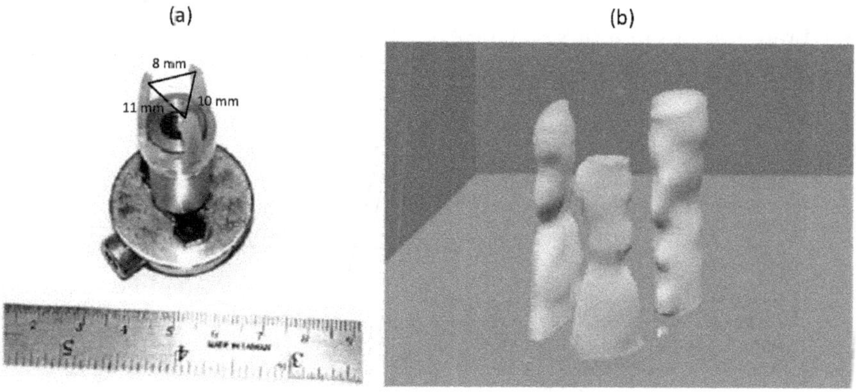

Fig. 7.17. (a) A test structure consists of three rectangular polyethylene bars arranged on a circle concentric with the axis of rotation. The thickness of the bars is 1.5 mm, and the widths are 2, 3.5, and 2.5 mm. (b) Reconstructed 3-D image of the test object. (Reprinted from [219]. ©2007 IEEE)

7.3 Imaging with Continuous-Wave THz Radiation

As discussed in Chapter 4, several technological disciplines have been involved in the development of the continuous-wave (CW) THz sources and detectors. Photomixers are optoelectronic devices; THz parametric oscillators (TPOs) are based on optical technologies; heterodyne detectors utilize solid-state diodes, etc. Consequently, many different types of CW THz imaging systems have been developed for numerous applications. The design of an imaging device is determined by the system requirements for a specific application. For example, when very high sensitivity is required, cryogenic detectors must be employed; in order to capture images of a moving object, detector arrays operating at a video frame rate are necessary; long-range imaging systems are built on millimeter wave devices, etc.

Millimeter-wave technologies, initially developed for space programs and military applications (see section 6.2.3 for spaceborne THz imaging systems), take up the lion's share of CW THz imaging technologies. We will continue to discuss millimeter-wave imaging for security in the next section. In the current section, we survey other competing CW THz imaging methods, in particular, operating above 0.5 THz and imaging for relatively small objects.

7.3.1 Raster-Scan Imaging

Raster scanning is a universal scheme for T-ray imaging. Any combination of CW THz sources and detectors listed in Fig. 7.18 can be used for this method. The THz detectors measure the power of either transmitted, reflected, or scattered radiation from a target. One exception is systems based on photomixing (see section 4.1).

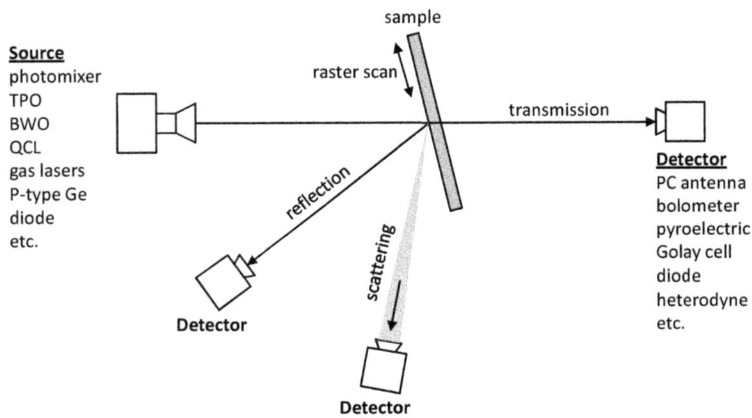

Fig. 7.18. Raster-scan imaging with CW THz radiation

Imaging System Based on Photomixing

A CW THz photomixing system uses two photomixers, both for emitters and coherent homodyne detectors. Its schematic arrangement is similar to that of a THz-TDS system using photoconductive antennas. Both photomixers are driven by the same dual-frequency laser source. Not only the amplitude but also the relative phase between the THz wave and the optical beat at the receiver is mapped out as the time delay between the pump and the probe beams is changed.

CW THz imaging with photomixers is demonstrated in Fig. 7.19 [220]. The imaging system utilizes two H-shaped photomixers with a 50-μm-long dipole and a 5×10 μm^2 photoconductive gap on LT-GaAs. The photomixers are excited by a dual-color CW Ti:sapphire laser. The sample is a biological specimen, a formalin-fixed, dehydrated, and wax-mounted slice through a canarys head. Its photograph is shown in Fig. 7.19(a). Figure 7.19(b) shows a logarithmic power transmission image taken with the CW THz imaging system at 1 THz. A pulsed THz image at 1 THz is shown in Fig. 7.19(c) for comparison. The spatial resolution and the contrast of the CW image are comparable with those of the pulsed image. Similar features are also shown in the CW phase image (Fig. 7.19(d)).

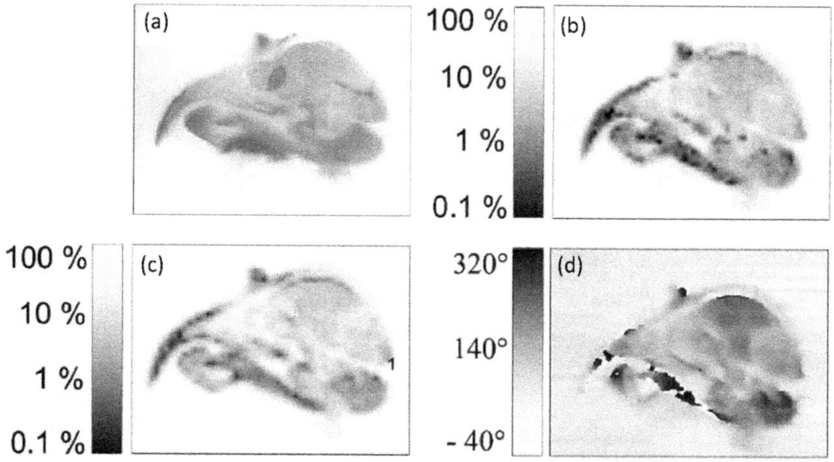

Fig. 7.19. THz images of a wax-mounted thin-cut canarys head at 1 THz. The sample dimension of 32 mm×24 mm×3 mm. (a) Photograph (b) CW THz power (c) Pulsed THz (d) CW THz phase (Reprinted with permission from [220]. ©2002, American Institute of Physics.)

A photomixing system can use diode lasers as its optical source. From a practical perspective, this is a huge advantage, because all of the main components are relatively small optoelectronic devices, and hence they are easily integrated into one compact system. A few commercial T-ray imaging systems are based on photomixing with diode lasers.

A carefully optimized photomixing system demonstrates a dynamic range of ~60 dB [221]. Figure 7.20 shows the photomixer used in the CW THz imaging system. The 220-μm-long antenna is resonant at 0.53 THz. Interdigitated electrodes with size of 8×8 μm^2 are placed at the center of the antenna. The emitter and receiver photomixers are driven by two diode lasers, operating at 830 nm. The laser linewidths are ~150 MHz, and the frequency difference is 0.53 THz.

Applications of the imaging system in security screening are demonstrated in Fig. 7.21. The optical photographs of a steel razor blade and a ceramic knife are shown in Fig. 7.21(a) and (d). The CW THz images of the objects, formed by peak-to-peak amplitude data, are taken from a distance of 20 cm in reflection geometry. The CW image of the razor blade (Fig. 7.21(c)) exhibits similar resolution and contrast with the pulsed THz image (Fig. 7.21(b)). The ceramic knife concealed underneath the denim cloth is clearly seen in the CW image of Fig. 7.21(e).

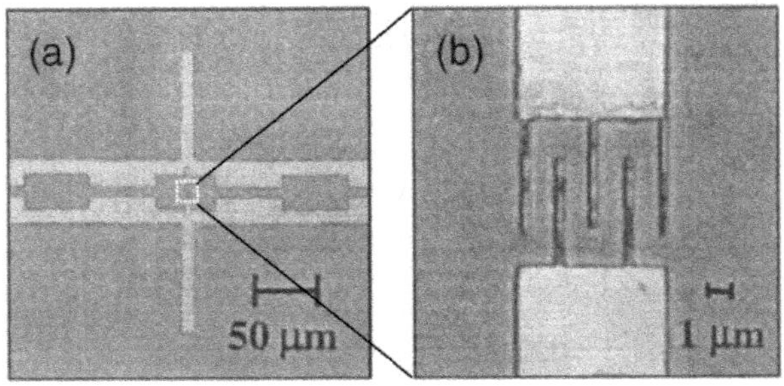

Fig. 7.20. Micrographs of the photomixer used in the CW THz imaging system: (a) 0.53-THz resonant antenna and (b) 8×8-μm^2 interdigitated electrodes. (Reprinted with permission from [221]. ©2005, American Institute of Physics.)

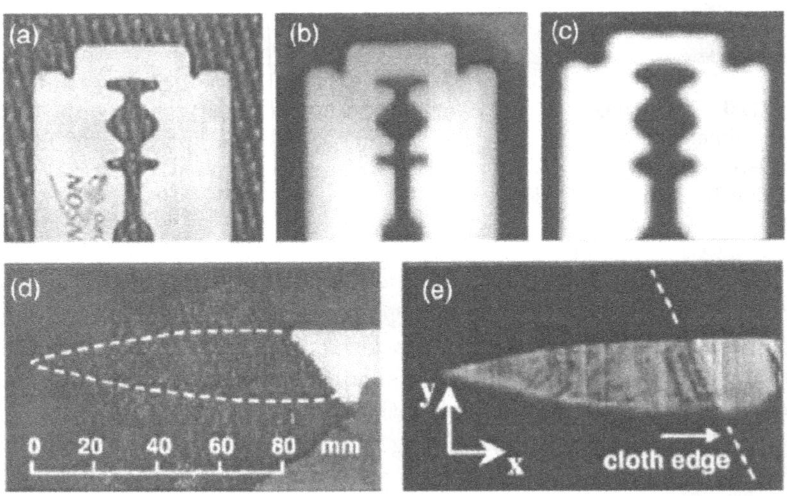

Fig. 7.21. Images of a razor blade, (a) photograph, (b) pulsed THz image, (c) CW THz image at 0.53 THz; Images of a ceramic knife, concealed underneath a denim cloth, (d) photograph, (e) CW THz image at 0.53 THz. (Reprinted with permission from [221]. ©2005, American Institute of Physics.)

Imaging with a Backward Wave Oscillator

A backward wave oscillator (BWO) is bulkier than optoelectronic and solid-state devices, but its high output power, 100 mW below 200 GHz and ~1 mW around 1 THz, has proven to be essential to acquire images of high quality. Furthermore, relatively insensitive detectors operating at room temperature such as pyroelectric detectors can be used for imaging applications.

A high-performance CW THz imaging system based on a BWO is demonstrated using a BWO of 15-mW output power at 0.6 THz and a pyroelectric detector [222]. The BWO output is modulated with an optical chopper, and the detector signal is measured with a lock-in amplifier.

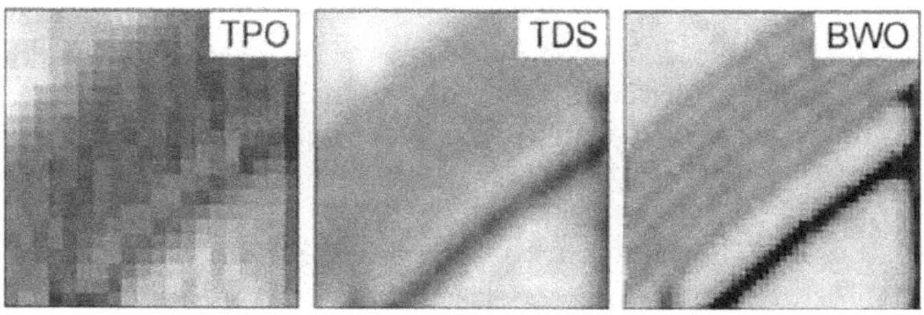

Fig. 7.22. THz images of metal wires embedded in a dielectric medium. The images are taken by three T-ray imaging systems based on TPO, TDS, and BWO. The TPO image was taken at 1.5 THz; the TDS and BWO images were taken at 0.59 THz. The size of the imaged area is 10×10 mm^2. (Reprinted from [222].)

Figure 7.22 compares the performance of the BWO system with those of other T-ray imaging systems based on THz-TDS and THz parametric oscillator (TPO). The TPO system operates at 1.5 THz and uses a Si bolometer as its detector. The sample is made of closely spaced metal wires embedded in a dielectric medium. Because of the wide tunability, the TPO system is useful for spectroscopic imaging, yet the spatial resolution is inferior to the BWO system. A direct comparison between the BWO and TDS images is not fair because the TDS image formed by the truncated spectral intensity at 0.59 THz does not exhibit the full capacity of the TDS system. Nevertheless, under this condition, the fine features of the sample are better resolved in the BWO image than in the TDS image.

The effects of detector sensitivity on image quality are demonstrated in Fig. 7.23. The sample is a 500-yen coin illuminated by a 1-mW THz beam. The images are taken slightly off the specular angle in order to avoid the bright specular reflection. Consequently, non-flat areas and edges emerge on top of the dark background formed by flat regions. The images of Fig. 7.23(b) and (c) are taken with a deuterated L-Alanine triglycine sulfate (DLATGS) sensor and a $LiTaO_3$ pyroelectric detector. The sensitivity of the DLATGS detector, a noise equivalent power of 6×10^{-10} W·Hz$^{-1/2}$, is about 3000 times higher than that of the $LiTaO_3$ detector, which accounts for the stark difference in image quality.

The high output power of a BWO can be exploited to measure weakly transmitted signals from a lossy sample. Figure 7.24 shows that the BWO

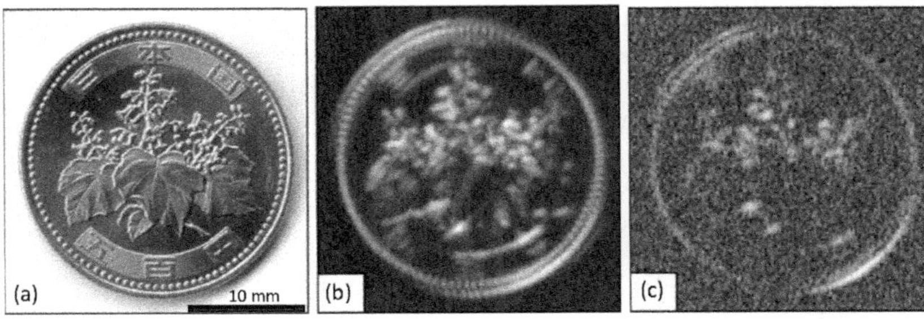

Fig. 7.23. THz images of a 500 yen coin captured in reflection geometry. A photograph of the sample is shown in (a) for comparison. The THz images are taken with (b) a DLATGS sensor and (c) a $LiTaO_3$ pyroelectric sensor.(Reprinted from [211].)

imaging system clearly identifies three needles embedded in 3-mm-thick milk powder. A highly sensitive detector based on superconducting tunnel junctions is employed to detect the weak signal. The noise equivalent power of the detector is of the order of 10^{-16} W·Hz$^{-1/2}$ at 0.3 K, lower than the noise equivalent power of a typical Si bolometer, $\sim 10^{-14}$ W·Hz$^{-1/2}$.

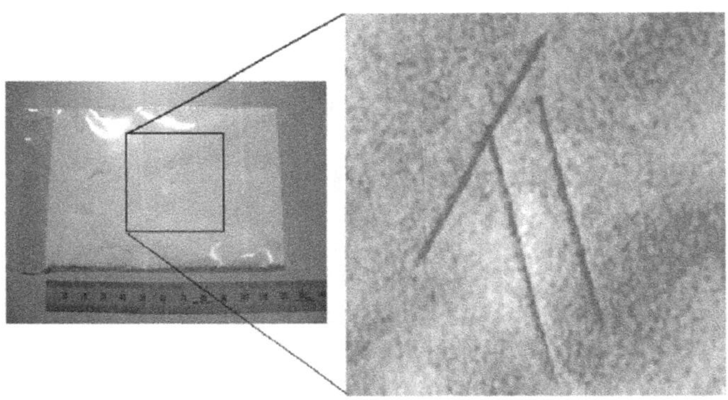

Fig. 7.24. THz Image of dehydrated milk powder and three needles contained in a nylon bag. The THz image is taken for the 5×5-cm^2 area indicated as the square box in the photograph. (Reprinted with permission from [223]. ©2006, American Institute of Physics.)

7.3.2 Real-Time Imaging with a Microbolometer Camera

In spite of the rapid advances in THz technology, real-time CW THz imaging, especially in the spectral range above 1 THz, is still technically challenging mainly due to the absence of high-power sources and sensitive detectors. A practical approach to this challenge is to take advantage of technologies developed in other adjacent fields. Microbolometer detector arrays, developed for thermal imaging (temperature sensitivity $\sim$ 100 mK), were originally designed for operating in the long wavelength IR band, 8–14 μm, and work at room temperature. Each pixel of the microbolometer camera is composed of a thin film of vanadium oxide (VO_x) and silicon nitride (Si_3N_4). Microbolometer cameras generating real-time images (30–60 frames per second) are commercially available. In the 8-14-μm band the noise equivalent power of the microbolometer sensors is in the range of 10^{-12} W·Hz$^{-1/2}$. The detector sensitivity at THz frequencies is unknown, yet it turns out that the camera is sensitive enough to take real-time THz images.

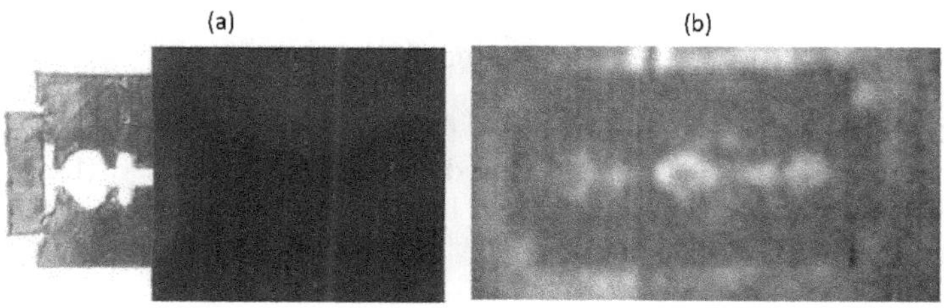

Fig. 7.25. (Images of a scaled razor blade partially covered by a black polyethylene sheet. (a) Photograph. (b) THz image taken by a 160×120 microbolometer camera. The object was illuminated by 2.52-THz, 10-mW radiation from a gas laser. (Reprinted from [224].)

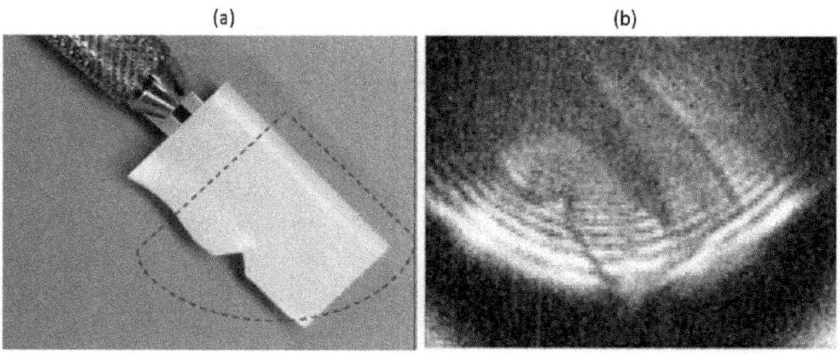

Fig. 7.26. Images of a knife blade wrapped in opaque plastic tape. (a) Photograph. (b) THz image taken by a 160×120 microbolometer camera. The sample was illuminated by a 2.8-THz QCL. (Reprinted from [225].)

Two examples of real-time CW THz imaging are shown in Fig. 7.25 and Fig. 7.26 [224, 225]. The THz images are taken by two different 160×120-pixel microbolometer cameras in transmission geometry. The cameras are designed for passive imaging in the 8-14-μm band, but due to the low sensitivity active illumination is necessary in the THz region. The 10-mW, 2.52-THz radiation generated by a gas laser was used to acquire the image of the scaled razor blade in Fig. 7.25. A 2.8-THz quantum cascade laser (QCL) was employed as an illuminator for the knife shown in Fig. 7.25. The QCL was operated at a 300-kHz repetition rate with an 8–15% duty cycle. The peak and average output power were 5–9 mW and 0.4–1.4 mW, respectively.

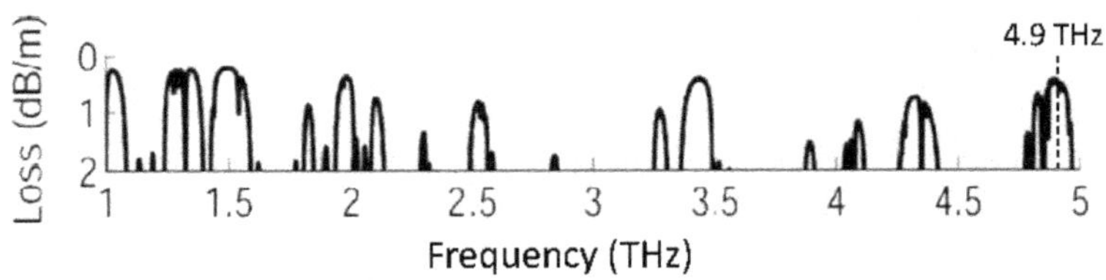

Fig. 7.27. Atmospheric transparency in the THz region. The loss in dB/m is calculated from HITRAN 2004 at 296 K and 40% relative humidity. (Reprinted with permission from [226]. ©2006, American Institute of Physics.)

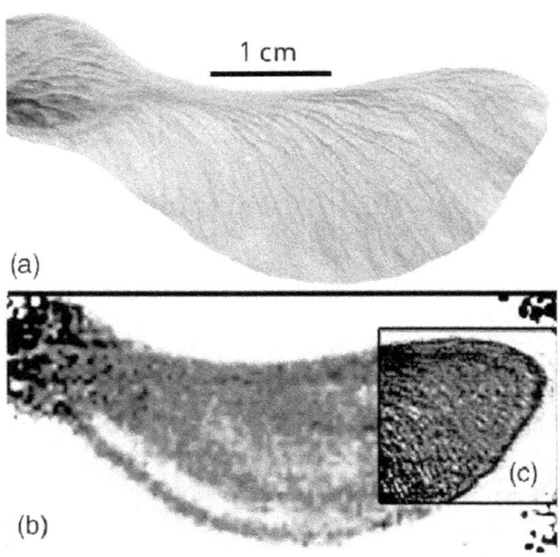

Fig. 7.28. Images of a dried seed pod. (a) Photograph. (b) Low- and (c) high-resolution THz images taken in transmission geometry at a standoff distance of 25 meters. The THz source and detector are a 4.9-THz QCL and a 320×240 microbolometer camera, respectively. (Reprinted with permission from [226]. ©2006, American Institute of Physics.)

CW imaging has an advantage over pulsed imaging: long-range imaging is possible at certain frequency bands where atmospheric attenuation is relatively low. Figure 7.27 shows the atmospheric transparency in the spectral region from 1 to 5 THz. The primary window is centered at 1.5 THz, and the bands near 3.4 and 4.9 THz are also usable.

Real-time THz imaging at a standoff distance of 25 meters has been demonstrated by use of a 4.9-THz QCL and a microbolometer camera [226]. Figure 7.28 shows the 20-frame average images of a dried seed pod taken in transmission geometry. The QCL produced a CW power of 38 mW at 9 K and a peak pulsed power of ~17 mW (13.5 ms pulse duration at 27% duty cycle) at 30 K. The microbolometer camera has 320×240 pixels spaced at a 46.25-μm pitch. The noise equivalent power of the detector array at 4.3 THz is $\sim 3 \times 10^{-10}$ W·Hz$^{-1/2}$. The image shown in Fig. 7.28 was taken with the object placed a few meters away from the camera. The spatial resolution is enhanced as the object is put closer to the camera (Fig. 7.28(c)).

7.4 Millimeter-Wave Imaging for Security

T-ray imaging and sensing has attracted a great deal of attention due to its applications in security. Materials of security interest show characteristic optical properties in the THz region. Metals are highly reflective and explosives and illicit drugs have spectral fingerprints in the THz range, whereas typical wrapping and packaging materials such as clothes, paper, and plastic are transparent to THz radiation. Furthermore, non-ionizing THz radiation appears to be unharmful to the human body.

Millimeter-wave technologies have been used in radio astronomy (see section 6.2.3) and military surveillance for many years. The accumulated technological capacities facilitate the quick adaptation of the technologies to security applications occurring in recent years. For example, the Transportation Security Administration (TSA) in the United States launched a pilot program in early 2008 to test millimeter-wave passenger imaging technology at several airports for detection of concealed weapons and explosives under layers of clothing without physical contact.

Millimeter-wave or sub-THz imaging in the range of 0.1-0.5 THz has some practical advantages over imaging at frequencies over 1 THz. Atmospheric opacity is much lower in the sub-THz region, and hence long-range imaging is available. Because attenuation lengths in common packaging materials also are relatively long, millimeter waves have an advantage over THz waves in the high-frequency side for inspection of bulky objects wrapped in a thick cover. Furthermore, the all-solid-state components for millimeter-wave applications are compact in size, operate at room temperature, generate powerful radiation, and are readily integrated into a single device.

Up to this point, all the T-ray imaging systems we have discussed in this chapter employ active imaging in which a radiation source illuminates an object and a sensor detects the transmitted, reflected, or scattered radiation from the object. On the other hand, passive imaging systems such as the spaceborne THz sensor arrays discussed in section 6.2.3 measure thermally generated radiation from objects. Millimeter imaging systems can be largely categorized by these two imaging schemes. The pros and cons of the two schemes are compared in Table 7.1. A notable characteristic of active imaging is that, if the illuminator produces highly coherent radiation, the scattered waves from the object interfere each other, and subsequently create coherent artifacts in its image.

Table 7.1. Active versus Passive Imaging

	Active Imaging	Passive Imaging
detector sensitivity	low	high
sensitivity to environment	relatively low	high
detection range	limited by source-object distance	relatively long
image interpretation	difficult due to coherent artifacts	easy
covert operation	no	yes
safety concern	unknown	no

7.4.1 Active Imaging

Many active imaging systems use a raster-scan method because of its simple architecture. A raster-scan imaging system demonstrated millimeter-wave imaging in transmission and reflection geometry utilizing a Gunn diode emitter and a Schottky diode detector [227]. The frequency-doubled Gunn diode oscillator produces an output power of 12 mW at 0.2 THz (1.5 mm in wavelength). The spatial resolution of this system is roughly 4 mm. Figure 7.29(c) shows a 0.2-THz reflective image of a space shuttle insulating foam sample embedded with intentional void defects; a photograph and a defect map of the sample are shown in (a) and (b), respectively. The foam in this sample is sprayed on an aluminium substrate reflecting most of the incoming THz radiation. The void defects are identified as dark circles in the THz image, which are formed by scattering losses and interferences at the void/foam interfaces.

The system capability of security screening is demonstrated in Fig. 7.30. The THz transmission images show a leather briefcase when it is empty and when it contains various objects, including a sword. The metal object is clearly identified as it blocks the THz radiation, while other non-metallic items are somewhat obscured by multiple reflections. The primary limitation of the single detection system is that it takes a few minutes to acquire a single THz image. It is desirable to reduce the acquisition time to 1-10 seconds for practical security applications.

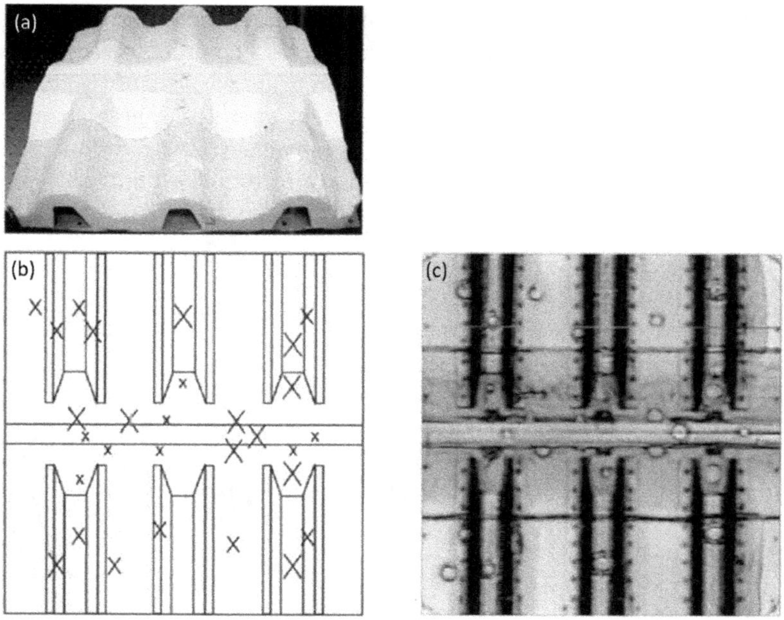

Fig. 7.29. Images of the space shuttle insulating foam sample. (a) Photograph (b) Defect map. Each void defect is marked with an "X". (c) 0.2-THz CW image in reflection geometry. (Reprinted with permission from [227]. ©2005, American Institute of Physics.)

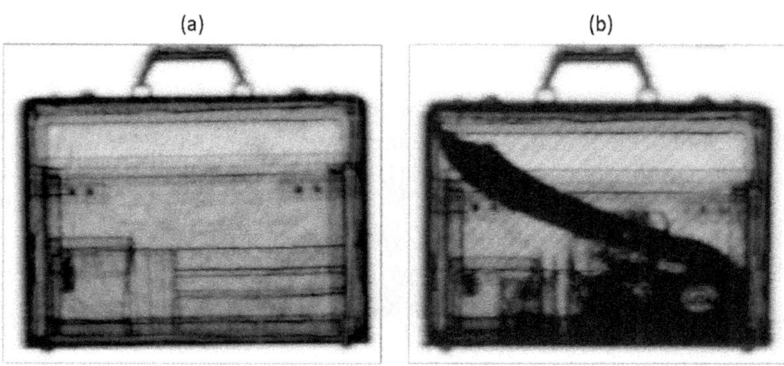

Fig. 7.30. 0.2-THz CW transmission images of (a) an empty leather briefcase and (b) the same briefcase containing a large knife and various harmless contents such as a compact disc, a video cassette, and audio cassette and pens. (Reprinted with permission from [227]. ©2005, American Institute of Physics.)

A straightforward strategy for lowering the image acquisition time is to increase the number of transmitters and detectors. The millimeter wave passenger imaging system used in the TSA pilot program consists of two transceiver arrays. Figure 7.31 illustrates how the system operates. The transmitters of the transceiver arrays emit millimeter waves as they rotate around the body and the receivers detect the reflected radiation from the body and other objects. It takes about one second for a single scan to acquire a whole body image.

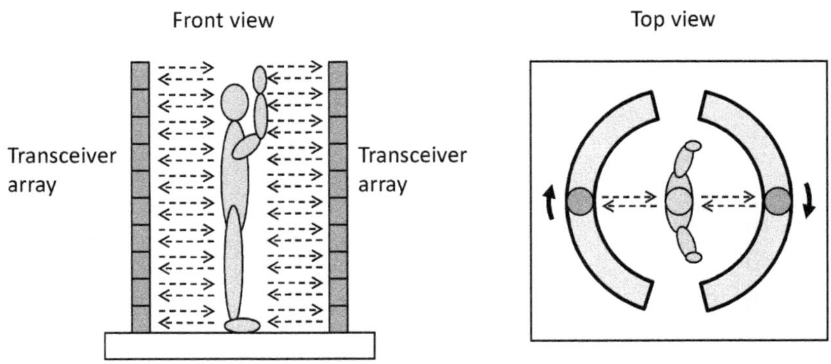

Fig. 7.31. Schematic diagram of the TSA millimeter-wave passenger imaging system

7.4.2 Passive Imaging

The dominant technology of passive millimeter-wave (PMMW) imaging is based on monolithic millimeter-wave integrated circuits (MMICs) incorporating high electron mobility transistors (HEMTs). Figure 7.32 sketches the basic structure of a MMIC receiver. A receiving antenna couples incoming millimeter waves into low noise amplifiers (LNAs) in which HEMTs are implemented. The LNAs preamplify the signal above the noise floor of the millimeter-wave detector such as a Schottky diode or a bolometer.

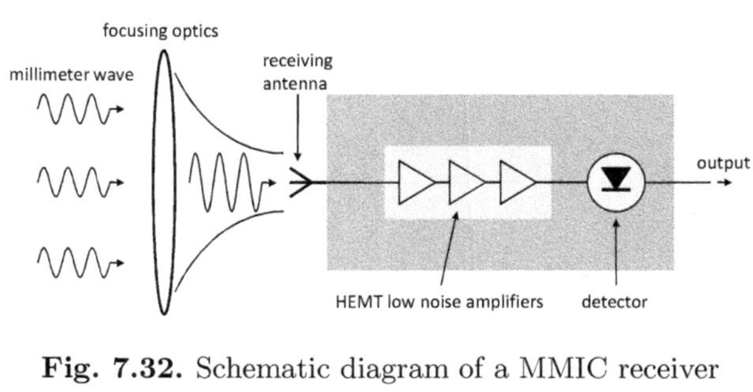

Fig. 7.32. Schematic diagram of a MMIC receiver

Early PMMW imaging systems employ GaAs-based HEMTs[1]. The performance of PMMW imaging systems is greatly improved by the recent adoption of InP-based HEMTs. Newly developed PMMW imaging systems take advantage of the superior characteristics of InP HEMTs in the high-frequency region: the cutoff frequency is very high (>1 THz) [2] and the noise figure is extremely low (<2 dB at 100 GHz).

Fig. 7.33. Images of Malvern Hills: (a) passive millimeter-wave image taken with an eight-channel 94-GHz instrument and (b) photograph. (Reprinted from [228].)

Figure 7.33(a) shows a millimeter-wave image taken by an eight-channel PMMW imaging system operating at 94 GHz [228]. Whereas the visible image shown in Fig. 7.33(b) is obscured by fog, the PMMW image exhibits much better visibility. In clear weather PMMW images and visible pictures show similar contrast, yet millimeter waves experience much less scattering in fog than visible lights because of their long wavelength. The attenuation in fog is less than 1 dB/km at 0.1 THz. Due to the superior visibility in poor-weather conditions such as fog, snow, clouds, smoke, sandstorms, etc. compared with visible imaging, PMMW imaging has been applied to aircraft navigation and landing.

---

[1] GaAs HEMTs are key components in microwave remote sensing and communication devices such as automotive radar systems, cellphones, and direct broadcast satellite (DBS) receivers.

[2] At present, an InP HEMT developed by Northrop Grumman Corporation holds the title of the world's fastest transistor, operating above 1 THz. The InP HEMT technology is relatively new, and its rapid progress has seen the operation frequency almost doubled since 2005.

The acquisition time of the PMMW image in Fig. 7.33(a) is substantially long (∼10 min), because the PMMW imaging system has only eight receiver channels and operates at a slow scanning speed. An ideal detector type for real-time passive imaging is a focal plane array (FPA) detector like a digital camera, yet the high cost of MMICs poses a considerable barrier to build a detector array comprising a large number of pixels. A few high-end imaging systems implement the FPA scheme. A compromising solution is to combine a detector array and a fast mechanical scanner. Figure 7.34 illustrates a schematic of a real-time mechanically scanned imaging system, which produces a conical scan pattern. The receiver array includes 32 detector modules. A PMMW image of people in an outdoor setting taken by the imaging system is shown in Fig. 7.35. Metal objects hidden under clothing are easily identified due to their high reflectivity.

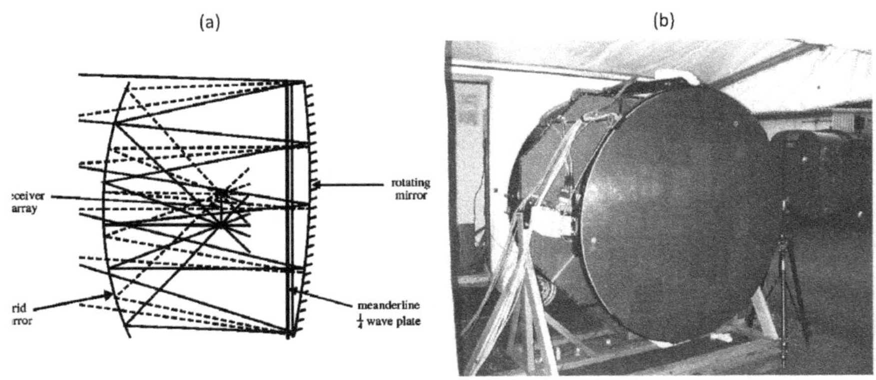

Fig. 7.34. Real-time mechanically scanned 35-GHz PMMW imaging system: (a) raytrace and (b) photograph. (Reprinted from [228].)

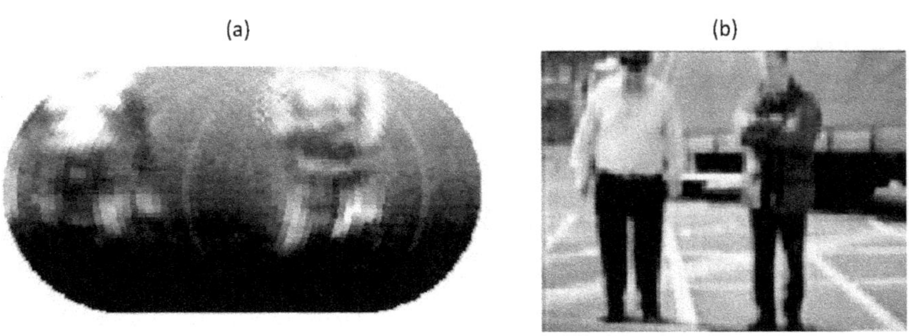

Fig. 7.35. Security scanning: (a) 35-GHz PMMW image and (b) photograph. (Reprinted from [228].)

Early PMMW imaging systems employ GaAs-based HEMTs[1]. The performance of PMMW imaging systems is greatly improved by the recent adoption of InP-based HEMTs. Newly developed PMMW imaging systems take advantage of the superior characteristics of InP HEMTs in the high-frequency region: the cutoff frequency is very high (>1 THz) [2] and the noise figure is extremely low (<2 dB at 100 GHz).

Fig. 7.33. Images of Malvern Hills: (a) passive millimeter-wave image taken with an eight-channel 94-GHz instrument and (b) photograph. (Reprinted from [228].)

Figure 7.33(a) shows a millimeter-wave image taken by an eight-channel PMMW imaging system operating at 94 GHz [228]. Whereas the visible image shown in Fig. 7.33(b) is obscured by fog, the PMMW image exhibits much better visibility. In clear weather PMMW images and visible pictures show similar contrast, yet millimeter waves experience much less scattering in fog than visible lights because of their long wavelength. The attenuation in fog is less than 1 dB/km at 0.1 THz. Due to the superior visibility in poor-weather conditions such as fog, snow, clouds, smoke, sandstorms, etc. compared with visible imaging, PMMW imaging has been applied to aircraft navigation and landing.

[1] GaAs HEMTs are key components in microwave remote sensing and communication devices such as automotive radar systems, cellphones, and direct broadcast satellite (DBS) receivers.

[2] At present, an InP HEMT developed by Northrop Grumman Corporation holds the title of the world's fastest transistor, operating above 1 THz. The InP HEMT technology is relatively new, and its rapid progress has seen the operation frequency almost doubled since 2005.

The acquisition time of the PMMW image in Fig. 7.33(a) is substantially long (~10 min), because the PMMW imaging system has only eight receiver channels and operates at a slow scanning speed. An ideal detector type for real-time passive imaging is a focal plane array (FPA) detector like a digital camera, yet the high cost of MMICs poses a considerable barrier to build a detector array comprising a large number of pixels. A few high-end imaging systems implement the FPA scheme. A compromising solution is to combine a detector array and a fast mechanical scanner. Figure 7.34 illustrates a schematic of a real-time mechanically scanned imaging system, which produces a conical scan pattern. The receiver array includes 32 detector modules. A PMMW image of people in an outdoor setting taken by the imaging system is shown in Fig. 7.35. Metal objects hidden under clothing are easily identified due to their high reflectivity.

(a) (b)

Fig. 7.34. Real-time mechanically scanned 35-GHz PMMW imaging system: (a) raytrace and (b) photograph. (Reprinted from [228].)

(a) (b)

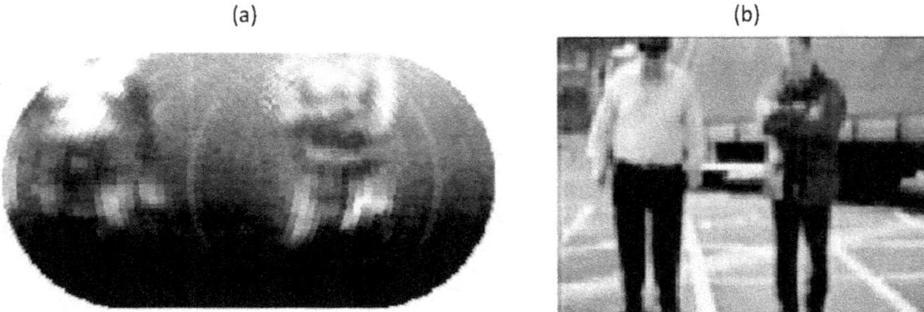

Fig. 7.35. Security scanning: (a) 35-GHz PMMW image and (b) photograph. (Reprinted from [228].)

Like the image in Fig. 7.35, most of the PMMW images are taken outdoors taking advantage of the large temperature difference between ground and sky (~200 K). The ground and sky are a dual-temperature radiation source at 300 K and 100 K for illumination, and the cold sky provides a stark contrast on outdoor scenery in the millimeter-wave band. Indoor PMMW imaging for security, however, must endure weak radiation power (~10% of outdoor) and small temperature contrasts between the body and other objects.

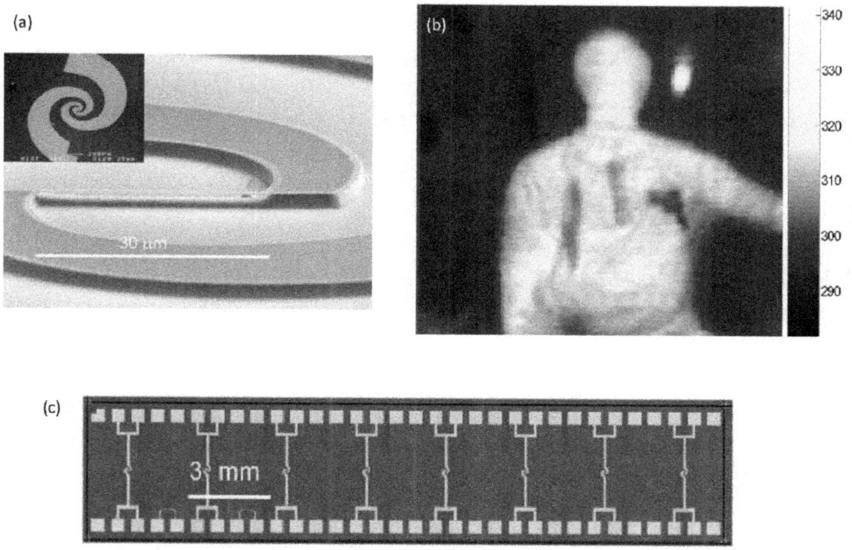

Fig. 7.36. (a) A SEM image of a superconducting NbN bolometer. The inset shows the lithographic log-spiral antenna. (b) A passive 0.1-1 THz image of a test subject carrying a concealed ceramic knife and a metal handgun. (c) An 8-pixel bolometer array module for video-rate passive imaging. (Reprinted from [229]. ©2006 EuMA)

In order to distinguish the small temperature differences, a highly sensitive detection system is necessary for passive imaging. The superconducting hot-electron bolometers discussed in section 4.10 are not only sensitive to THz radiation, but also compact in size and all-solid-state devices, hence it is possible to integrate a large number of bolometers into a detector array. Figure 7.36(a) shows a SEM image of a superconducting antenna-coupled bolometer. A suspended NbN bridge is coupled to a log-spiral antenna shown in the inset. The antenna has a broad bandwidth ranging from 0.1 to 1 THz. The noise equivalent temperature difference (NETD) of this detector is 125 mK in the region of 0.1-1 THz, meaning that it is capable of differentiating a temperature difference larger than ~0.1 K. Figure 7.36(b) shows a passive image of a subject carrying a ceramic knife and a metal handgun concealed underneath clothing. The image, scaled with temperature in Kelvin, demonstrates that the temperature difference of ~1 K between the body and the weapons are clearly discernible. The image was taken by raster scanning with a single bolometer, and the whole scanning took about 2 min. The first step toward video rate imaging is the fabrication of the 8-pixel module shown in Fig. 7.36(c).

7.5 Medical Applications of T-Ray Imaging

Radiation safety has always been an important issue concerning medical imaging procedures. Due to its low photon energy (1-10 meV), THz radiation is non-ionizing, thus it is generally considered to be unharmful to the human body. While no systematic clinical study has been conducted yet, the maximum permissible average beam power for skin exposure to radiation of wavelengths from 2.6 μm to 20 mm (0.15-115 THz) is conservatively estimated as 3 mW, which is still well above the output power of typical THz imaging systems [230].

Because of the strong attenuation by water, THz waves penetrate at best a few hundred microns into biological tissues. Nevertheless, it is deep enough to probe epithelial tissues. Furthermore, because of the long wavelength, images suffer little from blurring effects induced by scattering unlike other optical imaging techniques. To an extent, the strong interaction of THz radiation with water is useful for medical diagnosis, because THz sensing and imaging is sensitive to small changes in hydration levels that could be a critical measure in determining normalcy for biological systems.

The spatial resolution of THz imaging methods is usually in the submillimeter range, depending on the spectral region where the imaging is performed. THz-TDS imaging systems are also capable of resolving depth profiles with a resolution of ~10 μm. The spatial resolution is good enough for many medical imaging applications. T-ray medical imaging is still in its infancy, but some preliminary studies show its potential to provide a safe and efficient technique for medical diagnosis.

7.5.1 Optical Properties of Human Tissue

Interpretation of medical images sometimes poses daunting challenges unless the relevant optical properties of different tissue types are available. A recent study has established a catalogue of the optical properties of freshly excised healthy human tissue samples in the spectral range of 0.5-1.5 THz [231, 232]. Figure 7.37 shows the absorption coefficient of skin, adipose tissue, striated muscle, vein, and nerve as a function of frequency. The data are compared with those of deionized water. The broadband refractive index and absorption coefficient of ten biological samples are shown in Table 7.2. This is the first systematic effort to catalog the optical properties of human tissue at THz frequencies and little other data are available. Further studies are highly desirable.

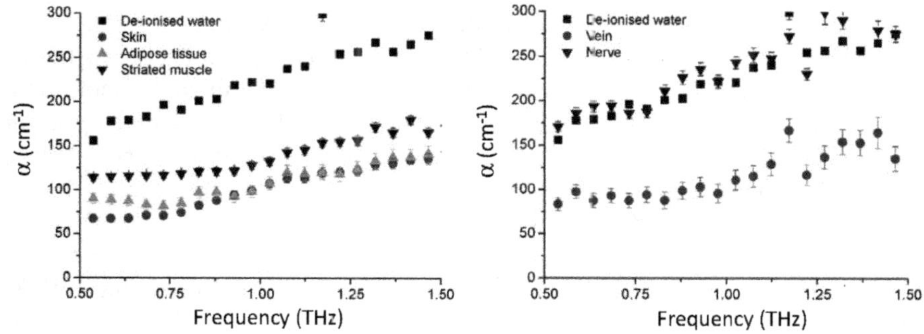

Fig. 7.37. Absorption coefficient α of different human tissue types in the spectral range of 0.5-1.5 THz: skin, adipose tissue, striated muscle, vein, and nerve. Data of deionized water are also shown for comparison. (Reprinted from [231].)

Table 7.2. Broadband optical coefficients of human tissue [232]

Tissue Type	Refractive Index n	Absorption Coefficient α (cm^{-1})
Deionized water	2.04±0.07	225±21
Tooth enamel	3.06±0.09	62±7
Tooth dentine	2.57±0.05	70±7
Skin	1.73±0.29	121±18
Adipose tissue	1.50±0.47	89±23
Striated muscle	2.00±0.35	164±17
Cortical bone	2.49±0.07	61±3
Vein	1.58±0.49	110±43
Artery	1.86±0.40	151±25
Nerve	1.95±0.46	246±27

7.5.2 Cancer Diagnostics

The main targets of THz medical imaging are epithelial diseases that several advanced imaging modalities are also competing to diagnose. In particular, confocal microscopy, optical coherence tomography, and ultrasound imaging are specialized techniques in the diagnosis of such diseases. No imaging technique, however, can readily differentiate benign from malignant lesions near the surface of skin.

Basal cell carcinoma (BCC) is the most common form of skin cancer that arises in the deepest layer of the epidermis. BCCs are diagnosed by a biopsy after a visual exam. The most effective treatment is Mohs surgery where the rate of cancer returning is about 1%. In this surgical treatment, samples of skin are cut out and immediately examined under a microscope. This process is repeated until no cancer is found in the skin sample. THz pulsed imaging is a promising technique to assess the size and depth of the invading tumor prior to surgery.

Figure 7.38 shows an *ex vivo* THz image of a BCC lesion, compared with a visible image and histology. In Fig. 7.38(b), THz waveforms reflected from diseased tissue and normal tissue are clearly distinguished from each other. Generally, the maximum amplitude E_{max} for the diseased tissue is much greater than that for normal tissue, especially if there is encrustation on the surface of the lesion. The THz pulse broadens more in the diseased tissue than in the normal tissue, which provides the contrast of the THz image in Fig. 7.38(c). The THz image shows that the diseased tissue is clearly differentiated from the surrounding healthy tissue. Figure 7.38(d) shows the histology that a BCC extends over the region indicated by the horizontal line, consistent with the contrasted area in the THz image.

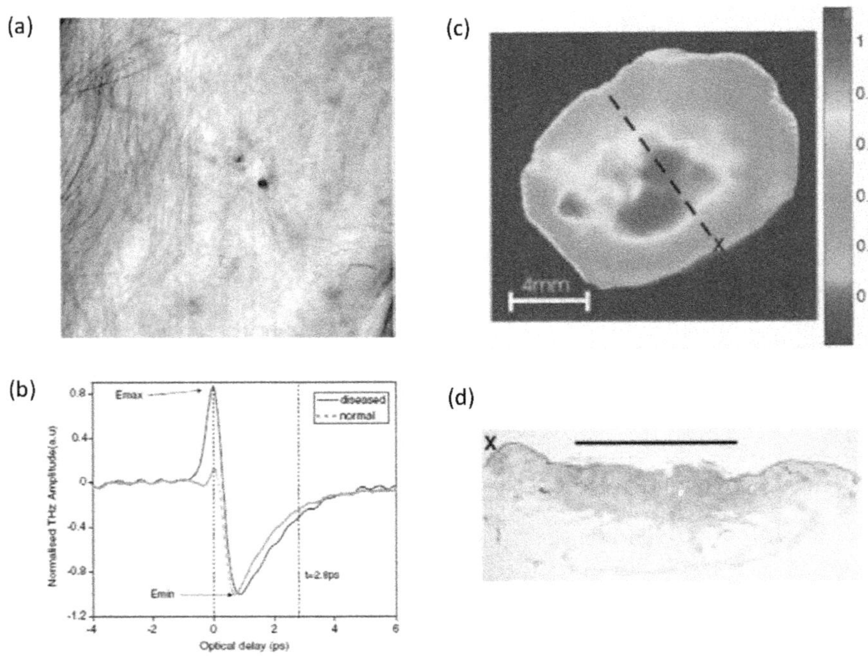

Fig. 7.38. THz pulsed imaging of basal cell carcinoma *ex vivo*. (a) Clinical photograph (b) Averaged THz waveforms from diseased tissue (solid line) and normal tissue (grey dotted line). The best contrast of the THz image is obtained by using the difference in the two waveforms at t=2.8 ps. (c) THz image acquired by imaging the excised tissue. The "X" indicates the suture location and the dotted line shows the axis of the vertical histology section. (d) Histology section. The suture position is marked with an "X" and extent of tumor indicated with the horizontal lines above the tissue. (Reprinted from [233].)

In vivo THz images of a BCC are are shown in Fig. 7.39, compared with a clinical photograph and histology. In Fig 7.39(b), the THz image, formed by the minimum amplitude E_{min}, shows surface features characterized by an inflammatory crust. The THz image in Fig.Fig. 7.39(c), formed by using the THz field amplitude E_t at t=2.8 ps normalized with E_{min}, provides depth information of the tumor with the maximum extent around 0.25 mm below the surface. The representative histology shown in Fig. 7.39(d) and (e) confirms the THz measurements.

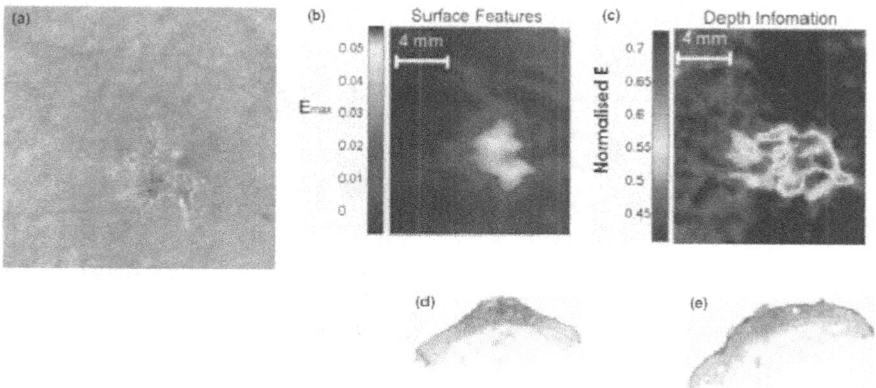

Fig. 7.39. THz pulsed imaging of basal cell carcinoma *in vivo*. (a) Clinical photograph of the lesion. (b) THz image formed by E_{min}, showing surface features. (c) Normalized THz image at t=2.8 ps, indicating the extent of the tumor at depth (~250 μm). (d) Representative histology section showing acute inflammatory crust corresponding to THz image (b). (e) Representative histology section showing lateral extent of tumor corresponding to normalized THz image (c). (Reprinted from [233].)

Breast cancer is by far the most common cancer amongst women. The first line of treatment against breast cancer is still surgery. Breast-conserving surgery, a.k.a. lumpectomy, is the most common form of breast cancer surgery. In this surgical procedure, only the part of the breast containing the tumor and some of the surrounding normal tissue is removed from the breast. The tissue in the margins is examined to check presence of cancer cells. It is, therefore, of great importance to accurately determine the margins, yet no technology exists to perform this procedure.

The feasibility of using THz pulsed imaging to map tumor margins has been tested in freshly excised human breast tissue [234]. Figure 7.40 compares the size and shape of tumor regions on a THz image with those on a photomicrograph. The THz image was formed by the minimum THz field E_{min}. There is good correlation between the tumor regions identified by the two methods. In this study, 22 specimens were examined, and all 22 samples show good correlation for the size and shape of the tumor area on a THz image and a photomicrograph.

In general, the absorption coefficient and refractive index of cancerous tissue at THz frequencies are higher than those of normal tissue. This may be accounted for by higher water content in cancerous tissue. Water concentration, however, should not be the sole factor, because cancerous tissue is still distinguishable from normal tissue on THz images of dehydrated samples. Structural changes such as increased cell and protein density may also give rise to the changes in optical properties.

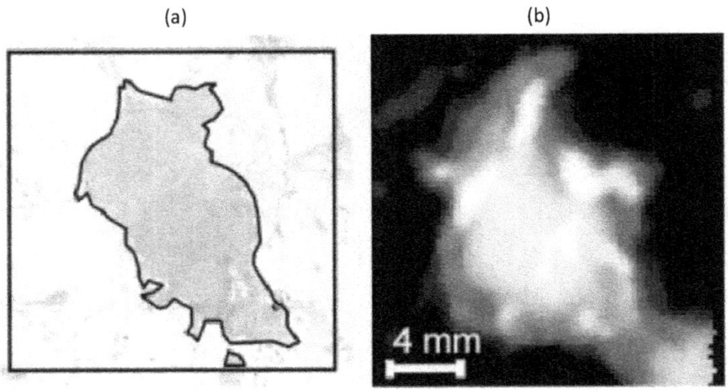

Fig. 7.40. Images of a human breast tissue specimen with carcinoma: (a) photomicrograph (Tumor regions are marked with black outline.) and (b) THz image formed by E_{min}. (Reprinted from [234].)

7.5.3 Reflective Imaging of Skin Burns

Burns are classified as first, second, or third degree, depending on the depth and severity: first-degree burns are limited to the epidermis, second-degree burns involve the dermis, and third-degree burns involve all the layers of the skin and underlying tissue. In order for efficient treatment of burn injuries, an accurate diagnosis of burn depth is desirable. Clinical assessment of burn depth, however, is often inaccurate and ambiguous. Currently, there is no non-invasive technology to make reliable assessments.

Burn injuries result in fluid accumulation and inflammation in and around the wound, and hence T-ray imaging has a potential to make an accurate diagnosis of burn depth and extent. A few preliminary studies have been conducted on animal skins. Figure 7.41(a) shows a THz pulsed image of a burn on a chicken breast, acquired in reflection geometry. An argon ion laser was used to produce the wound on the sample skin. The THz waveforms reflected from damaged and undamaged tissue show different features. The THz image indicates that the damaged region is less reflective than the undamaged area. Figure 7.41(b) and (c) show the differences in amplitude and phase of the THz

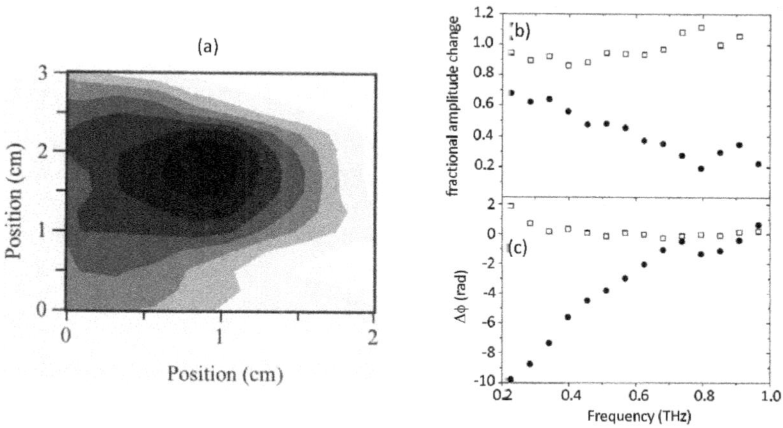

Fig. 7.41. (a) THz image of a circular burned region on chicken breast, formed by the reflected THz amplitude in the frequency range 0.5-1.0 THz. Darker shades indicate lower reflectivity. (b) Fractional amplitude and (c) phase difference between the waveforms reflected from damaged and undamaged tissue. The solid circles represent $E_{\text{damaged}}/E_{\text{undamaged}}$. For comparison, the open circles represent the measurement on two undamaged areas. (Reprinted from [213].)

waveforms as a function of frequency. The fractional amplitude of damaged tissue is less than unity in the broad spectral range, and decreases as frequency is increased. The phase difference also shows a strong frequency dependence. The spectral characteristics may be useful for a precise classification of burn injuries. Change in hydration level may be a primary source of the different optical properties between damaged and undamaged tissue, but other factors such as chemical and morphological modifications should also be considered. A similar result is obtained for burned pig skin (Fig. 7.42).

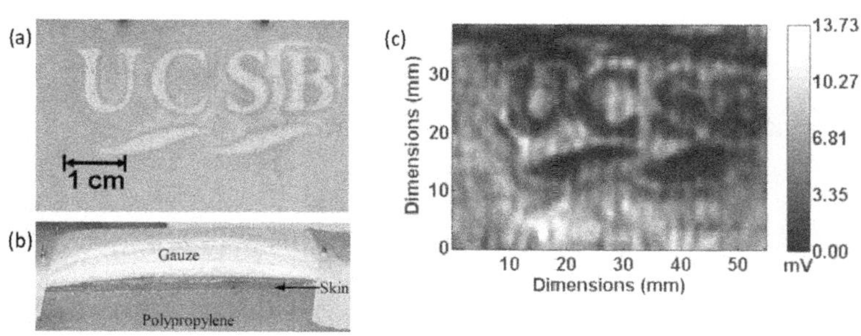

Fig. 7.42. (a) Burned pig skin sample. (b) Side view of mounted pig skin with ten layers of gauze. (c) THz image of burned pig skin beneath five layers of gauze, formed by THz signals centered at 0.5 THz with ~0.125 THz of bandwidth. Darker shades indicate lower reflectivity. (Reprinted from [235].)

7.5.4 Detection of Dental Caries

Traditional caries detection methods rely largely on visual examination based on subjective data such as color, translucency, and hardness. The current caries detection rates indicate that a large number of lesions are being missed, in particular, those in early stages. Dental caries is a dynamic disease process from the earliest stages through the cavitation stage. Early lesions go through many cycles of demineralization and remineralization. If early caries are accurately detected before cavitation, therapeutic intervention can stop and even reverse the disease process by remineralization of the non-cavitated lesion. Preliminary studies have been conducted to test the feasibility of THz imaging for caries detection [236].

8

Terahertz Spectroscopy of Condensed Matter

The interaction of THz radiation with condensed matter is a highly diverse subject. The wide spectrum of phenomena ranges from lattice vibrations and intraband transitions in semiconductors to the dynamics of complex fluids, electron spins, and strongly correlated electrons. The unique and advanced techniques of THz spectroscopy are a powerful tool to explore the material properties which have been inaccessible until recently. Many of these relatively unexplored subjects have the potential for next-generation technologies such as ultrahigh speed and ultralarge capacity information processing. In this chapter we shall focus on the two major topics of contemporary condensed matter physics, dealing with carrier dynamics in semiconductors and low-energy excitations in strongly correlated electron systems.

8.1 Intraband Transitions in Semiconductors

The bandgap energy of usual semiconductors, which is in the range of 1-10 eV, is far greater than the THz photon energy; hence THz excitations in semiconductors involve transitions within an energy band. The intraband carrier dynamics and associated nonlinear optical effects are of great interest because they not only directly manifest fundamental physical processes such as many-body interactions and Coulomb correlations, but also have broad applications in ultrahigh speed optoelectronic devices beyond the signal switching rates of 100 Gigabits/sec.

The electronic structure of intrinsic semiconductors is modified by doping impurities, by applying external electric and magnetic fields, and by lowering dimensionality. These manipulations lead to the localization of carriers in nanometer scale, and the consequent energy levels of the quantized bound states coincide with photon energies in the THz region. The elementary excitations of semiconductors involve impurity states, Landau levels, excitons, and subbands of semiconductor nanostructures.

8.1.1 Band Structure of Intrinsic Semiconductors

Semiconductors are characterized by their bandgap, referring to the energy difference (bandgap energy E_g) between the top of the highest valence band and the bottom of the lowest conduction band. The electronic band structure of a homogeneous, three-dimensional semiconductor can be described by parabolic bands in the effective mass approximation. In the parabolic approximation, the conduction and the valence band energies (E_C and E_V) are expressed as

$$E_C(k) = E_g + \frac{\hbar^2 k^2}{2m_e^*}, \tag{8.1}$$

$$E_V(k) = -\frac{\hbar^2 k^2}{2m_h^*}, \tag{8.2}$$

where $\hbar k$ is the momentum, and m_e^* and m_h^* are the electron and the hole effective masses, respectively. Here we consider a direct bandgap semiconductor.

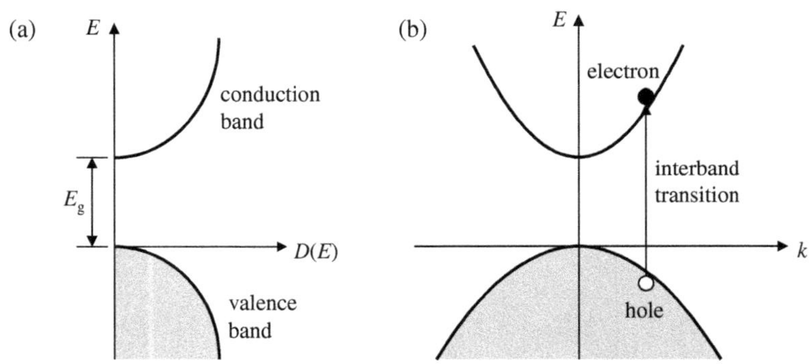

Fig. 8.1. Schematic diagram of the conduction and valence bands of a direct bandgap semiconductor in the effective mass approximation. (a) Density of states, $D(E) \propto \sqrt{E}$. (b) The band energies are parabolic functions of the momentum.

The wave function of an energy eigenstate is a plane wave:

$$\Psi_{\mathbf{k}}(\mathbf{r}) = \frac{1}{\sqrt{V}} e^{i\mathbf{k}\cdot\mathbf{r}} = \frac{1}{\sqrt{V}} e^{i(k_x x + k_y y + k_z z)}, \tag{8.3}$$

where V is the volume of the material. The density of states is proportional to the square root of the energy:

$$D_C(E) = \frac{V}{2\pi^2} \left(\frac{2m_e^*}{\hbar^2}\right)^{3/2} \sqrt{E - E_g}, \tag{8.4}$$

$$D_V(E) = \frac{V}{2\pi^2} \left(\frac{2m_h^*}{\hbar^2}\right)^{3/2} \sqrt{-E}. \tag{8.5}$$

Figure 8.1 illustrates the band structure of a direct bandgap semiconductor. An interband transition excites an electron from the valence band to the conduction band, and leaves a hole in the valence band.

8.1.2 Photocarrier Dynamics

THz spectroscopy of carrier dynamics in semiconductors is of great interest because it has the unique capability to resolve some fundamental physical processes, such as many-body Coulomb interactions, and carrier-phonon scattering. It sheds light on how optoelectronic devices would behave under ultrahigh speed modulation at THz frequencies.

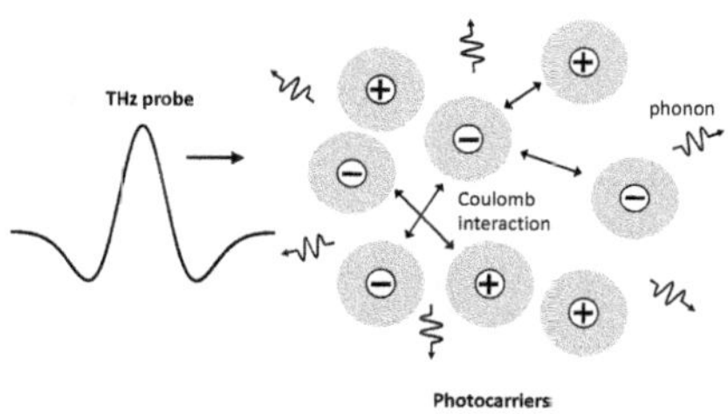

Fig. 8.2. Interaction of broadband THz pulses with optically generated electrons and holes in a semiconductor.

An ultrashort optical pulse with a photon energy greater than the bandgap generates electrons and holes in a semiconductor via interband excitations. The optical transitions are almost instantaneous, and the interactions among the optically excited carriers evolve in time depending on the environment, characterized by physical parameters such as carrier density and lattice temperature. At the moment of carrier excitation the bare Coulomb potential describes the interaction between the charged carriers. Subsequently, the Coulomb interaction forces the charged particles to be rearranged by a process called dressing or screening. The dressed particle is called a *quasiparticle.*

In momentum space, the bare Coulomb potential and the dynamically screened Coulomb potential are expressed as

$$V_q = \frac{e^2}{\epsilon_0 q^2}, \tag{8.6}$$

$$V_q^s(\omega, t_D) = \frac{V_q}{\epsilon_q(\omega, t_D)}, \tag{8.7}$$

where $\epsilon_q(\omega, t_D)$ is the dynamic dielectric function and t_D is the delay time from the moment of carrier generation. $\epsilon_q(\omega, t_D)$ is a macroscopic physical quantity which can be determined by THz conductivity measurements. The temporal evolution of the screening process is observed by an optical-pump and THz-probe experiment with ultrashort optical and THz pulses. Figure 8.3 shows the experimentally measured spectra of the imaginary and real parts of the inverse dielectric function of GaAs in the long-wavelength limit for several delay times [237]. The time-resolved dielectric function indicates that a sharp resonance builds up at the plasma frequency $\omega_p/2\pi$=14.4 THz within 150 fs. This demonstrates the ultrafast dynamics of the nonequilibrium many-body system of electron-hole plasma. When the many-body system reaches equilibrium, the experimental data agree well with the inverse of the Drude dielectric function

$$\frac{1}{\epsilon_q(\omega)} = \frac{\omega^2}{\omega^2 - \omega_p^2 - i\gamma\omega} \tag{8.8}$$

in the high frequency limit, $\omega \gg \gamma$, where γ is the scattering rate.

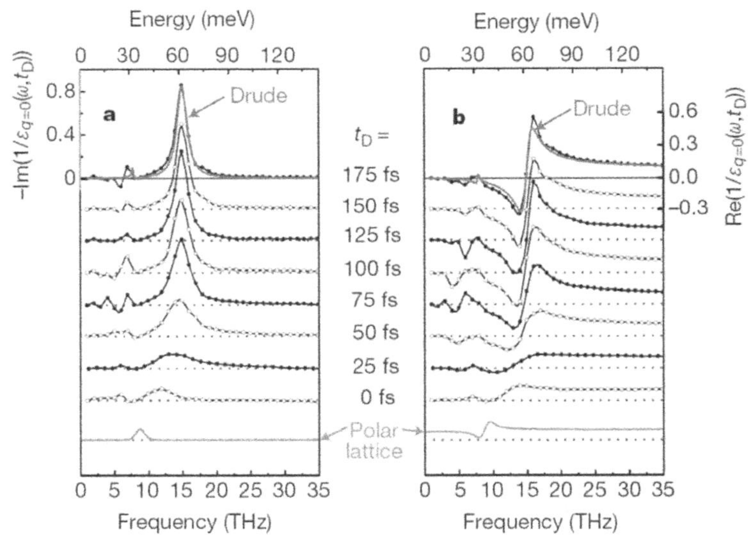

Fig. 8.3. Temporal evolution of Coulomb screening and plasmon scattering. Spectra of (a) the imaginary and (b) the real parts of the long-wavelength limit of the inverse dielectric function of GaAs for several relative delay times between optical and THz pulses. The Drude curves fit the data at 175 fs. (Reprinted by permission from Macmillan Publishers Ltd: Nature [237], ©2001.)

The Drude-type frequency dependence of the electron-hole plasma in equilibrium is an excellent measure to probe the scattering processes of charged carriers in a photoexcited semiconductor. The real and the imaginary part of the Drude conductivity, $\sigma = \sigma_1 + i\sigma_2$, are given as

$$\sigma_1(\omega) = \frac{\sigma_0}{1+\omega^2\tau^2}, \tag{8.9}$$

$$\sigma_2(\omega) = \frac{\omega\tau\sigma_0}{1+\omega^2\tau^2}, \tag{8.10}$$

where σ_0 is the DC conductivity, and $\tau = \gamma^{-1}$ is the relaxation time. Figure 8.4(a) shows the spectra of the real and the imaginary part of the conductivity of a photoexcited sapphire crystal in equilibrium, obtained by an optical-pump and THz-probe experiment [238]. The solid lines represent a fit to the Drude model. The fit results in the scattering rate at room temperature, $\gamma/2\pi \approx 1.6$ THz. $\gamma/2\pi$ is larger than the THz frequencies of the measured spectra (0-1.6 THz), and accordingly σ_1 decreases gradually as the frequency is increased, while σ_2 increases. The temperature dependence of the scattering rate is shown in Figure 8.4(b). The solid line is fit to a transport model including LO-phonon and acoustic phonon scattering contributions, which indicates that carrier scattering is dominated by interactions with LO-phonons at high temperature and with acoustic phonons at low temperature.

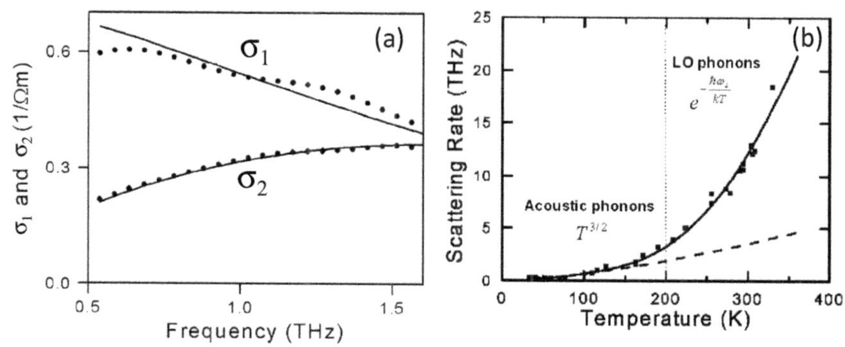

Fig. 8.4. (a) Spectra of the real and the imaginary part of the conductivity of a photoexcited sapphire crystal at room temperature. The solid lines are fit to the Drude model. (b) Scattering rate versus temperature. The solid line is fit to a transport model including LO-phonon and acoustic phonon scattering. The dashed line represents the acoustic phonon contribution. (Reprinted with permission from [238]. ©2003, American Physical Society.)

8.1.3 Impurity States

An impurity atom that replaces an original atom of a semiconductor can supply or take away electrons depending on the electronegativity of the atoms. The point defects supplying extra electrons and holes are called *donors* and *acceptors*, respectively. Consider a GaAs crystal doped with Si atoms, shown in Fig. 8.5. Each Ga atom provides three valence electrons and As provides five to form the valence band of GaAs. A Si atom, having four valence electrons, supplies an additional electron at a Ga site and takes away one from its surroundings at a As site. The Si atom and the extra electron or hole forms a hydrogen-like system via the Coulomb interaction. The theoretical description of the impurity states is identical to that for a hydrogen atom (see

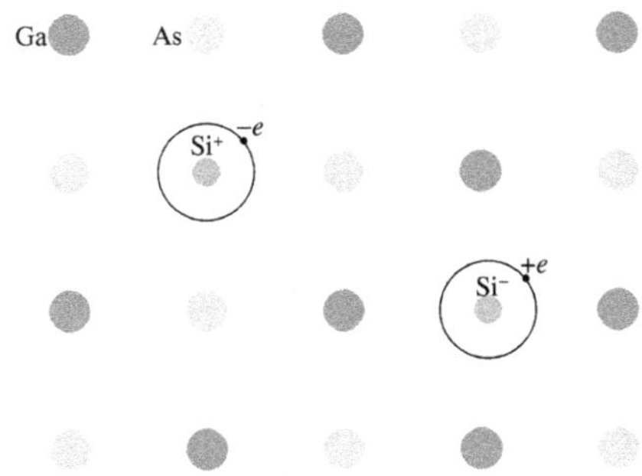

Fig. 8.5. Silicon atoms substitute a gallium and an arsenic atoms in a GaAs crystal.

section 2.2.2), except that the screening of the Coulomb potential is factored with the dielectric constant ϵ_r and the electron mass m_e is replaced with the effective mass m^*. Consequently, the radius of the ground state becomes

$$r_0 = \frac{m_e}{m^*}\epsilon_r a_0, \tag{8.11}$$

and the energy levels of the bound states are expressed as

$$E_n = -\frac{m^*}{m_e}\frac{1}{\epsilon_r^2}\frac{\mathcal{R}}{n^2} \tag{8.12}$$

with the principal quantum number n. Given that m_e^* of GaAs is $0.067m_e$ and ϵ_r is 12.9, we obtain that the radius r_0 is 10 nm, and the binding energy of the ground state is 5.5 meV. It is important to note that the radius is much larger than the lattice constant, 0.565 nm, because the hydrogen model is valid only if the electron distribution covers a large enough volume containing many atoms so that we can apply the macroscopic dielectric constant to account for the effective field screening.

Occasionally a uniform magnetic field is applied to a doped semiconductor to control shifting and splitting of the impurity levels using the Zeeman effect (see Fig. 8.6). The magnetic field also induces the collapse of the conduction band into Landau levels. The Zeeman effect makes the $2p$ level of the hydrogenic impurity system split into three levels: $2p^+$, $2p^0$, and $2p^-$. If the magnetic field is strong enough, the $2p^+$ level goes above the bottom of the lowest Landau level.

These impurity states can be coherently manipulated by intense pulses of THz radiation. Figure 8.7(a) illustrates the schematics of the coherent manipulation of $1s$ and $2p^+$ impurity levels in Si-doped GaAs using THz pulses. When the magnetic field is stronger than ~2 T, the $2p^+$ level exceeds the bottom of the lowest Landau level [239]. The incident THz pulse induces Rabi flopping between the $1s$ and the $2p^+$ states (see section 2.2.1). At the end of the THz pulse some electrons may remain in the $2p^+$ state depending on the pulse duration and the field amplitude. The population of the excited state ρ_{11}

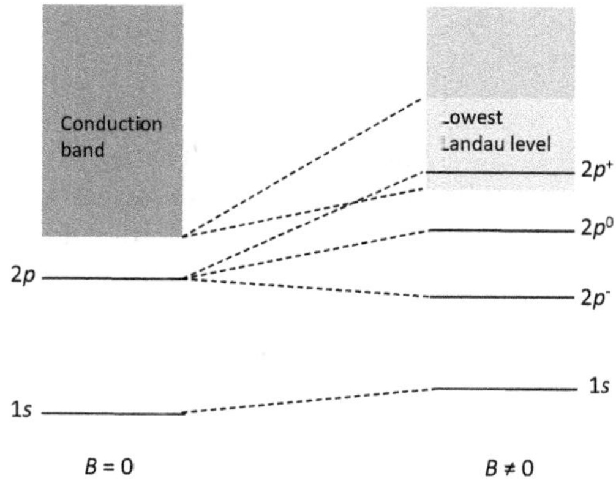

Fig. 8.6. Level shifting and splitting of impurity states and conduction band

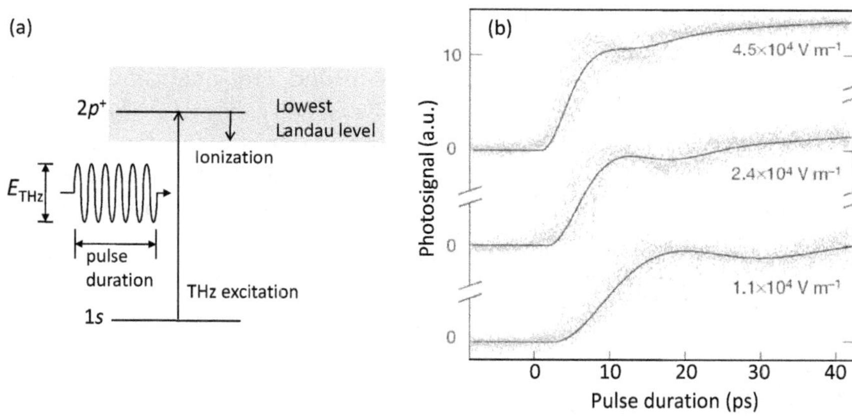

Fig. 8.7. Rabi oscillations of $1s$ and $2p^+$ impurity states in Si-doped GaAs. (a) Schematic diagram and (b) photocurrent versus THz pulse duration for THz fields $E_{THz} = 0.45$, 0.24, and 0.11 kV/cm. (Reprinted by permission from Macmillan Publishers Ltd: Nature [4], ©2001.)

is the physical quantity of interest. Eventually the electrons in the $2p^+$ state are ionized and get into the conduction band. Since the conductivity is proportional to the number of free carriers, the $2p^+$ population can be obtained by measuring the photocurrent of free carriers. Figure 8.7(b) shows that the photocurrent oscillates as a function of pulse duration for several THz field amplitudes [4]. The Rabi frequency increases as the field intensity increases. The Rabi oscillations are damped because of dephasing. The observed dephasing time is 20-30 ps. The theoretical description of Rabi oscillations is presented in section 2.2.1. It is useful to understand the coherent manipulation of impurity states. An important result of the theoretical analysis is that the Rabi frequency is proportional to the field amplitude:

$$\omega_R = \frac{2d_{12}}{\hbar} E_\omega. \tag{8.13}$$

Using a rough estimation of the dipole moment $d_{12} \sim er_0$, we can calculate the Rabi frequencies as $\omega_R = 0.3$, 0.7, and 1.4 THz for $E_{THz} = 0.45$, 0.24, and 0.11 kV/cm, respectively. These numbers are consistent with the experimental results within the empirical uncertainty of the THz field amplitude.

8.1.4 Semiconductor Nanostructures: Quantum Wells, Quantum Wires, and Quantum Dots

While a discrete energy gap exists between the conduction and valence bands in a semiconductor, the energy levels within a band are continuous for a bulk material. The continuity of the band structure, however, is transfigured in lower dimensional systems. The modification of the electronic structure arises from the quantum mechanical premise that, if a particle is spatially confined, its energy levels are discrete. A practical system used to accomplish such spatial confinement is an epitaxially-grown semiconductor quantum well (QW). A QW is a quasi two-dimensional electron system, consisting of a thin semiconductor sandwiched between two layers of a material with a larger bandgap. The thickness of a semiconductor QW is usually in the nanometer scale, which corresponds to the de Broglie wavelength of an electron in semiconductors.

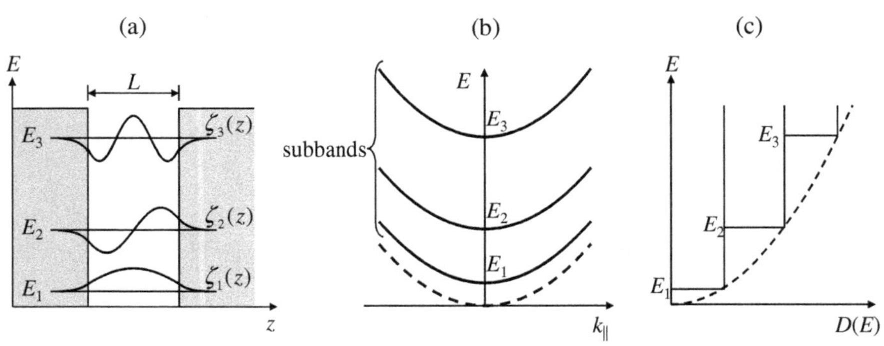

Fig. 8.8. Electronic structure of the conduction band in a semiconductor quantum well. (a) Discrete energy levels of a square potential well and quantized wave functions. (b) Parabolic energy levels of the subbands. (c) Density of states.

Figure 8.8(a) shows a one-dimensional rectangular potential well for the conduction band in a QW. An electron can move freely in the QW plane, but the motion is confined within the potential well. The wave functions of the energy eigenstates have the form

$$\Psi_{\mathbf{k}_\parallel, n_z}(\mathbf{r}) = e^{i\mathbf{k}_\parallel \cdot \mathbf{r}} \zeta_{n_z}(z), \qquad n_z = 1, 2, 3, \ldots, \tag{8.14}$$

where $\mathbf{k}_{\parallel} (= k_x \mathbf{e}_x + k_y \mathbf{e}_y)$ is an in-plane wave vector, and $\zeta_{n_z}(z)$ is a quantized standing wave. $\zeta_1(z)$, $\zeta_2(z)$, and $\zeta_3(z)$ of the square potential well are also shown in Fig. 8.8(a). The energy eigenvalues of this two-dimensional system are expressed as

$$E_{n_z}(k_{\parallel}) = \frac{\hbar^2 k_{\parallel}^2}{2m_e^*} + E_{n_z}. \tag{8.15}$$

A subband refers to the energy levels with a specific quantum number n_z, and hence the lowest energy for the subband is given as E_{n_z} with $\mathbf{k}_{\parallel} = 0$. Figure 8.8(b) shows the energy dispersion for the first few subbands. The density of states depends on the dimensionality of a system: for a d-dimensional space it has the form

$$D_C(E) \propto E^{\frac{d}{2}-1}. \tag{8.16}$$

Therefore, the density of states for a QW subband is constant, as shown in Fig. 8.8(c). If the potential well is deep, the discrete energies are approximately

$$E_{n_z} \approx \frac{\hbar^2 \pi^2}{2m_e^* L^2} n_z^2. \tag{8.17}$$

This equation implies that the energy levels can be tuned by varying the QW thickness L. A typical inter-subband excitation energy ranges from 10 to 100 meV, which falls within the THz region.

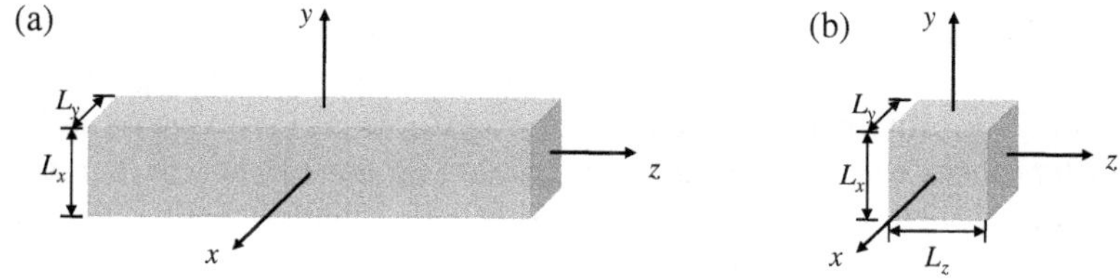

Fig. 8.9. (a) Quantum wire and (b) quantum dot

The analysis can be easily extended to one- and zero-dimensional systems, which are called a quantum-wire and a quantum-dot, respectively (Fig. 8.9). The energy eigenvalues of a quantum wire, in which electrons are confined to the x-y plane, but move freely along the z-axis, have the form

$$E_{n_1,n_2}(k_z) = \frac{\hbar^2 k_z^2}{2m_e^*} + E_{n_1,n_2}, \tag{8.18}$$

where the two quantum numbers, n_1 and n_2, are integers. For example, if the system has an infinite square potential,

$$E_{n_x,n_y} = \frac{\hbar^2 \pi^2}{2m_e^* L^2}(n_x^2 + n_y^2). \tag{8.19}$$

The density of states for the one-dimensional system, shown in Fig. 8.10, is proportional to $1/\sqrt{E}$ according to Eq. 8.16. The three-dimensional confinement of a particle in a quantum dot leads to discrete energy eigenvalues associated with three quantum numbers. Accordingly, the density of states is a group of delta functions not vanishing only at the energy eigenvalues, $D(E) \propto \sum_{n_1,n_2,n_3} \delta(E - E_{n_1,n_2,n_3})$.

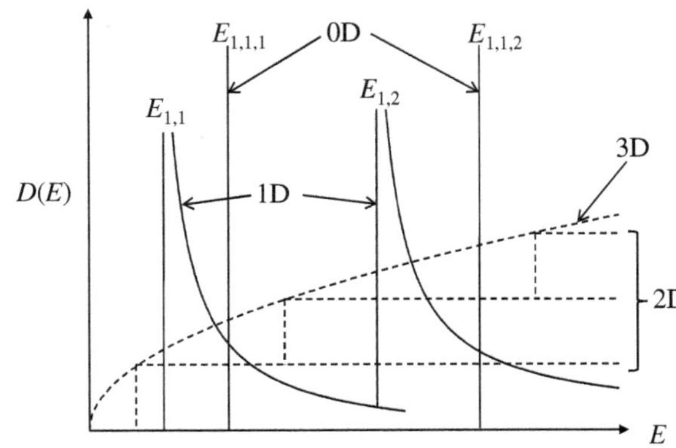

Fig. 8.10. Density of states of a quantum wire and quantum dot.

Now we consider electron-hole pair excitations including the Coulomb interaction between the quasiparticles. The parabolic band model provides a simple and intuitive picture. As we have seen in Fig. 8.1(b), an interband transition in an intrinsic semiconductor creates an electron and a hole simultaneously. These quasiparticles are attracted to each other by the Coulomb interaction, which leads to the formation of a hydrogen-like system. The hydrogen-like system of an electron-hole pair is called an *exciton.*

We can apply the same theoretical framework of the hydrogen problem (see section 2.2.2) to excitons in a bulk semiconductor. We only need to replace a couple of parameters: the reduced mass and the material permittivity instead of the electron mass and the vacuum permittivity. Then, the corresponding exciton energy has the form

$$E_n^{ex}(\mathbf{K}) = E_g + \frac{\hbar^2 K^2}{2M} - \frac{\mathcal{R}_{ex}}{n^2}. \tag{8.20}$$

The second term is merely the center-of-mass kinetic energy with the total mass, $M = m_e^* + m_h^*$, and the center-of-mass wave vector, $\mathbf{K} = \mathbf{k}_e + \mathbf{k}_h$. The last term represents the binding energy of the excitonic states. The excitonic Rydberg energy,

$$\mathcal{R}_{ex} = \frac{m_\mu}{m_e}\frac{1}{\epsilon_r^2}\mathcal{R} \tag{8.21}$$

with the reduced mass m_μ and dielectric constant ϵ_r, is only a fraction of the hydrogen Rydberg energy, $\mathcal{R} = 13.6$ eV, because m_μ is relatively smaller than

m_e and ϵ_r is substantially larger than unity in semiconductors. For example, in GaAs $\mathcal{R}_{ex}$=4.7 meV with m_μ/m_e=0.058 and ϵ_r=12.9. You can see that the spectrum of inter-excitonic transitions falls right into the THz region.

Excitons in a semiconductor QW have a few notable attributes associated with dimensionality. The exciton energy in two-dimensional space has the form,

$$E_n^{ex}(n_z, \mathbf{K}_{\|}) = E_g + E_{n_z} + \frac{\hbar^2 K_{\|}^2}{2M} - \frac{\mathcal{R}_{ex}}{\left(n - \frac{1}{2}\right)^2}. \tag{8.22}$$

The binding energy of the exciton in the ground state, $4\mathcal{R}_{ex}$, is larger than that in three dimensions, $\mathcal{R}_{ex}$. The spatial confinement in the two-dimensional system also leads to the smaller Bohr radius and larger oscillator strength,

$$a_0^{ex}(2\text{D}) = \frac{1}{2} a_0^{ex}(3\text{D}) \quad \text{and} \quad f^{ex}(2\text{D}) = 8 f^{ex}(3\text{D}), \tag{8.23}$$

for the ground state. Semiconductor QWs have been rigorously studied to understand exciton dynamics in semiconductors, partly because the large binding energy and oscillator strength of excitons and the flat density of states near the band edge in a two-dimensional system are all favorable properties for optical measurements on excitons.

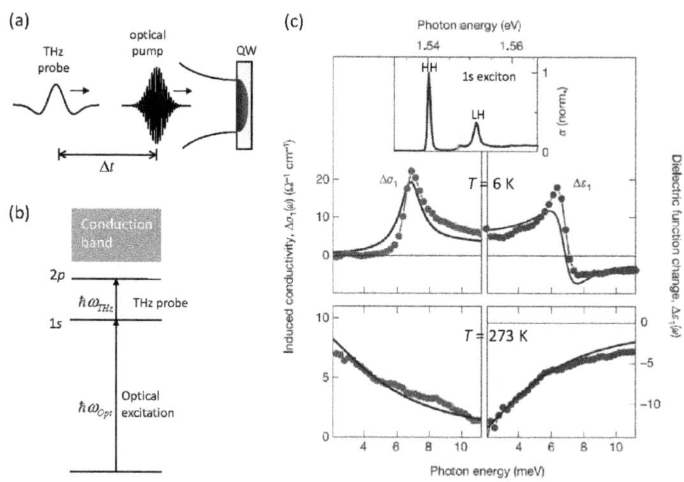

Fig. 8.11. (a) Optical-pump and THz-probe experiment on a semiconductor QW. (b) Energy level diagram for the optical excitation and the THz probe. (c) Spectra of the THz conductivity ($\Delta\sigma_1$) and dielectric constant ($\Delta\varepsilon_1$) when the optical pump is tuned to the heavy-hole 1s exciton resonance at lattice temperature of 6 K (top panels) and when the optical pump excites free carriers in the conduction band at room temperature (bottom panels). The inset is the low temperature optical spectrum of the heavy-hole and light-hole 1s exciton resonance lines. (Reprinted by permission from Macmillan Publishers Ltd: Nature [240], ©2003.)

The energy levels of excitons in semiconductor QWs can be directly probed by the optical-pump and THz-probe technique [240]. Figure 8.11(a) and (b) illustrate the experimental scheme. An ultrashort optical pulse tuned to the $1s$ exciton resonance generates coherent polarization waves in a semiconductor QW. They transform into a $1s$ exciton population within a few picosecond while they lose their coherence by scattering with lattice vibrations and defects. A broadband THz pulse arrives afterwards, and detects the THz response of the excitons. The THz conductivity and dielectric constant of the excitonic system are obtained by analyzing the time-resolved waveforms of the transmitted THz pulses. The low temperature optical spectrum of heavy-hole (HH) and light-hole (LH) $1s$ exciton resonance lines of GaAs QWs with $Al_{0.3}Ga_{0.7}As$ barriers is shown in the inset of Fig. 8.11(c). The low temperature spectra of the THz conductivity and dielectric constant shown in Fig. 8.11(c) shows a clear resonance near 7 meV, which is attributed to the transition from the $1s$ to the $2p$ level of the HH exciton [240]. The solid lines, calculations based on quasi-two-dimensional exciton wavefunctions, fit well with the measurements. As a comparison, the spectra of the free carrier excitation at room temperature are also shown. The measured THz conductivity and dielectric constant are consistent with calculations based on the Drude model (solid lines).

Now we turn our attention to coherent manipulation of excitonic states in QWs by intense THz radiation. We consider a three-level system consisting of the ground state, the $1s$ exciton state in the first subband (h1X), and the $1s$ exciton state in the second subband (h2X). The subband levels are illustrated in Fig. 8.8. According to Eq. 8.22, the energy of the two exciton levels of different subbands (n_z=1 and 2) are written as

$$\text{h1X:} \quad E^{ex}_{n=1}(n_z = 1, \mathbf{K}_{\|} = 0) = E_1 - 4\mathcal{R}_{ex}, \tag{8.24}$$

$$\text{h2X:} \quad E^{ex}_{n=1}(n_z = 2, \mathbf{K}_{\|} = 0) = E_2 - 4\mathcal{R}_{ex}. \tag{8.25}$$

The two excitonic states can be strongly coupled by an intense THz field perpendicularly polarized to the plane of QW and tuned near the resonant intersubband transition.

Figure 8.12(a) and (b) illustrate the experimental scheme for an optical measurement of the strong coupling between the excitonic states mediated by a intense THz field. The strong light-matter interaction induces an alteration of the two energy levels such that a weak optical probe detects energy level splitting. This phenomenon is called the Autler-Townes effect. The experimental observation and theoretical calculation of the Autler-Townes effect in a semiconductor QW system are shown in Fig. 8.12(c) [6]. The sample consists of $In_{0.06}Ga_{0.94}As$ QWs with $Al_{0.3}Ga_{0.7}As$ barriers. The source of the THz radiation is the UCSB Free-Electron laser. A weak optical probe scans near the $1s$ exciton resonance of the first subband for several THz field intensities. The THz frequency is tuned at, below, and above the resonant intersubband transition. The Autler-Townes splitting increases linearly with the applied filed amplitude at resonance. In fact, the magnitude of the splitting is given by the generalized Rabi frequency,

$$\Omega_R = \sqrt{\omega_R^2 + \Delta\omega^2}, \tag{8.26}$$

where $\Delta\omega = \omega_{THz} - \omega_{12}$. The experimental data agree well with the theoretical calculations shown on the right-hand side of Fig. 8.12(c).

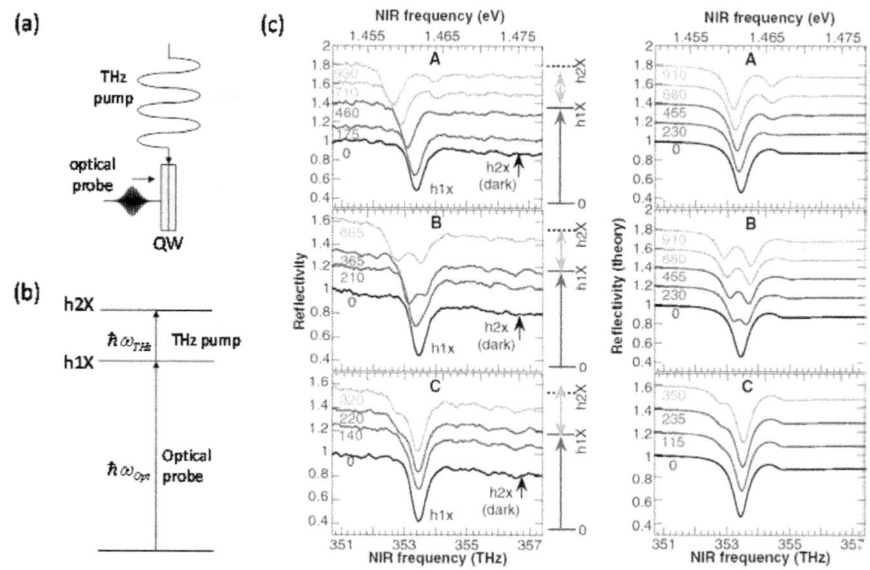

Fig. 8.12. Autler-Townes splitting of an excitonic state induced by a strong THz field (a) Experimental scheme (b) Energy level diagram (c) Reflectivity spectra of the excitonic resonance of $In_{0.06}Ga_{0.94}As/Al_{0.3}Ga_{0.7}As$ QWs for various THz intensities at $\omega_{THz}/2\pi$ = 2.52, 3,42, and 3.90 THz. Experimental data are shown on the left-hand side, and theoretical calculations on the right-hand side. The THz frequency is tuned below (top panel), at (middle panel), and above (bottom pane) the resonant intersubband transition. (From [6]. Reprinted with permission from AAAS.)

A THz field polarized in the plane of a QW induces internal transitions of confined excitons as we have seen in Fig. 8.11. If the field intensity reaches a certain point, the dynamics of the excitonic states enters a unique regime of extreme nonlinear effects. In this regime, three energy scales have comparable quantities: the THz photon energy, the Rabi energy, and the ponderomotive energy. They have the following expressions:

$$\text{Photon energy:} \quad \hbar\omega_{THz}, \tag{8.27}$$

$$\text{Rabi energy:} \quad \hbar\omega_R = 2d_{1s\to 2p}E_{THz}, \tag{8.28}$$

$$\text{Ponderomotive energy:} \quad \frac{e^2 E_{THz}^2}{4m_e^*\omega_{THz}^2}. \tag{8.29}$$

The ponderomotive energy refers to the average quivering energy of a free electron in an oscillating electric field. These three energies can have similar values with a moderate field intensity (E_{THz} ~1-10 kV/cm) in the THz region due to the relatively low photon energy, while in the optical region the photon energy is significantly larger than the Rabi energy and the ponderomotive energy is, in general, negligible. Subsequently, several distinctive nonlinear effects are observed in the interaction of intense THz fields with excitonic states.

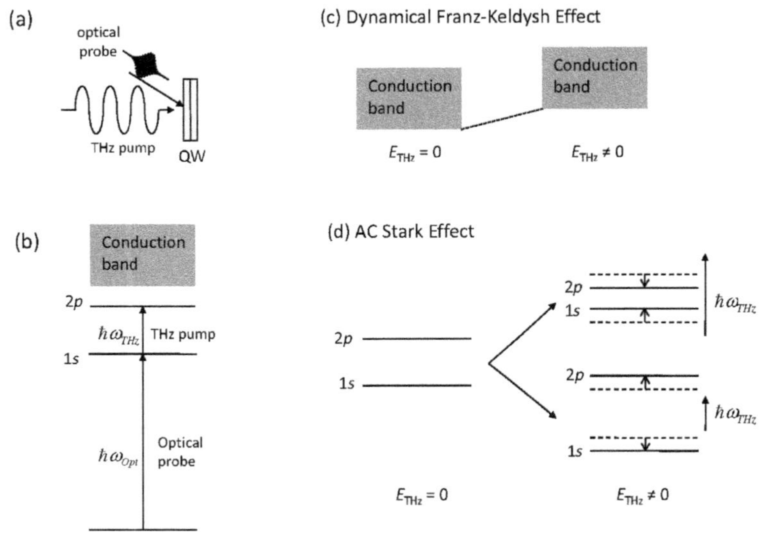

Fig. 8.13. (a) Schematic diagram of the THz-pump and optical-probe experiment with the THz field polarized in the plane of QW. (b) Energy diagram for the optical probe of the 1*s* exciton state and the THz-induced intra-excitonic transition from the 1*s* to the 2*p* state. (c) The dynamical Franz-Keldysh effect. (d) The AC Stark effect.

Figure 8.13(a) illustrates the THz-pump and optical-probe experiment to measure the THz-induced nonlinear effects of an excitonic system in semiconductor QWs. The THz field is polarized in the plane of QW to induce inter-excitonic transitions such as the transition from the 1*s* to the 2*p* state as shown in Fig. 8.13(b). The optical probe scans the spectral region near the 1*s* exciton resonance to observe the excitonic nonlinear effects. The intense THz radiation gives rise to two distinctive effects on the QW system: the dynamical Franz-Keldysh effect and the AC Stark effect. The dynamical Franz-Keldysh effect refers to the phenomenon that the THz field pushes up the conduction band edge because of the positive ponderomotive energy as shown in Fig. 8.13(c). Consequently, the excitonic states also shift to the higher energy side. The AC Stark effect (Fig. 8.13(d)) is the energy level shifts of a two-level system induced by an off-resonant oscillating field: the energy levels are pushed out for a radiation frequency below the resonant transition, and they are squeezed in for a radiation above the resonance. It is notable that the AC Stark shift increases linearly with the field amplitude, while the dynamical Franz-Keldysh effect is quadratic.

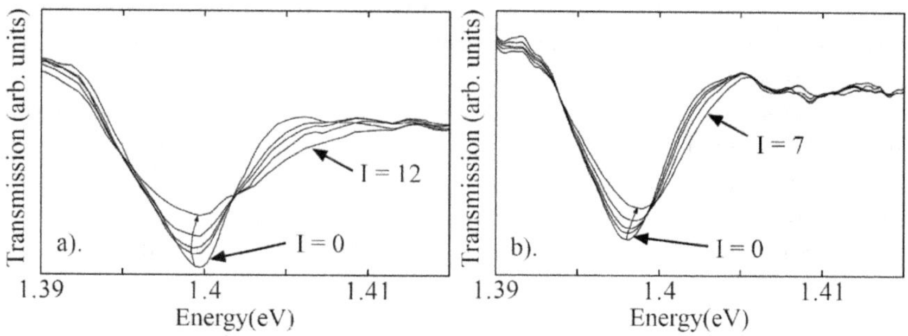

Fig. 8.14. Experimental transmission spectra of the 1s exciton resonance of $In_{0.2}Ga_{0.8}As/GaAs$ QWs with (a) $\hbar\omega_{THz} = 2.5$ meV ($< \hbar\omega_{12}$) at $I_{THz} = 0, 1, 2, 4, 12$ (arbitrary units) and (b) $\hbar\omega_{THz} = 14$ meV ($> \hbar\omega_{12}$) at $I_{THz} = 0, 1, 2, 4, 7$ (arbitrary units). (Reprinted with permission from [5]. ©1998, American Physical Society.)

Figure 8.14 shows the experimental transmission spectra of the 1s exciton resonance of $In_{0.2}Ga_{0.8}As/GaAs$ QWs when the THz photon energy, $\hbar\omega_{THz}$, is (a) smaller and (b) larger than the exciton $1s \rightarrow 2p$ transition energy, $\hbar\omega_{12}$ ~8 meV [5]. The source of THz radiation is the UCSB free-electron laser. When $\omega_{THz} < \omega_{12}$, the exciton line shifts to the lower energy side at low THz intensities, where the AC stark effect pushing down the 1s level is dominant. The direction of the level shift is reversed as the THz intensity increases due to the quadratic increase of the ponderomotive energy with the field amplitude. The results indicate that the ponderomotive energy is on a similar scale as the Rabi energy in this experimental condition. When $\omega_{THz} > \omega_{12}$, the AC Stark effect induces a blueshift of the exciton line so that the level shift increases monotonically to the higher energy side as the THz intensity increases.

When strong single-cycle THz pulses are applied to a QW system, time-resolved optical measurements reveal some interesting aspects of THz nonlinear effects [241]. The schematic diagram of the THz-pump and optical-probe experiment is shown in Fig. 8.15(a). The dotted line in Fig. 8.15(b) is the unperturbed optical transmission spectrum of the LH and HH excitons of $GaAs/Al_{0.3}Ga_{0.7}As$ QWs at 5 K. The solid line shows that the spectrum is strongly modified by the applied THz fields, $E_{THz} \simeq 10$ kV/cm, for the relative time delay between the optical and THz pulses Δt=0.0 ps. The spectral modulation is so large that the normalized differential transmission spectrum shown in the inset reaches up to 0.6. Figure 8.15(c) shows the ultrafast dynamics of the pronounced spectral modulations as a two-dimensional contour plot of the differential transmission $\Delta T(\hbar\omega_{opt}, \Delta t)$. The results of microscopic many-body calculations identify the dominant physical mechanisms contributing to the spectral modulations. The broad peak centered at 1.542 eV in the differential spectra at Δt=-0.22 and 0.78 ps shown in Fig. 8.15(d) and (e) is attributed to the dynamical Franz-Keldysh effect. This feature is short-lived because the ponderomotive energy is quadratic in E_{THz}. Another competing process is the AC Stark effect, which is manifest in the relatively long-lived peak near 1.537 eV. A distinctive extreme-nonlinear effect is shown in Fig. 8.15(f), (g), and (h) displaying the temporal evolution of ΔT at $\hbar\omega$=1.533, 1.540, and 1.548 eV. Because of the ultrashort waveform of the single-cycle THz pulses, the rotating-wave-approximation breaks down. The fast oscillations in the time-resolved ΔT arise from the non-rotating-wave-approximation parts. Another interesting feature is the spectral modulation on the high energy side near 1.548 eV, which results from THz third harmonic generation.

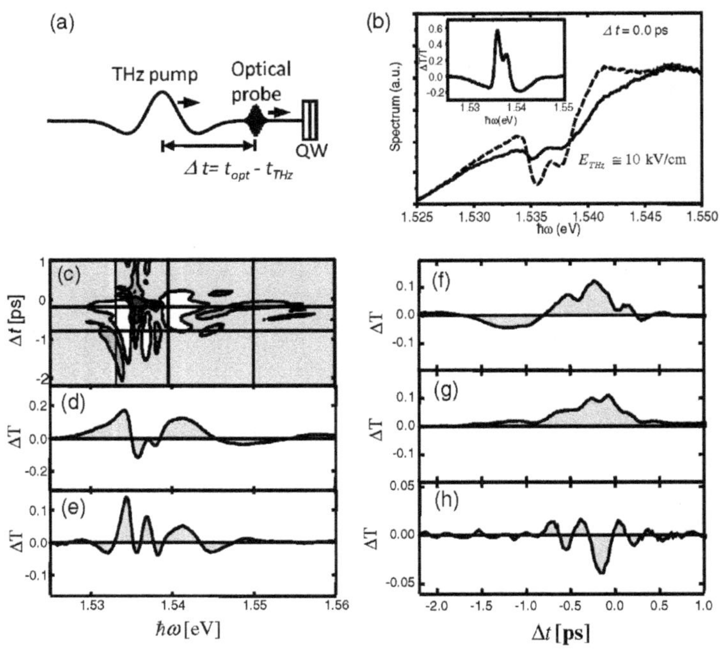

Fig. 8.15. (a) Schematic of the THz-pump and optical-probe experiment. (b) The solid-line indicates the optical transmission spectrum of LH and HH 1s excitons at 5 K, for Δt=0.0 ps. The unperturbed spectrum is also shown for comparison (dotted-line). The inset shows the normalized differential transmission, $\Delta T/T$. (c) Contour plot of $\Delta T(\hbar\omega_{opt}, \Delta t)$ (d), (e) ΔT spectra at Δt = -0.22 and 0.78 ps (f), (g), and (h) Time dependent Δt at $\hbar\omega$ = 1.533, 1.540, and 1.548 eV [241].

8.2 Strongly Correlated Electron Systems

A solid material is, in principle, a many-body system which contains myriad ions and electrons mutually interacting with each other. It is an impossible task to solve the many-body problem with the full Hamiltonian including all the interactions between the particles. A powerful approach to handle this problem is to replace all interactions with any one electron with an averaged potential energy, which is known as the independent particle approximation or the mean-field approximation. The basic concept of the approximation is portrayed in the following simple, classical picture:

$$\begin{aligned} H &= \sum_i h_i(x_i) + \frac{1}{2} \sum_{i,j(\neq i)} v_{i,j}(x_i, x_j) \\ \rightarrow H_{MF} &= \sum_i \left[h_i(x_i) + v_{MF,i}(x_i) \right], \end{aligned} \tag{8.30}$$

where $h_i(x_i)$ and $v_{ij}(x_i, x_j)$ are non-interacting and interacting Hamiltonians, respectively, and

$$v_{MF,i}(x_i) \approx \frac{1}{2} \sum_{j(\neq i)} v_{i,j}(x_i, x_j) \tag{8.31}$$

is a single-particle effective potential energy. The independent particle approximation is hugely practical, because it converts the intangible many-body problem to a single-body problem which can be treated systematically. It is valid only if the electronic correlation beyond the mean-field theory, $\Delta H = H - H_{MF}$, is relatively small compared to the total Hamiltonian. If this condition is satisfied, the perturbation theory is applied to handle the remaining interactions between the dressed particles.

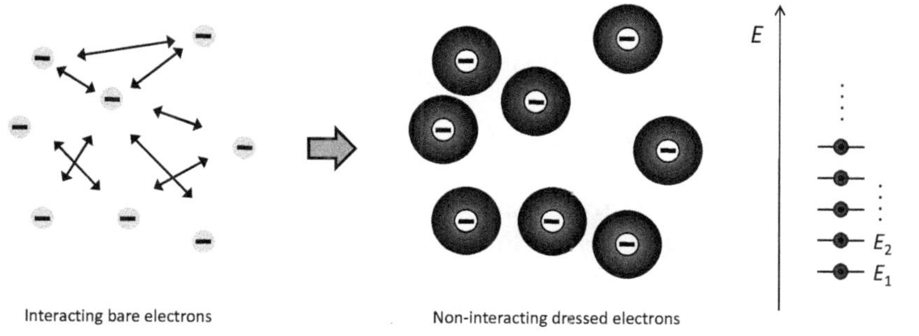

Fig. 8.16. Independent particle approximation

As Fig. 8.16 illustrates, the independent particle approximation treats the mutually interacting bare electrons as non-interacting dressed electrons. The energy levels of the non-interacting system are determined by the single particle Hamiltonian. The Pauli exclusion principle prevents two fermions from occupying the same quantum state simultaneously, and hence the quasiparticles fill up the levels, starting from the lowest energy state, to form the ground state of the many-body system.

It is fortunate that the independent particle approximation and the perturbation theory are suitable to describing the electron systems of typical metals, semiconductors, and insulators. In large part, modern solid state physics is founded on this extraordinary blessing of Nature. The electron system of a crystalline solid with a periodic effective potential is described by a band structure as discussed in section 8.1.1. The energy levels of the conduction and valence band (Eq. 8.1 and 8.2) indicate that the electrons are free particles and independent of each other in the effective mass approximation. In this picture, the electrons and holes are quasiparticles with the free dressed-state energies.

Not all materials, however, are as tame as the usual metals and insulators. The independent particle approximation fails drastically to account for the exotic properties of so-called strongly correlated electron systems, because, in such systems, mutual interactions between electrons dominate their kinetic energy, thus screening is too weak to form non-interacting quasiparticles. The strong correlation of electrons gives rise to metal-insulator transitions, high-temperature superconductivity, superconductivity in organic materials, ferromagnetism, colossal magnetoresistance, heavy fermions, Kondo effects, charge density waves, and fractional quantum hall effects, among many others. These remarkable attributes are caused by delicate and subtle interplay between electron, lattice, orbital, and spin degrees of freedom.

THz spectroscopy is a powerful method for investigating the electrodynamics of strongly correlated electron systems, because crucial elementary excitations in strongly correlated electron systems have the same energy scale as THz photons. A conspicuous example is the superconducting gap, which is a manifestation of Cooper pair binding energy. In this section we shall focus on the THz spectroscopy of electrodynamics in superconductors.

8.2.1 Quasiparticle Dynamics in Conventional Superconductors

Superconductivity refers to the fascinating phenomenon in which electrical resistance drops down to zero below the critical temperature T_c in some materials. Temperature-dependent resistivity of a conventional or type-I superconductor is shown in Fig. 8.17(a). Conventional superconductors such as aluminium, tin, and mercury are metals, and have critical temperatures less than $\sim$20 K. For example, the critical temperatures of Al, Sn, and Hg are 1.19, 3.72, and 4.15 K, respectively. The BCS theory, developed by Bardeen, Cooper, and Schrieffer, elucidates the microscopic mechanism of how the delicate interactions between the electron and lattice bring about superconductivity. The key physical process of the theory is the formation of binding pairs of electrons with opposite spin known as Cooper pairs. The formation of a Cooper pair is illustrated in Fig. 8.17(b). As a free electron close to the Fermi level travels through a lattice with the velocity $v_F \sim 10^6$ m/s, the attractive Coulomb interaction between the electron and positive ions induces a deformation of the lattice. Because the lattice response is not instantaneous[1] and the electron moves fast, the electron leaves a positively charged area behind, and the second electron is attracted by the net positive charge. Since the size of the positively charged area is roughly $v_F \cdot \frac{2\pi}{\omega_D} \sim 100$ nm and the Coulomb interaction between the two electrons are completely screened at this distance, the two electrons are weakly bound by the excessive charge induced by the lattice distortion. The quantum mechanical analysis of the BCS theory requires that the two electrons have opposite spin and that lattice vibrations are quantized. From the quantum mechanical point of view, phonons mediate the pairing of two electrons in a spin-singlet state.

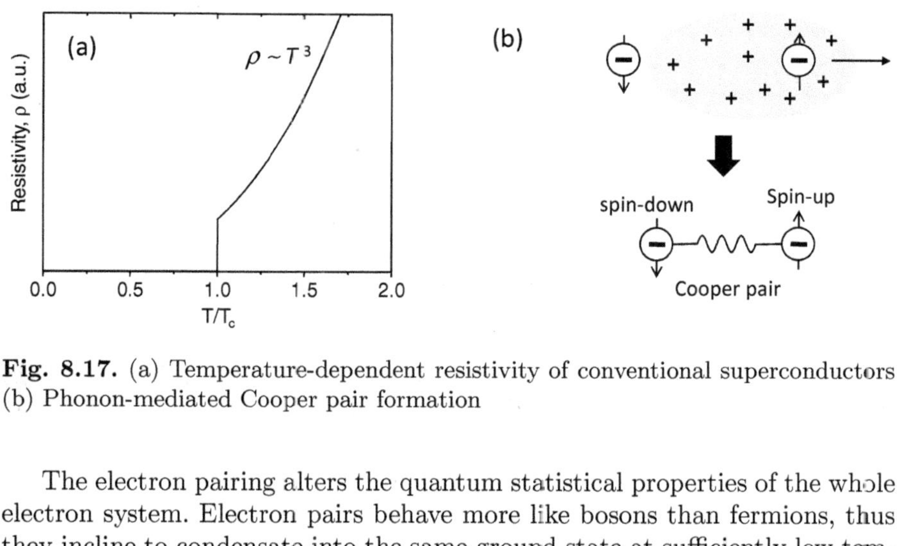

Fig. 8.17. (a) Temperature-dependent resistivity of conventional superconductors (b) Phonon-mediated Cooper pair formation

The electron pairing alters the quantum statistical properties of the whole electron system. Electron pairs behave more like bosons than fermions, thus they incline to condensate into the same ground state at sufficiently low temperatures. The many electrons in the superconducting ground state form a macroscopic quantum system expressed as

$$\psi(\mathbf{r}) = |\psi(\mathbf{r})|e^{i\phi(\mathbf{r})}, \tag{8.32}$$

where $|\psi(\mathbf{r})|^2$ is the Cooper pair density and $\phi(\mathbf{r})$ is the macroscopic phase. The most striking result emerging from the BCS theory is that the formation of the BCS ground state opens an energy gap $\Delta(T)$ at the Fermi level. The energy gap is the minimum energy for exciting a single electron from the ground state so that it needs $2\Delta(T)$ to break a binding pair. The BCS theory predicts that the zero-temperature energy gap is

[1] The characteristic time scale is $\frac{2\pi}{\omega_D} \sim 100$ fs, where ω_D is the Debye frequency which is the maximum frequency of lattice vibration.

$$\Delta_0 = 1.76 k_B T_c \tag{8.33}$$

in the weak coupling limit and

$$\frac{\Delta(T)}{\Delta_0} = 1.74 \left(1 - \frac{T}{T_c}\right)^{1/2} \tag{8.34}$$

for $T \approx T_c$. The superconducting gap is a subject of THz spectroscopy. For example, $T_c = 10$ K corresponds to the pair binding energy $2\Delta_0 = 3$ meV. Superconductivity is accounted for as a direct consequence of the existence of the energy gap, because scattering at an energy scale smaller than the pair binding energy is completely suppressed. Consequently, resistivity drops to zero when a material enters into the superconducting state.

The optical properties of a superconductor are governed by its complex conductivity, and its superconducting gap energy is a determining parameter. The BCS theory being applied, the real and imaginary parts of the conductivity, σ_1 and σ_2, are expressed by the Mattis-Bardeen equations [242]

$$\begin{aligned}\frac{\sigma_1(\omega)}{\sigma_N(\omega)} &= \frac{2}{\hbar\omega}\int_{\Delta_0}^{\infty}\left[f(E) - f(E+\hbar\omega)\right] g(E) dE \\ &\quad + \frac{1}{\hbar\omega}\int_{\Delta_0-\hbar\omega}^{-\Delta_0}\left[1 - 2f(E+\hbar\omega)\right] g(E) dE,\end{aligned} \tag{8.35}$$

$$\frac{\sigma_2(\omega)}{\sigma_N(\omega)} = -\frac{1}{\hbar\omega}\int_{\Delta_0-\hbar\omega,-\Delta_0}^{\Delta_0} \frac{\left[1-2f(E+\hbar\omega)\right]\left(E^2+\Delta_0^2+\hbar\omega E\right)}{\left(\Delta_0^2 - E^2\right)^{\frac{1}{2}}\left[(E+\hbar\omega)^2 - \Delta_0^2\right]^{\frac{1}{2}}} dE, \tag{8.36}$$

where $\sigma_N(\omega)$ is the conductivity of the normal metallic state, the Fermi-Dirac function $f(E)$ and the density of states $g(E)$ have the forms

$$f(E) = \frac{1}{e^{(E-E_F)/k_B T} + 1}, \tag{8.37}$$

and

$$g(E) = \frac{E^2 + \Delta_0^2 + \hbar\omega E}{\left(E^2 - \Delta_0^2\right)^{\frac{1}{2}}\left[(E+\hbar\omega)^2 - \Delta_0^2\right]^{\frac{1}{2}}}. \tag{8.38}$$

The lower limit of the integral in Eq. 8.36 is either $-\Delta_0$ for $\hbar\omega > 2\Delta_0$ or $\Delta_0 - \hbar\omega$ for $\hbar\omega < 2\Delta_0$.

The superconducting energy gap can be directly observed by measuring the optical properties of superconductors. Niobium (Nb), a basic element with atomic number 41, is a transition metal in its normal state. Technically, Nb is a type-II superconductor because its penetration depth is much larger than coherence length, yet its superconducting mechanism is the conventional phonon-mediated pairing. The energy gap of Nb was observed by experimental measurements using a backward wave oscillator and a Golay cell [243]. Figure 8.18 shows the frequency-dependent real and imaginary part of the conductivity for several temperatures. The sample is a 150-nm thick film on a sapphire substrate, and the critical temperature is measured as 8.31 K. The real part of the conductivity σ_1 (Figure 8.18(a)) shows a drastic change from the flat frequency response of Drude-type behavior to the development of the sharp dip at the pair binding energy. The imaginary part of the conductivity (Figure 8.18(b)) is inversely proportional to the frequency at low frequencies and low temperatures, which indicates the δ-function-like behavior of the DC conductivity. The inset shows the temperature-dependent binding energy and the BCS fit to the data with T_c=8.31 K and $2\Delta_0/k_BT_c$=4.1. The binding energy is larger than the BCS prediction of Eq. 8.33, which indicates that the electron-phonon coupling in Nb may exceed the weak-coupling limit of the BCS theory.

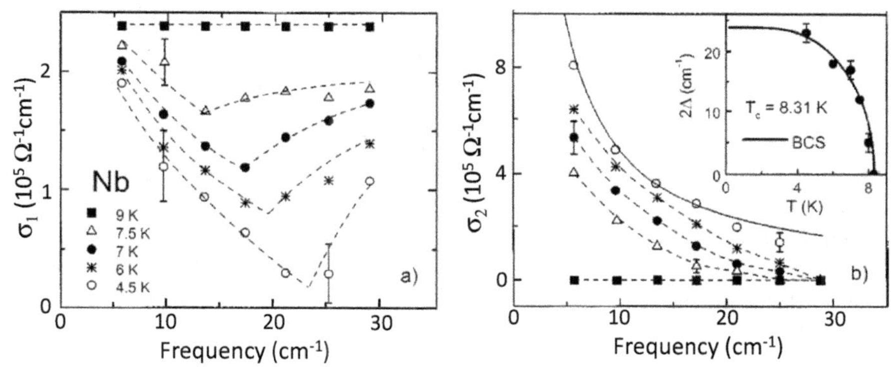

Fig. 8.18. (a) Real and (b) imaginary part of the conductivity of Nb versus frequency at T = 9, 7.5, 7, 5, and 4.5 K. The dashed lines are guides for the eye. The inset shows the temperature-dependent binding energy $2\Delta(T)$. The line indicates the BCS binding energy with T_c=8.31 K and $2\Delta_0/k_BT_c$=4.1. (Reprinted with permission from [243]. ©1998, American Physical Society.)

The recent discovery of superconductivity in magnesium diboride (MgB_2) has attracted a lot of attention because its critical temperature, T_c = 39 K, is remarkably high for a superconductor characterized by the conventional phonon-mediated pairing. Its phonon-mediated mechanism, however, is not yet fully understood. The opening of a superconducting gap in MgB_2 has been observed in a study of THz time-domain spectroscopy (THz-TDS) [244]. Figure 8.19 shows that a spectral dip corresponding to the pair binding energy 2Δ develops in the real part of the frequency-dependent conductivity as the temperature cools down below the critical temperature. The sample is an epitaxially-grown 100-nm film, and the critical temperature is measured as 30 K. The zero-temperature pair binding energy $2\Delta_0$ is estimated as 5 meV, which corresponds to the gap ratio, $\Delta_0/k_BT_c = 0.95$. The ratio is much smaller than the BCS prediction of the weak-coupling limit, 1.76 (Eq. 8.33). This demonstrates the complex nature of the electron-phonon coupling mechanism in MgB_2.

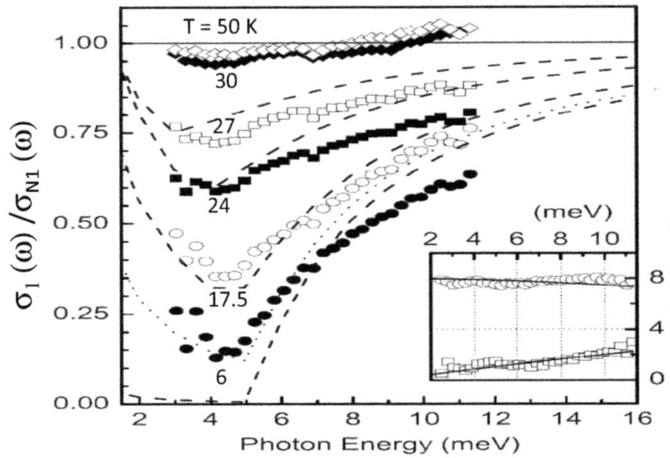

Fig. 8.19. Spectra of real part of conductivity σ_1 for the 100-nm MgB_2 film normalized to its normal state value σ_{N1}(40 K) for T = 6, 17.5, 24, 27, 30, and 50 K. The dashed and dotted lines are Mattis-Bardeen calculations for $2\Delta_0 = 5$ meV and $T_c = 30$ K (T = 6, 12, 17.5, 24, and 27 K). Inset: real (circles) and imaginary (squares) part of normal state conductivity at 40 K, along with a Drude calculation. (Reprinted with permission from [244]. ©2002, American Physical Society.)

While the THz spectroscopy techniques used in the above studies measure the linear optical properties of superconductors in thermal equilibrium, pump-probe experiments can resolve the dynamics of excited quasiparticles driven far-from equilibrium. Optical radiation on a superconductor breaks apart the weakly bound electron pairs and generates high energy electrons. Subsequently, the excited electrons get paired again after a series of relaxation processes. The pair breaking and relaxation dynamics have been observed in MgB_2 by a time-resolved study using the optical-pump and THz-probe technique [245].

Figure 8.20 shows the imaginary and the real parts of the frequency-dependent conductivity of MgB_2 at several delay times after an optical excitation with 800-nm central wavelength, 150-fs pulse duration, and 3-μJ/cm^2 fluence. The modulation of the $1/\omega$ dependence of the imaginary conductivity after the optical excitation is a good indicator of how the pair breaking and recovery evolves in time, as shown in Fig. 8.20(a) and the inset of Fig. 8.20(b). The pair-breaking process continues for the first 10 ps followed by a sub-nanosecond relaxation process. An analysis of the data, together with optical-pump and optical-probe measurements, affirms the following scenario. Because of the very strong coupling between electrons and high-frequency optical phonons in MgB_2, electron-electron scattering contributes little to the relaxation process in contrast to its usual dominance. Thus the optically-excited electrons lose their energy mostly by emitting high frequency optical phonons. These high-frequency phonons induce the continuing pair-breaking for the initial 10 ps. When the electron energy becomes lower than the optical phonon energy, electrons cannot emit optical phonons anymore, and lose their energy by much slower relaxation processes involving acoustic phonons. The slow recovery dynamics after 10 ps are strongly influenced by this phonon bottle-neck effect. The sub-nanosecond relaxation time corresponds to the lifetime of acoustic phonons with their energy $\hbar\omega_{ph} > 2\Delta$.

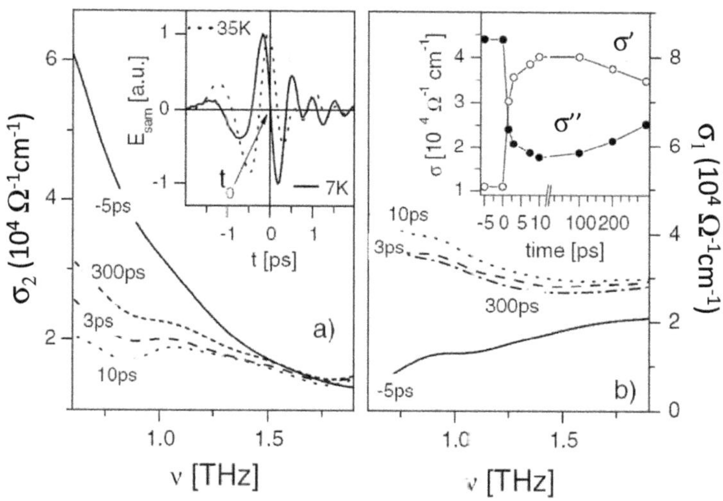

Fig. 8.20. (a) Imaginary and (b) real part of conductivity of MgB_2 versus frequency at various delay times following optical excitation with a fluence $\sim$3 J/cm^2 at 7 K. Inset of (a): $E_{sam}(t)$ at 7 and 35 K. Inset of (b): the time evolution of σ_1 and σ_2 taken at $\nu = 0.8$ THz. (Reprinted with permission from [245]. ©2003, American Physical Society.)

8.2.2 Low Energy Excitations in High Temperature Superconductors

High temperature (high-T_c) superconductivity is an astonishing, yet much more complicated phenomenon than conventional superconductivity. High-T_c superconductors are surely the most widely studied strongly correlated electron systems. High-T_c superconductivity is one of the most remarkable scientific discoveries in the last few decades. It has overturned some conventional wisdoms in one way or another. As the name suggests, their critical temperatures are extraordinarily high while conventional superconductors have

T_c less than 10 K under ambient pressure.[2] Furthermore, superconductivity is found in ceramic compounds unlike conventional metal or metal-alloy superconductors. All high-T_c superconductors are similar in that they contain lightly doped CuO_2 planes. The carrier transport in the CuO_2 planes is believed to be closely associated with the microscopic mechanism of the high-T_c superconductivity. The doping concentration is also a significant factor governing the electrical properties of the high-T_c superconductors. A superconductor is converted from an antiferromagnetic insulator by introducing a moderate density of electrons and holes.

A representative high-T_c superconductor is yttrium barium copper oxide ($YBa_2Cu_3O_7$). This compound was the first superconductor to be discovered whose $T_c(= 92$ K) is higher than the boiling point of nitrogen, 77.2 K. The crystalline structure of $YBa_2Cu_3O_7$ is shown in Fig. 8.21. The basic structure of this compound is described as an oxygen-deficient perovskite structure with triple layers. In a unit cell, an yttrium atom is in the center layer, while two barium atoms occupy the other two. The layers are separated by two CuO_2 planes. Other high-T_c superconductors have similar structural features. Some contain one CuO_2 plane per unit cell, and others have two or three. For example, $Bi_2Sr_2CaCu_2O_8$ (T_c=92 K) contains two CuO_2 planes like $YBa_2Cu_3O_7$ while $La_{1.85}Ba_{0.15}CuO_4$ (T_c=30 K) has only one. For a given cuprate family, compounds containing two or three CuO_2 planes tend to have higher T_c than those with a single CuO_2 plane.

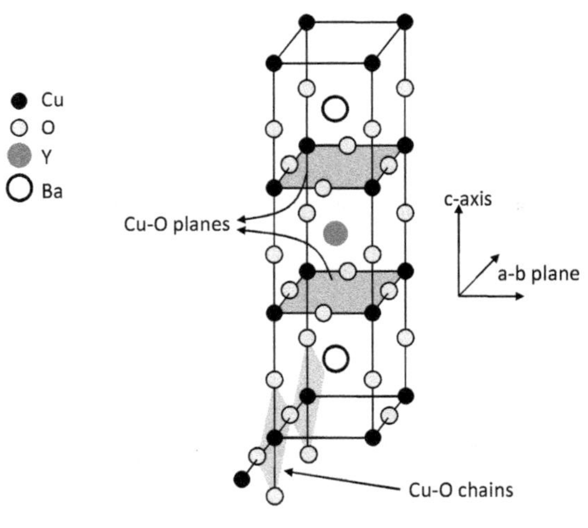

Fig. 8.21. Crystal structure of $YBa_2Cu_3O_7$

[2] At present the highest T_c of stoichiometrically formed materials under ambient pressure is 138 K attained in a mercury-based cuprate: $Hg_{0.8}Tl_{0.2}Ba_2Ca_2Cu_3O_{8+\delta}$.

The crystal anisotropy gives rise to significantly different carrier transport properties depending on the crystal axis. Carriers move freely in the CuO_2 planes while the interlayer direction is nearly insulating. Although its microscopic mechanism remains to be understood, the formation of binding pairs of charge carriers is again the key process of the high-T_c superconductivity. A profound difference between the conventional and high-T_c superconductors lies in the detailed electronic structures of the paired states. The total orbital angular momentum of a binding pair in conventional superconductors is zero so that it is called an s-wave state. On the other hand, the binding pairs in high-T_c superconductors are in a so-called d-wave state having non-zero angular momentum.

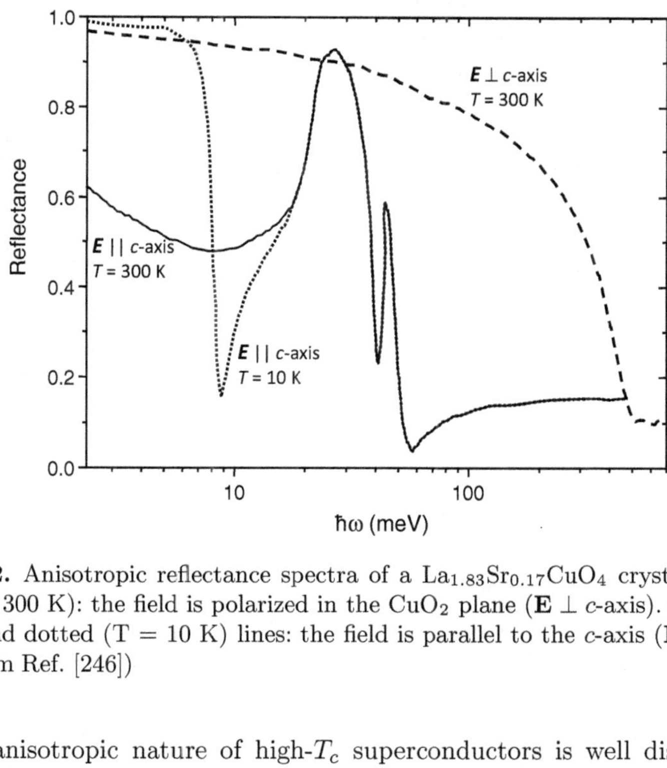

Fig. 8.22. Anisotropic reflectance spectra of a $La_{1.83}Sr_{0.17}CuO_4$ crystal. Dashed line (T = 300 K): the field is polarized in the CuO_2 plane (**E** $\perp$ c-axis). Solid (T = 300 K) and dotted (T = 10 K) lines: the field is parallel to the c-axis (**E** $\parallel$ c-axis). (Data from Ref. [246])

The anisotropic nature of high-T_c superconductors is well displayed in Fig. 8.22 depicting the reflectance measurements on a $La_{1.83}Sr_{0.17}CuO_4$ crystal [246]. The CuO_2 plane behaves like a metal surface at room temperature so that the reflectance with **E** $\perp$ c-axis is close to unity in the THz region and drops sharply at the plasma frequency in the infrared region. On the contrary, the c-axis reflectance (**E** $\parallel$ c-axis) shows some strong features of interlayer lattice vibrations in the infrared region. The gradual decrease of the room-temperature reflectance with an increase of frequency in the low-frequency region (0-7 meV) signifies the contribution of c-axis carrier transport. Nevertheless, the overall c-axis response is characterized as insulator-like behavior.

As the material enters into a superconducting state, the c-axis spectrum in the THz region undergoes a drastic transformation. The reflectance becomes near unity in the low-frequency region and drops sharply around 7 meV. It gradually recovers to the normal state reflectance in the THz-infrared boundary region. This feature arises from a pronounced phenomenon of high-T_c superconductors, known as the Josephson plasma resonance.

We shall make a brief overview of the Josephson effect to get an insight into the Josephson plasma resonance. It is not only a very interesting phenomenon, but has also turned out to be an effective mechanism for generating coherent THz radiation. A solid-state THz source exploits the intrinsic Josephson junctions of $Bi_2Sr_2CaCu_2O_8$ to produce continuous-wave THz radiation with decent emission power [247]. The Josephson effect is a quantum tunneling phenomenon in which Cooper pairs pass through a thin insulating barrier separating two superconducting layers without resistance. The macroscopic coherence of the superconducting states reaches over the thin barrier preserving their long-range order. The phase difference between the two superconducting states is the driving force for the tunneling current. Figure 8.23 illustrates the physical system and the key electrical properties of the Josephson effect. The superconductor-insulator-superconductor tunnel junction is called a Josephson junction. The Josephson current is the quantum-tunneling induced supercurrent at zero bias. Assuming no magnetic field is present, the Josephson current density is written as

$$J_z = J_0 \sin \Delta\phi, \tag{8.39}$$

where $\Delta\phi = \phi_1 - \phi_2$

$$J_0 \sim \frac{e\hbar}{m^* d} |\psi|^2 \tag{8.40}$$

is the maximum current density.

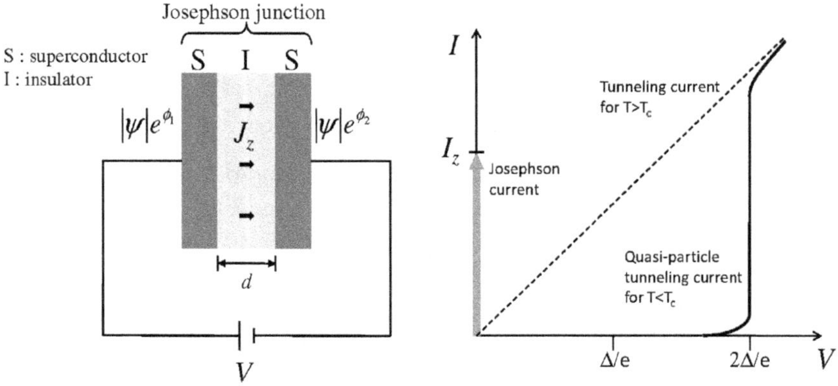

Fig. 8.23. Josephson junction and its I-V curve.

If a DC bias voltage V is applied across a Josephson junction, the relative phase difference varies in time, depending on the energy difference across the barrier. The time evolution of the quantum states is written as

$$\psi_1 = |\psi| e^{\frac{iE_1 t}{\hbar}} \quad \text{and} \quad \psi_2 = |\psi| e^{\frac{iE_2 t}{\hbar}}, \tag{8.41}$$

where $E_1 - E_2 = 2eV$. Consequently, the phase difference is given as

$$\Delta\phi = \frac{E_1 - E_2}{\hbar} t = \frac{2eV}{\hbar} t, \tag{8.42}$$

and the tunneling current $J_0 \sin \Delta\phi$ oscillates with the frequency $\nu = \frac{2eV}{\hbar}$.

If an oscillating field,

$$E_z(t) = \frac{V(t)}{d} = E_z^0 e^{-i\omega t}, \tag{8.43}$$

is applied, the phase difference also evolves in time accordingly. The rate of change in the phase difference is given as

$$\frac{\partial}{\partial t} \Delta\varphi = \frac{2edE_z}{\hbar}. \tag{8.44}$$

Assuming the phase variation is small ($\Delta\phi \ll 1$), we can write the current density as

$$J_z(t) = J_0 \Delta\phi(t). \tag{8.45}$$

The time derivative of Eq. 8.45 leads to the conductivity

$$\sigma(\omega) = \frac{2edJ_0}{-i\hbar\omega} = i\epsilon_0 \frac{\omega_{JP}^2}{\omega}, \tag{8.46}$$

where

$$\omega_{JP} = \sqrt{\frac{2eJ_0 d}{\hbar\epsilon_0}} \sim \sqrt{\frac{2e^2|\psi|^2}{\epsilon_0 m^*}} \tag{8.47}$$

is the Josephson plasma resonance frequency. The dielectric function of the Josephson junction is expressed as

$$\epsilon(\omega) = \epsilon_0 \left[1 - \frac{\omega_{JP}^2}{\omega^2} \right], \tag{8.48}$$

having the same form as the dielectric function of a plasma.

The layered structure of high-T_c superconductors can be considered as a stack of Josephson junctions consisting of superconducting CuO_2 planes and insulating barrier layers. The sharp drop in the c-axis reflectance of the superconducting state shown in Fig. 8.22 is caused by the Josephson plasma resonance in the $La_{1.83}Sr_{0.17}CuO_4$ crystal. The resonance frequency is significantly smaller than the in-plane plasma frequency because the pair concentration $|\psi|^2$ is smaller than the total carrier concentration and the c-axis effective mass m_c^* is larger than the in-plane effective mass m_{ab}^*.

Probing the Josephson plasmon mode is a sensitive measure for investigating the microscopic density distribution of Cooper pairs in CuO_2 planes [248]. Figure 8.24 shows the spectra of the real conductivity $\sigma_1(\omega)$ (top panels) and the imaginary part of the inverse dielectric function $-\Im\left[\frac{1}{\epsilon(\omega)}\right]$ (bottom panels) of the $La_{2-x}Sr_xCuO_4$ crystals for $x = 0.08$, 0.125, and 0.17, at $T < T_c$. An anomalous feature appears in the conductivity spectra for the crystal with $x = 0.125$: an additional absorption resonance develops near the plasma edge below T_c. The spectra of the inverse dielectric function are compared with the fits using a two fluid model. The two fluid model is a phenomenological description of carrier dynamics in a superconductor containing both paired and unpaired carriers. The c-axis dielectric function based on the two fluid model has the form,

$$\epsilon(\omega) = \epsilon_0 \left[\epsilon_\infty - \frac{\omega_{JP}^2}{\omega^2} - \frac{\omega_p^2}{\omega^2 + i\gamma\omega} \right], \tag{8.49}$$

where ϵ_∞ is the high frequency dielectric constant, ω_p is the plasma frequency of unpaired carriers, and γ is their scattering rate. In this model, $-\Im\left[\frac{1}{\epsilon(\omega)}\right]$ has the Josephson resonance peak at $\omega_0 = \omega_{JP}/\sqrt{\epsilon_\infty}$ with a Lorentzian lineshape. The spectrum of the inverse dielectric function for $x = 0.125$, however, is largely asymmetric, which is inconsistent with the two fluid model. It has been known that the normal-state charge distribution in CuO_2 planes is not

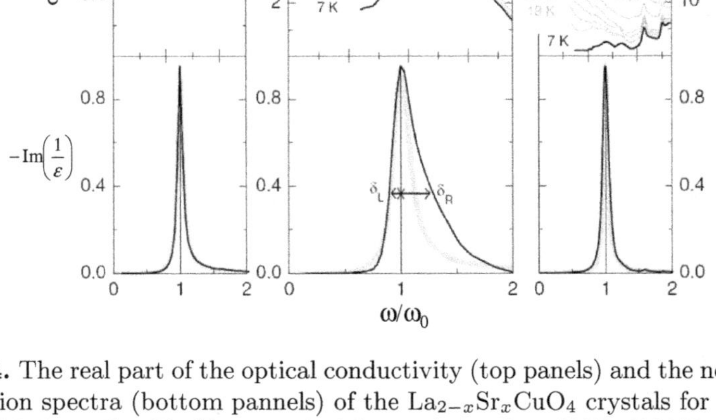

Fig. 8.24. The real part of the optical conductivity (top panels) and the normalized loss function spectra (bottom pannels) of the $La_{2-x}Sr_xCuO_4$ crystals for $x = 0.08$, 0.125, and 0.17, at $T < T_c$. (Reprinted with permission from [248]. ©2003, American Physical Society.

uniform, but no observation has been made on the superconducting carrier distribution. The pronounced asymmetry in the spectrum indicates that the in-plane superconducting carrier density is also spatially heterogeneous. A theoretical analysis shows that the spectral asymmetry is an effect of the presence of normal state or weak superconducting regions with a characteristic length scale of 10-20 nm in the CuO_2 planes.

A conspicuous and puzzling property of high-T_c superconductors is the formation of a pseudogap at temperature T* well above T_c. The pseudogap is a kind of partial energy gap whose existence depends on direction in momentum space. Between T_c and T^*, electron pairs and short-range phase correlations still do exist, yet the long-range phase coherence among the pairs vanishes. Complex conductivity measurements in the THz region reveal the presence of the short-range phase correlations in the pseudogap state [249]. The conductivity of the superconducting state has the form,

$$\sigma(\omega) = i\epsilon_0 \frac{\omega_p^2}{\omega} = i\frac{n_s e^2}{m^* \omega} = i\sigma_Q \frac{k_B T_\theta}{\hbar\omega}, \tag{8.50}$$

where n_s is the density of paired electrons,

$$\sigma_Q = \frac{e^2}{\hbar d} \tag{8.51}$$

is the quantum conductivity of multi-layer conductors with a layer spacing d, and

$$k_b T_\theta = \frac{\hbar^2 n_s d}{m^*} \tag{8.52}$$

is the phase-stiffness energy. The phase stiffness indicates that long-range coherence is imperative to keep the superconducting state. When no current is present, the phase of a superconducting state tends to keep a spatial uniformity. The phase-stiffness energy corresponds to the amount of energy required to force spatial fluctuations in the phase.

Eq. 8.50 indicates that the phase stiffness can be directly probed by conductivity measurements. Figure 8.25 shows the temperature-dependent phase stiffness of two $Bi_2Sr_2CaCu_2O_{8+\delta}$ samples (T_c = 33 K and 74 K) at several frequencies in the THz region. The phase stiffness is independent of frequency at low temperatures. A general tendency is that it weakens as temperature is increased, because the pair concentration n_s reduces. At high temperatures, it splits off for different frequencies, and its weakening becomes more severe at lower frequencies. Above the temperature at which the splitting occurs, the phase correlation time becomes finite and gets shorter as temperature is increased. Consequently, it becomes easier to exert phase fluctuations at a lower frequency. The frequency dependence of the pseudogap conductivity in the high-frequency region is consistent with Eq. 8.50, which indicates that there is no distinguishable difference between the phase fluctuations in the pseudogap state and in the superconducting state. On the contrary, the phase

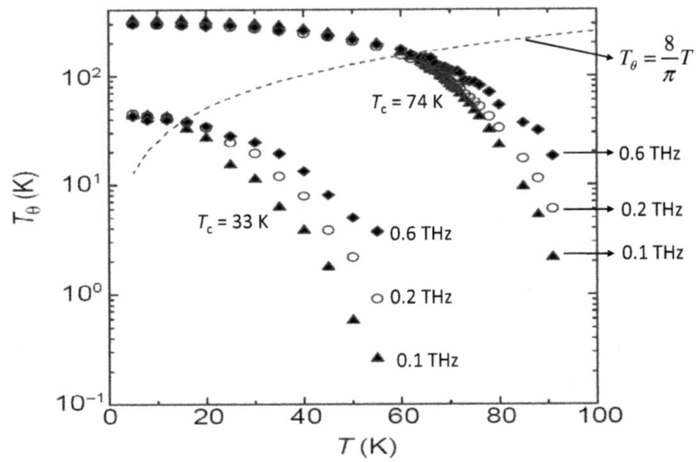

Fig. 8.25. Phase-stiffness temperature of two $Bi_2Sr_2CaCu_2O_{8+\delta}$ samples with $T_c =$ 33 K and 74 K at $\nu = 0.1$, 0.2, and 0.6 THz as a function of temperature. (Reprinted by permission from Macmillan Publishers Ltd: Nature [249], ©1999.)

correlations in the pseudogap state are not discernible at low frequencies as the conductivity of the pseudogap state converges to the normal-state DC conductivity. The dashed curve corresponds to the simple relation

$$T_\theta = \frac{8}{\pi}T, \tag{8.53}$$

which fits well with the points where the splittings occur. This is consistent with the Kosterlitz-Thouless-Berezinskii theory of two-dimensional melting, where the thermal fluctuations of unbound vortices, governing the phase correlation times, start to appear when temperature reaches $(\pi/8)T_\theta$.

Vortex dynamics is one of the principal microscopic phenomena of high-T_c superconductivity. A purely superconducting state excludes magnetic fields, which is called the Meissner effect. As the magnetic field increases, a type-II superconductor undergoes two transitions: (i) the onset of a mixed superconducting and normal state by the formation of vortices and (ii) the complete wipeout of the superconducting state. A vortex is a topological singularity of swirling electrical current carrying a quantized magnetic flux,

$$\Phi_0 = \frac{\hbar}{2e}. \tag{8.54}$$

Figure 8.26 shows the formation of a haxagonal vortex lattice in an $YBa_2Cu_3O_7$ crystal [250]. The size of the vortices is on a mesoscopic scale, $\sim$1 μm.

The characteristic time scale of vortex dynamics, such as the vortex relaxation time, is in the range of a picosecond, thus THz spectroscopy is a powerful method for investigating the dynamical properties of the vortices. A

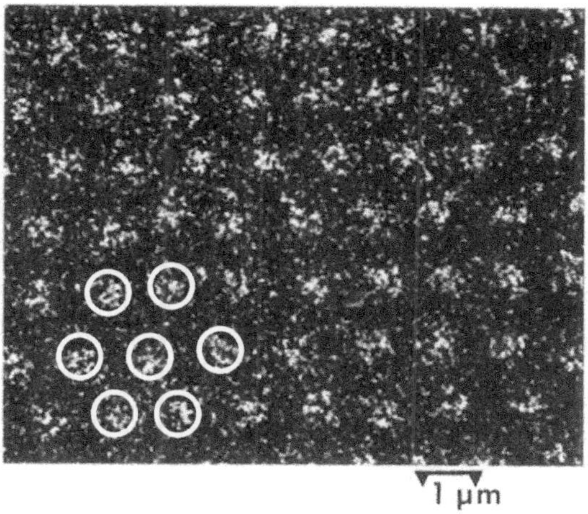

Fig. 8.26. Hexagonal vortex structure in $YBa_2Cu_3O_7$. (Reprinted with permission from [250]. ©1987, American Physical Society.)

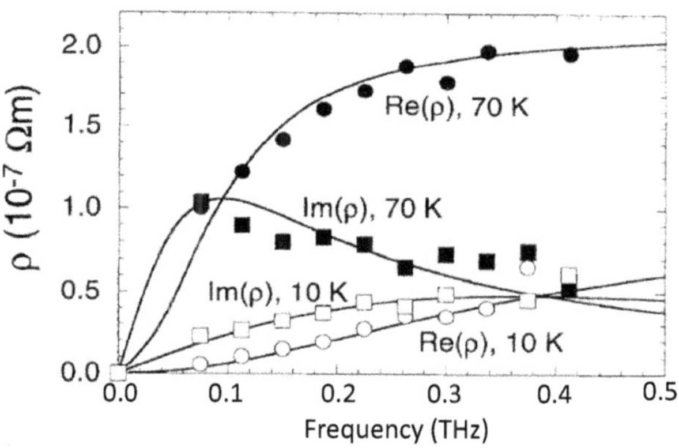

Fig. 8.27. Resistivity tensor component ρ_v^{xx} of an $YBa_2Cu_3O_{7-\delta}$ crystal at $T = 10$ and 70 K for $B = 6$ T versus frequency. The lines are fits to the data with Eq. 8.56. (Reprinted with permission from [251]. ©1995, American Physical Society.)

phenomenological model of the vortex system is that the vortices form a lattice of damped harmonic oscillators. In the presence of electric and magnetic fields, the motion of the vortices is balanced by several forces such as restoring, damping, dragging, and Lorentz forces. The equation of the momentum balance has the form

$$\eta \mathbf{v}_L + \frac{\kappa}{i\omega}\mathbf{v}_L = \left[\frac{n_s h}{2}\mathbf{v}_s - \alpha \mathbf{v}_L\right] \times \mathbf{e}_z, \tag{8.55}$$

where η is the damping coefficient, κ is the spring constant, α is the Magnus parameter, $\mathbf{v}_L$ is the velocity of the vortex with respect to the crystal lattice, $\mathbf{v}_s$ is the superconducting carrier velocity, and n_s is the density of paired electrons [251]. The vortex resistivity ρ_v is defined as $\mathbf{E}_v = \rho_v \mathbf{J}_s$, where $\mathbf{E}_v = \mathbf{B} \times \mathbf{v}_L$ is the vortex induced electric field and $\mathbf{J}_s = en_s\mathbf{v}_s$ is the superconducting current density. Due to the magnetic field, the superconducting carrier transport is anisotropic, and the resistivity becomes a second-rank tensor. Algebraic manipulations of the equations for $\mathbf{v}_L$, $\mathbf{v}_s$, and $\mathbf{J}_s$ yield the vortex resistivity tensor components:

$$\rho_v^{xx} = iX_v \frac{1 + i\omega/\Gamma}{(1 + i\omega/\Gamma)^2 - (\alpha\omega/\kappa)^2}, \tag{8.56}$$

$$\rho_v^{xy} = X_v \frac{\alpha\omega/\kappa}{(1 + i\omega/\Gamma)^2 - (\alpha\omega/\kappa)^2}, \tag{8.57}$$

where $X_v = \omega\Phi_0 B/\kappa$ and $\Gamma = \kappa/\eta$. Figure 8.27 shows the spectra of the vortex resistivity tensor component $\rho_v^{xx}(\omega)$ of an $YBa_2Cu_3O_{7-\delta}$ crystal at 10 and 70 K for $B = 6$ T. The solid lines are fits to the data with Eq. 8.56. The fitting results suggest that the strong anisotropy of the superconducting gap is an important factor in explaining the experimental measurements.

References

1. N. E. Tielking and R. R. Jones, Phys. Rev. A **52**, 1371 (1995).
2. J. Ahn, T. C. Weinacht, and P. H. Bucksbaum, Science **287**, 463 (2000).
3. J. Ahn, D. N. Hutchinson, C. Rangan, and P. H. Bucksbaum, Phys. Rev. Lett. **86**, 1179 (2001).
4. B. E. Cole, J. B. Williams, B. T. King, M. S. Sherwin, and C. R. Stanley, Nature **410**, 60 (2001).
5. K. B. Nordstrom, K. Johnsen, S. J. Allen, A.-P. Jauho, B. Birnir, J. Kono, T. Noda, H. Akiyama, and H. Sakaki, Phys. Rev. Lett. **81**, 457 (1998).
6. S. G. Carter, V. Birkedal, C. S. Wang, L. A. Coldren, A. V. Maslov, D. S. Citrin, and M. S. Sherwin, Science **310**, 651 (2005).
7. N. Kida, H. Murakami, and M. Tonouchi, in *Terahertz Optoelectronics*, edited by K. Sakai (Springer-Verlag, Berlin, 2005), Chap. 6, pp. 275–334.
8. M. Schall, M. Walther, and P. Uhd Jepsen, Phys. Rev. B **64**, 094301 (2001).
9. H. Harde, S. Keiding, and D. Grischkowsky, Phys. Rev. Lett. **66**, 1834 (1991).
10. T. Globus, D. Woolard, T. Khromova, T. Crowe, M. Bykhovskaia, B. Gelmont, J. Hesler, and A. Samuels, J. Bio. Phys. **29**, 89 (2003).
11. N. W. Ashcroft and N. D. Mermin, in *Solid State Physics* (Holt, Rinehart and Winston, New Yoik, 1976), Chap. 27, p. 553.
12. F. E. Doany, D. Grischkowsky, and C.-C. Chi, Appl. Phys. Lett. **50**, 460 (1987).
13. K. A. McIntosh, K. B. Nichols, S. Verghese, and E. R. Brown, Appl. Phys. Lett. **70**, 354 (1997).
14. S. Gupta, M. Y. Frankel, J. A. Valdmanis, J. F. Whitaker, and G. A. Mourou, Appl. Phys. Lett. **59**, 3276 (1991).
15. D. H. Auston, in *Ultrashort Laser Pulses: Generation and Applications*, 2nd ed., edited by W. Kaiser (Springer, Berlin, 1993), p. 188.
16. D. C. Look, D. C. Walters, , G. D. Robinson, M. G. M. J. R. Sizelove, and C. E. Stutz, J. Appl. Phys. **74**, 306 (1993).
17. L. Duvillaret, F. Garet, J.-F. Roux, and J.-L. Coutaz, IEEE J. Sel. Top. Quantum Electron. **7**, 615 (2001).
18. P. U. Jepsen, R. H. Jacobsen, and S. R. Keiding, J. Opt. Soc. Am B **13**, 2424 (1996).
19. M. Tani, S. Matsuura, K. Sakai, and S. Nakashima, Appl. Opt. **36**, 7853 (1997).
20. Y. Cai, I. Brener, J. Lopata, J. Wynn, L. Pfeiffer, and J. Federici, Appl. Phys. Lett. **71**, 2076 (1997).

21. P. R. Smith, D. H. Auston, and M. C. Nuss, IEEE J. Quantum Electron. **24**, 255 (1988).
22. J. V. Rudd and D. M. Mittleman, J. Opt. Soc. Am B **19**, 319 (2002).
23. D. Grischkowsky, S. Keiding, M. van Exter, and C. Fattinger, J. Opt. Soc. Am B **7**, 2006 (1990).
24. J. D. Jackson, in *Classical Electrodynamics*, 2nd ed. (Wiley, New York, 1975), pp. 278–282.
25. P. U. Jepsen, R. H. Jacobsen, and S. R. Keiding, Opt. Lett. **20**, 807 (1995).
26. J. T. Darrow, X.-C. Zhang, and D. H. Auston, IEEE J. Quantum Electron. **28**, 1607 (1992).
27. J. D. Jackson, in *Classical Electrodynamics*, 2nd ed. (Wiley, New York, 1975), pp. 17–22.
28. S. L. Teitel and J. W. Wilkins, IEEE Trans. Electron. Devices **ED-30**, 150 (1983).
29. A. J. Taylor, P. K. Benicewicz, and S. M. Young, Opt. Lett. **18**, 11340 (1993).
30. D. You, R. R. Jones, P. H. Bucksbaum, and D. R. Dykaar, Opt. Lett. **18**, 290 (1993).
31. S.-G. Park, M. R. Melloch, and A. M. Weiner, Appl. Phys. Lett. **73**, 3184 (1998).
32. R. W. Boyd, in *Nonlinear Optics* (Academic Press, Inc., San Diego, 1992), Chap. 1.
33. L. Xu, X.-C. Zhang, and D. H. Auston, Appl. Phys. Lett. **61**, 1784 (1992).
34. Y.-S. Lee, T. Meade, V. Perlin, H. Winful, T. B. Norris, and A. Galvanauskas, Appl. Phys. Lett. **76**, 2505 (2000).
35. A. Nahata, A. S. Weling, and T. F. Heinz, Appl. Phys. Lett. **69**, 2321 (1996).
36. M. Nagaia, K. Tanaka, H. Ohtake, T. Bessho, T. Sugiura, T. Hirosumi, and M. Yoshida, Appl. Phys. Lett. **85**, 3974 (2004).
37. G. Gallot, J. Zhang, R. W. McGowan, T.-I. Jeon, and D. Grischkowsky, Appl. Phys. Lett. **74**, 3450 (1999).
38. Y. J. Ding and I. B. Zotova, Opt. Quant. Electron. **32**, 531 (2000).
39. A. Leitenstorfer, S. Hunsche, J. Shah, M. C. Nuss, and W. H. Knox, Appl. Phys. Lett. **74**, 1516 (1999).
40. A. Debernardi, Phys. Rev. B **57**, 12847 (1998).
41. N. Kuroda, O. Ueno, and Y. Nishina, Phys. Rev. B **35**, 3860 (1987).
42. S. Kojima, N. Tsumura, H. Kitahara, M. W. Takeda, and S. Nishizawa, Jpn. J. Appl. Phys. **41**, 7033 (2002).
43. S. Kojima, H. Kitahara, S. Nishizawa, and M. W. Takeda, Jpn. J. Appl. Phys. **42**, 6238 (2002).
44. R. W. Boyd, in *Nonlinear Optics* (Academic Press, Inc., San Diego, 1992), Chap. 10.
45. P. Y. Han and X.-C. Zhang, Appl. Phys. Lett. **73**, 3049 (1998).
46. R. Huber, A. Brodschelm, F. Tauser, and A. Leitenstorfer, Appl. Phys. Lett. **76**, 3191 (2000).
47. K. Liu, J. Xu, and X.-C. Zhang, Appl. Phys. Lett. **85**, 863 (2004).
48. S. Kono, M. Tani, P. Gu, and K. Sakai, Appl. Phys. Lett. **77**, 4104 (2000).
49. Y. C. Shen, P. C. Upadhya, H. E. Beere, E. H. Linfield, A. G. Davies, I. S. Gregory, C. Baker, W. R. Tribe, and M. J. Evans, Appl. Phys. Lett. **85**, 164 (2004).
50. Y. Shen, T. Watanabe, D. A. Arena, C.-C. Kao, J. B. Murphy, T. Y. Tsang, X. J. Wang, and G. L. Carr, Phys. Rev. Lett. **99**, 043901 (2007).

51. G. L. Carr, M. C. Martin, W. R. McKinney, K. Jordan, G. R. Neil, and G. P. Williams, Nature **420**, 153 (2002).
52. M. Abo-Bakr, J. Feikes, K. Holldack, G. Wüstefeld, and H.-W. Hübers, Phys Rev. Lett. **88**, 254801 (2002).
53. F. Wang, D. Cheever, M. Farkhondeh, W. Franklin, E. Ihloff, J. van der Laan, B. McAllister, R. Milner, C. Tschalaer, D. Wang, D. F. Wang, A. Zolfaghari, T. Zwart, G. L. Carr, B. Podobedov, and F. Sannibale, Phys. Rev. Lett. **96**, 064801 (2006).
54. S. Gu and M. Tani, in *Terahertz Optoelectronics*, edited by K. Sakai (Springer, Berlin, 2005), pp. 63–97.
55. H. Ohtake, S. Ono, and N. Sarukura, in *Terahertz Optoelectronics*, edited by K. Sakai (Springer, Berlin, 2005), pp. 99–116.
56. A. G. Stepanov, J. Hebling, and J. Kuhl, Appl. Phys. Lett. **83**, 3000 (2003).
57. K.-L. Yeh, M. C. Hoffmann, J. Hebling, and K. A. Nelson, Appl. Phys. Lett. **90**, 1121 (2007).
58. H. Hamster, A. Sullivan, S. Gordon, W. White, and R. W. Falcone, Phys. Rev. Lett. **71**, 2725 (1993).
59. T. Löffler, F. Jacob, and H. G. Roskos, Appl. Phys. Lett. **77**, 453 (2000).
60. D. J. Cook and R. M. Hochstrasser, Opt. Lett. **25**, 1210 (2000).
61. M. Kress, T. Löffler, S. Eden, M. Thomson, and H. G. Roskos, Opt. Lett. **29**, 1120 (2004).
62. T. Bartel, P. Gaal, K. Reimann, M. Woerner, and T. Elsaesser, Opt. Lett. **30**, 2805 (2005).
63. X. Xie, J. Dai, and X.-C. Zhang, Phys. Rev. Lett. **96**, 075005 (2006).
64. K.-Y. Kim, J. H. Glownia, A. J. Taylor, and G. Rodriguez, Opt. Express **15**, 4577 (2007).
65. Y. S. Lee, T. Meade, T. B. Norris, and A. Galvanauskas, Appl. Phys. Lett. **78**, 3583 (2001).
66. W. C. Hurlbut, B. J. Norton, N. Amer, and Y.-S. Lee, J. Opt. Soc. Am B **23**, 90 (2006).
67. Y.-S. Lee, W. C. Hurlbut, K. L. Vodopyanov, M. M. Fejer, and V. G. Kozlov, Appl. Phys. Lett. **89**, 1104 (2006).
68. K. L. Vodopyanov, M. M. Fejer, X. Yu, J. S. Harris, Y.-S. Lee, W. C. Hurlbut, V. G. Kozlov, D. Bliss, and C. Lynch, Appl. Phys. Lett. **89**, 1119 (2006).
69. K. L. Vodopyanov, Opt. Express **14**, 2263 (2006).
70. Y.-S. Lee, N. Amer, and W. C. Hurlbut, Appl. Phys. Lett. **82**, 170 (2003).
71. Y.-S. Lee, T. Meade, M. Decamp, T. B. Norris, and A. Galvanauskas, Appl. Phys. Lett. **77**, 1244 (2000).
72. J. R. Danielson, N. Amer, and Y.-S. Lee, Appl. Phys. Lett. **89**, 1118 (2006).
73. Y. Liu, S.-G. Park, and A. M. Weiner, IEEE J. Sel. Top. Quantum Electron. **2**, 709 (1996).
74. J. Y. Sohn, Y. H. Ahn, D. J. Park, E. Oh, and D. S. Kim, Appl. Phys. Lett. **81**, 13 (2002).
75. J. Ahn, A. V. Efimov, R. D. Averitt, and A. T. Taylor, Opt. Express **11**, 2486 (2003).
76. A. G. Stepanov, J. Hebling, and J. Kuhl, Opt. Express **12**, 4650 (2004).
77. E. R. Brown, F. W. Smith, and K. A. McIntosh, J. Appl. Phys. **73**, 1480 (1993).
78. S. Verghese, K. A. McIntosh, and E. R. Brown, IEEE Trans. Microwave Theory and Tech. **45**, 1301 (1997).

79. S. Matsuura, M. Tani, and K. Sakai, Appl. Phys. Lett. **70**, 559 (1997).
80. M. Kominami, D. M. Pozar, and D. H. Schaubert, IEEE Trans. Antennas Propag. **33**, 600 (1985).
81. S. M. Duffy, S. Verghese, K. A. McIntosh, A. Jackson, A. C. Gossard, and S. Matsuura, IEEE Trans. Microwave Theory and Tech. **49**, 1032 (2001).
82. S. Matsuura and H. Ito, in *Terahertz Optoelectronics*, edited by K. Sakai (Springer, Berlin, 2005), pp. 173–179.
83. S. Matsuura and H. Ito, in *Terahertz Optoelectronics*, edited by K. Sakai (Springer, Berlin, 2005), pp. 179–186.
84. R. W. Boyd, in *Nonlinear Optics* (Academic Press, Inc., San Diego, 1992), Chap. 2.
85. N. B. Singh, T. B. Norris, T. Buma, R. N. Sing, M. Gottlieb, D. Suhre, and J. J. Hawkins, Opt. Engin. **45**, 094002 (2006).
86. K. L. Vodopyanov and L. A. Kulevskii, Opt. Commun. **118**, 375 (1995).
87. W. Shi, Y. J. Ding, N. Fernelius, and K. L. Vodopyanov, Opt. Lett. **27**, 1454 (2002).
88. J. E. Schaar, K. L. Vodopyanov, and M. M. Fejer, Opt. Lett. **32**, 1284 (2007).
89. K. Imaia, K. Kawase, J. ichi Shikata, , H. Minamide, and H. Ito, Appl. Phys. Lett. **78**, 1026 (2001).
90. K. Kawase, J. Shikata, K. Imai, and H. Ito, Appl. Phys. Lett. **78**, 2819 (2001).
91. T. J. Edwards, D. Walsh, M. B. Spurr, C. F. Rae, M. H. Dunn, and P. G. Browne, Opt. Express **14**, 1582 (2006).
92. G. W. Chantry, in *Long-wave Optics* (Academic Press, Inc., London, 1984), Vol. 2, Chap. 6, pp. 574–576.
93. E. Bründermann, D. Chamberlin, and E. Haller, Infrared Phys. Tech. **40**, 141 (1999).
94. A. Bergner, U. Heugen, E. Bündermann, G. Schwaab, M. Havenith, D. R. Chamberlin, and E. E. Haller, Rev. Sci. Instrum. **76**, 063110 (2005).
95. A. Maestrini, J. S. Ward, J. J. Gill, H. S. Javadi, E. Schlecht, C. Tripon-Canseliet, G. Chattopadhyay, and I. Mehdi, IEEE Trans. Microw. Theoy Tech. **53**, 2835 (2005).
96. M. A. Belkin, J. A. Fan, S. Hormoz, F. Capasso, S. P. Khanna, M. Lachab, A. G. Davies, and E. H. Linfield, Optics Express **16**, 3242 (2008).
97. B. S. Williams, H. Callebaut, S. Kumar, Q. Hu, and J. L. Reno, Appl. Phys. Lett. **82**, 1015 (2003).
98. R. Köhler, A. Tredicucci, F. Beltram, H. E. Beere, E. H. Linfield, A. G. Davies, D. A. Ritchie, R. C. Iotti, and F. Rossi, Nature **417**, 156 (2002).
99. S. E. Tsimring, in *Electron Beams and Microwave Vacuum Electronics* (John Wiley & Sons, Inc., Hoboken, New Jersey, 2007), Chap. 8.
100. G. Kozlov and A. Volkov, in *Millimeter and Submillimeter Wave Spectroscopy of Solids*, edited by G. Grüner (Springer-Verlag, Berlin, 1998), Chap. 3.
101. C. A. Brau, Science **239**, 1115 (1988).
102. P. L. Richards, J. Appl. Phys. **76**, 1 (1994).
103. A. Hossain and M. H. Rashid, IEEE Trans on Industrial Applications **27**, 825 (1991).
104. H. P. Beerman, IEEE Trans on Electron Devices **16**, 554 (1969).
105. C. Côté and A. W. DeSilva, Rev. Sci. Inst. **67**, 4146 (1996).
106. M. J. E. Golay, Rev. Sci. Inst. **20**, 816 (1949).
107. J.-B. Chevrier, K. Baert, T. Slater, and A. Verbist, Microsyst. Technol. **1**, 71 (1995).

108. T. W. Kenny, J. K. Reynolds, J. A. Podosek, E. C. Vote, L. M. Miller, H. K. Rockstad, and W. J. Kaiser, Rev. Sci. Inst. **67**, 112 (1996).
109. O. Ajakaiye, J. Grade, C. Shin, and T. Kenny, Sensors and Actuators A **134**, 575 (2007).
110. P. Yagoubov, M. Kroug, H. Merkel, E. Kollberg, G. Gol'tsman, S. Svechnikov, and E. Gershenzon, Appl. Phys. Lett. **73**, 2814 (1998).
111. E. H. Putley, Appl. Opt. **4**, 649 (1965).
112. A. Skalare, W. R. McGrath, B. Bumble, H. G. Leduc, P. J. Burke, A. A. Verheijen, R. J. Schoelkopf, and D. E. Prober, Appl. Phys. Lett. **68**, 1558 (1996).
113. J. D. D. J. R. Birch and J. Lesurf, Infrared Physics **21**, 225 (1981).
114. J. R. Birch, Infrared Physics **24**, 573 (1984).
115. J. R. Birch, Infrared Physics **30**, 195 (1990).
116. J. R. Birch, Infrared Physics **33**, 33 (1992).
117. M. Naftaly and R. E. Miles, Proceedings of the IEEE **95**, 1658 (2007).
118. Y.-S. Jin, G.-J. Kim, and S.-G. Jeon, J. Korean Phys. Soc. **49**, 513 (2006).
119. J. Dai, J. Zhang, W. Zhang, and D. Grischkowsky, J. Opt. Soc. Am B **21**, 1379 (2004).
120. K. Winer and M. Cardona, Phys. Rev. B **35**, 8189 (1987).
121. E. V. Loewenstein, D. R. Smith, and R. L. Morgan, Appl. Opt. **12**, 398 (1973).
122. R. H. Stolen, Phys. Rev. B **11**, 767 (1975).
123. L. Harris and P. Fowler, J. Opt. Soc. Am **51**, 164 (1961).
124. K. Bane and G. Stupakov, in *Proceeding of EPAC 2006* (JACoW, Edinburgh, 2006), pp. 2955–2957.
125. E. D. Walsby, S. Wang, J. Xu, T. Yuan, R. Blaikie, S. M. Durbin, X.-C. Zhang, and C. D. R. S., J. Vac. Sci. Technol. B **20**, 2780 (2002).
126. J. C. Wiltse, in *Proceedings of the SPIE, Volume 5411*, edited by R. J. Hwu and D. L. Woolard (SPIE, Bellingham, 2004), Vol. 5411, pp. 127–135.
127. K. R. Armstrong and F. J. Low, Appl. Opt. **13**, 425 (1974).
128. K. Kawase and N. Hiromoto, Appl. Opt. **37**, 1862 (1998).
129. I. Hosako, Appl. Opt. **42**, 4045 (2003).
130. I. Hosako, Appl. Opt. **44**, 3769 (2005).
131. R. Ulrich, Infrared Physics **7**, 37 (1967).
132. D. W. Porterfield, J. L. Hesler, R. Densing, E. R. Mueller, T. W. Crowe, and R. M. Weikle II, Appl. Opt. **33**, 6046 (1994).
133. J.-B. Masson and G. Gallot, Opt. Lett. **31**, 265 (2006).
134. C.-F. Hsieh, R.-P. Pan, T.-T. Tang, H.-L. Chen, and C.-L. Pan, Opt. Lett. **31**, 1112 (2006).
135. N. Amer, W. C. Hurlbut, B. J. Norton, Y.-S. Lee, and T. B. Norris, Appl. Phys. Lett. **87**, 1111 (2005).
136. G. Gallot, S. P. Jamison, R. W. McGowan, and D. Grischkowsky, J. Opt. Soc. Am B **17**, 851 (2000).
137. S. P. Jamison, R. W. McGowan, and D. Grischkowsky, Appl. Phys. Lett. **76**, 1987 (2000).
138. L.-J. Chen, H.-W. Chen, T.-F. Kao, J.-Y. Lu, and C.-K. Sun, Opt. Lett. **31**, 308 (2006).
139. R. Mendis and D. Grischkowsky, Opt. Lett. **26**, 846 (2001).
140. K. Wang and D. M. Mittleman, Nature **432**, 376 (2004).
141. T.-I. Jeon, J. Zhang, and D. Grischkowsky, Appl. Phys. Lett. **86**, 1904 (2005).

142. K. Wang and D. M. Mittleman, J. Opt. Soc. Am B **22**, 2001 (2005).
143. J. A. Deibel, K. Wang, M. D. Escarra, and D. Mittleman, Opt. Express **14**, 279 (2006).
144. V. G. Veselago, Sov. Phys. Usp. **10**, 509 (1968).
145. D. R. Smith, W. J. Padilla, D. C. Vier, S. C. Nemat-Nasser, and S. Schultz, Phys. Rev. Lett. **84**, 4184 (2000).
146. M. C. K. Wiltshire, J. B. Pendry, I. R. Young, D. J. Larkman, D. J. Gilderdale, and J. V. Hajnal, Science **291**, 849 (2001).
147. C. Enkrich, M. Wegener, S. Linden, S. Burger, L. Zschiedrich, F. Schmidt, J. F. Zhou, T. Koschny, and C. M. Soukoulis, Phys. Rev. Lett. **95**, 203901 (2005).
148. T. J. Yen, W. J. Padilla, N. Fang, D. C. Vier, D. R. Smith, J. B. Pendry, D. N. Basov, and X. Zhang, Science **303**, 1494 (2004).
149. J. B. Pendry, Phys. Rev. Lett. **85**, 3966 (2000).
150. R. A. Shelby, D. R. Smith, and S. Schultz, Science **292**, 77 (2001).
151. H.-T. Chen, W. J. Padilla, J. M. O. Zide, A. C. Gossard, A. J. Taylor, and R. D. Averitt, Nature **444**, 597 (2006).
152. N. Amer, W. C. Hurlbut, B. J. Norton, Y.-S. Lee, S. L. Etringer, and B. K. Paul, Appl. Opt. **45**, 1857 (2006).
153. H. Nemec, P. Kuzel, F. Garet, and L. Duvillaret, Appl. Opt. **43**, 1965 (2004).
154. N. Jukam and M. S. Sherwin, Appl. Phys. Lett. **83**, 21 (2003).
155. Y. Zhao and D. Grischkowsky, Opt. Lett. **31**, 1534 (2006).
156. Y. Zhao and D. Grischkowsky, IEEE Trans. Microw. Theory Tech. **55**, 656 (2007).
157. T. Prasad, V. L. Colvin, Z. Jian, and D. M. Mittleman, Opt. Lett. **32**, 683 (2007).
158. H. Han, H. Park, M. Cho, and J. Kim, Appl. Phys. Lett. **80**, 2634 (2002).
159. E. Özbay, E. Michel, G. Tuttle, R. Biswas, K. M. Ho, J. Bostak, and D. M. Bloom, Opt. Lett. **19**, 1155 (1994).
160. M. C. Wanke, O. Lehmann, K. Müller, Q. Wen, and M. Stuke, Science **275**, 1284 (1997).
161. K. Takagi, K. Seno, and A. Kawasaki, Appl. Phys. Lett. **85**, 3681 (2004).
162. W. L. Barnes, A. Dereux, and T. W. Ebbesen, Nature **424**, 824 (2003).
163. J. F. O'Hara, R. D. Averitt, and A. J. Taylor, Opt. Express **13**, 6117 (2005).
164. T.-I. Jeon and D. Grischkowsky, Appl. Phys. Lett. **88**, 1113 (2006).
165. J. B. Pendry, L. Martín-Moreno, and F. J. Garcia-Vidal, Science **305**, 847 (2004).
166. T. W. Ebbesen, H. J. Lezec, H. F. Ghaemi, T. Thio, and P. A. Wolff, Nature **391**, 667 (1998).
167. D. Qu and D. Grischkowsky, Phys. Rev. Lett. **93**, 196804 (2004).
168. H. Cao and A. Nahata, Opt. Express **12**, 1004 (2004).
169. T. Matsui, A. Agrawal, A. Nahata, and Z. V. Vardeny, Nature **446**, 517 (2007).
170. J. W. Lee, M. A. Seo, D. H. Kang, K. S. Khim, S. C. Jeoung, and D. S. Kim, Phys. Rev. Lett. **99**, 137401 (2007).
171. F. J. García-Vidal, E. Moreno, J. A. Porto, and L. Martín-Moreno, Phys. Rev. Lett. **95**, 103901 (2005).
172. K. Ishihara, K. Ohashi, T. Ikari, H. Minamide, H. Yokoyama, J.-I. Shikata, and H. Ito, Appl. Phys. Lett. **89**, 1120 (2006).
173. S. A. Maier, S. R. Andrews, L. Martín-Moreno, and F. J. García-Vidal, Phys. Rev. Lett. **97**, 176805 (2006).

174. T. Feurer, J. C. Vaughan, and K. A. Nelson, Science **299**, 374 (2003).
175. N. S. Stoyanov, T. Feurer, D. W. Ward, and K. A. Nelson, Appl. Phys. Lett. **82**, 674 (2003).
176. H. Maeda, D. V. L. Norum, and T. F. Gallagher, Science **307**, 1757 (2005).
177. R. R. Jones, D. You, and P. H. Bucksbaum, Phys. Rev. Lett. **70**, 1236 (1993).
178. C. Raman, C. W. S. Conover, C. I. Sukenik, and P. H. Bucksbaum, Phys. Rev. Lett. **76**, 2436 (1996).
179. A. Wetzels, A. Gürtler, H. G. Muller, and L. D. Noordam, Eur. Phys. J. D **14**, 157 (2001).
180. M. J. Molina and F. S. Rowland, Nature **249**, 810 (1974).
181. J. C. Farman, B. G. Gardiner, and J. D. Shanklin, Nature **315**, 207 (1985).
182. S. Solomon, R. R. Garcia, F. S. Rowland, and D. J. Wuebbles, Nature **321**, 755 (1986).
183. S. Albert, D. T. Petkie, R. P. A. Bettens, S. P. Belov, and F. C. De Lucia, Anal. Chem. News & Features **70**, 719A (1998).
184. C. D. Ball and F. C. De Lucia, Phys. Rev. Lett. **81**, 305 (1998).
185. H. Rabitz, Annu. Rev. Phys. Chem. **25**, 155 (1974).
186. J. C. Pearson, L. C. Oesterling, E. Herbst, and F. C. de Lucia, Phys. Rev. Lett. **75**, 2940 (1995).
187. J. W. Waters, L. Froidevaux, R. S. Harwood, R. F. Jarnot, H. M. Pickett, W. G. Read, P. H. Siegel, R. E. Cofield, M. J. Filipiak, D. A. Flower, J. R. Holden, G. K. Lau, N. J. Livesey, G. L. Manney, H. C. Pumphrey, M. L. Santee, D. L. Wu, D. T. Cuddy, R. R. Lay, M. S. Loo, V. S. Perun, M. J. Schwartz, P. C. Stek, R. P. Thurstans, M. A. Boyles, K. M. Chandra, M. C. Chavez, G.-S. Chen, B. V. Chudasama, R. Dodge, R. A. Fuller, M. A. Girard, J. H. Jiang, Y. Jiang, B. W. Knosp, R. C. Labelle, J. C. Lam, A. K. Lee, D. Miller, J. E. Oswald, N. C. Patel, D. M. Pukala, O. Quintero, D. M. Scaff, W. Vansnyder, M. C. Tope, P. A. Wagner, and M. J. Walch, IEEE Transactions on Geoscience and Remote Sensing **44**, 1075 (2006).
188. *Herschel Observers' Manual, Version 2.1* (ESA, Herschel Space Center, 2007), Chap. 1.
189. *HIFI Observers' Manual, Version 1.1* (ESA, Herschel Space Center, 2007), Chap. 2.
190. *PACS Observers' Manual, Version 1.5* (ESA, Herschel Space Center, 2007), Chap. 2.
191. *SPIRE Observers' Manual, Version 1.2* (ESA, Herschel Space Center, 2007), Chap. 2.
192. M. Nagai, H. Yada, T. Arikawa, and K. Tanaka, Int. J. Infrared Millimet. Waves **27**, 505 (2006).
193. J. Xu, K. W. Plaxco, and S. J. Allen, J. Chem. Phys. **124**, 036101 (2006).
194. L. Thrane, R. H. Jacobsen, P. U. Jepsen, and S. R. Keiding, Chem. Phys. Lett. **240**, 330 (1995).
195. O. Esenturk, A. Evans, and E. J. Heilweil, Chem. Phys. Lett. **442**, 71 (2007).
196. B. M. Fischer, M. Walther, and P. U. Jepsen, Phys. Med. Biol. **47**, 3807 (2002).
197. Y. C. Shen, P. C. Upadhya, E. H. Linfield, and A. G. Davies, Appl. Phys. Lett. **82**, 2350 (2003).
198. M. Heyden, E. Bründermann, U. Heugen, G. Niehues, U., D. M. Leitner, and M. Havenith, J. Am. Chem. Soc. **130**, 5773 (2008).
199. M. R. Leahy-Hoppa, M. J. Fitch, X. Zheng, L. M. Hayden, and R. Osiander, Chem. Phys. Lett. **434**, 227 (2007).

200. H.-B. Liu, Z. H., N. Karpowicz, Y. Chen, and X.-C. Zhang, Proc. IEEE **95**, 1514 (2007).
201. C. Strachan, T. Rades, D. A. Newnham, K. C. Gordon, M. Pepper, and P. F. Taday, Chem. Phys. Lett. **390**, 20 (2004).
202. P. F. Taday, Phil. Trans. R. Soc. Lond. A **362**, 351 (2004).
203. K. Kawase, Y. Ogawa, Y. Watanabe, and H. Inoue, Opt. Express **11**, 2549 (2003).
204. A. Markelz, S. Whitmire, J. Hillebrecht, and R. Birge, Phys. Med. Biol. **47**, 3797 (2002).
205. M. Havenith-Newen, Nachrichten aus der Chemie **55**, 1090 (2007).
206. C. Zhang and S. M. Durbin, J. Phys. Chem. B **110**, 23607 (2006).
207. C. Zhang, T. E., A. K. Ramdas, A. M. Weiner, and S. M. Durbin, J. Phys. Chem. B **108**, 10077 (2004).
208. S. Ebbinghaus, S. J. Kim, M. Heyden, X. Yu, U. Heugen, M. Gruebele, D. M. Leitner, and M. Havenith, PNAS **104**, 20749 (2007).
209. S. Ebbinghaus, S. J. Kim, M. Heyden, X. Yu, M. Gruebele, D. M. Leitner, and M. Havenith, J. Am. Chem. Soc. **130**, 2374 (2008).
210. P. Haring Bolivar, M. Brucherseifer, M. Nagel, H. Kurz, A. Bosserhoff, and R. Büttner, Phys. Med. Biol. **47**, 3815 (2002).
211. A. Dobroiu, C. Otani, and K. Kawase, Meas. Sci. Technol. **17**, R161 (2006).
212. D. M. Mittleman, R. H. Jacobsen, and M. C. Nuss, IEEE J. Sel. Top. Quantum Electron. **2**, 679 (1996).
213. D. M. Mittleman, M. Gupta, R. Neelamani, R. G. Baraniuk, J. V. Rudd, and M. Koch, Appl. Phys. B **68**, 1085 (1999).
214. Q. Wu, T. D. Hewitt, and X.-C. Zhang, Appl. Phys. Lett. **69**, 1026 (1996).
215. M. Usami, T. Iwamoto, R. Fukasawa, M. Tani, M. Watanabe, and K. Sakai, Phys. Med. Biol. **47**, 3749 (2002).
216. X.-C. Zhang, Phil. Trans. R. Soc. Lond. A **362**, 283 (2004).
217. D. M. Mittleman, S. Hunsche, L. Boivin, and M. C. Nuss, Opt. Lett. **22**, 904 (1997).
218. H. Zhong, J. Xu, X. Xie, T. Yuan, R. Reightler, E. Eric Madaras, and X.-C. Zhang, IEEE Sensors J. **5**, 203 (2005).
219. W. Withayachumnankul, G. M. Png, X. Yin, S. Atakaramians, I. Jones, H. Lin, B. S. Y. Ung, J. Balakrishnan, B. W.-H. Ng, B. Ferguson, S. P. Mickan, B. M. Fischer, and D. Abbott, Proc. IEEE **95**, 1528 (2007).
220. K. J. Siebert, H. Quast, R. Leonhardt, T. Löffler, M. Thomson, T. Bauer, H. G. Roskos, and S. Czasch, Appl. Phys. Lett. **80**, 3003 (2002).
221. I. S. Gregory, W. R. Tribe, C. Baker, B. E. Cole, M. J. Evans, L. Spencer, M. Pepper, and M. Missous, Appl. Phys. Lett. **86**, 204104 (2005).
222. A. Dobroiu, M. Yamashita, Y. N. Ohshima, Y. Morita, C. Otani, and K. Kawase, Appl. Opt. **43**, 5637 (2004).
223. S. Ariyoshi, C. Otani, A. Dobroiu, H. Sato, K. Kawase, H. M. Shimizu, T. Taino, and H. Matsuo, Appl. Phys. Lett. **88**, 203503 (2006).
224. A. W. M. Lee and Q. Hu, Opt. Lett. **30**, 2563 (2005).
225. B. N. Behnken, G. Karunasiri, D. R. Chamberlin, P. R. Robrish, and J. Faist, Opt. Lett. **33**, 440 (2008).
226. A. W. M. Lee, Q. Qin, S. Kumar, B. S. Williams, Q. Hu, and J. L. Reno, Appl. Phys. Lett. **89**, 141125 (2006).
227. N. Karpowicz, H. Zhong, C. Zhang, K.-I. Lin, J.-S. Hwang, J. Xu, and X.-C. Zhang, Appl. Phys. Lett. **86**, 054105 (2005).

228. R. Appleby, Phil. Trans. R. Soc. Lond. A **362**, 379 (2004).
229. P. Helistö, A. Luukanen, L. Grönberg, J. S. Penttilä, H. Seppä, H. Sipola, C Dietlein, and E. Grossman, in *Proceedings of the 1st European Microwave Integrated Circuits Conference* (Horizon House Publications Ltd, London, 2006), pp. 35–38.
230. E. Berry, J. Laser Appl. **15**, 192 (2003).
231. A. J. Fitzgerald, E. Berry, N. N. Zinov'ev, S. Homer-Vanniasinkam, R. E. Miles, J. M. Chamberlain, and M. A. Smith, J. Biol. Phys. **29**, 123 (2003).
232. E. Berry, A. J. Fitzgerald, N. N. Zinov'ev, G. C. Walker, S. Homer-Vanniasinkam, C. D. Sudworth, R. E. Miles, J. M. Chamberlain, and M. A. Smith, in *Proceedings of the SPIE* (SPIE, Bellingham, 2003), Vol. 5030, pp. 459–470.
233. V. P. Wallace, A. J. Fitzgerald, S. Shankar, N. Flangan, R. J. Pye, J. Cluff, and D. D. Arnone, British J. Dermatol. **151**, 424 (2004).
234. A. J. Fitzgerald, V. P. Wallace, M. Jimenez-Linan, L. Bobrow, R. J. Pye, A. D. Purushotha, and D. D. Arnone, Radiology **239**, 533 (2006).
235. Z. D. Taylor, R. S. Singh, C. D. Culjat, J. Y. Suen, W. S. Grundfest, H. Lee, and E. R. Brown, Opt. Lett. **33**, 1258 (2008).
236. D. A. Crawley, C. Longbottom, B. E. Cole, C. M. Ciesla, D. Arnone, V. P. Wallace, and M. Pepper, Caries Res. **37**, 352 (2003).
237. R. Huber, F. Tauser, A. Brodschelm, M. Bichler, G. Abstreiter, and A. Leitenstorfer, Nature **414**, 286 (2001).
238. J. Shan, F. Wang, E. Knoesel, M. Bonn, and T. F. Heinz, Phys. Rev. Lett. **90**, 247401 (2003).
239. M. F. Doty, B. T. King, M. S. Sherwin, and C. R. Stanley, Phys. Rev. B **71**, 201201 (2005).
240. R. A. Kaindl, M. A. Carnahan, D. Hägele, R. Lövenich, and D. S. Chemla, Nature **423**, 734 (2003).
241. J. R. Danielson, Y.-S. Lee, J. P. Prineas, J. T. Steiner, M. Kira, and S. W. Koch, Phys. Rev. Lett. **99**, 237401 (2007).
242. D. C. Mattis and J. Bardeen, Phys. Rev. **111**, 412 (1958).
243. A. V. Pronin, M. Dressel, A. Pimenov, A. Loidl, I. V. Roshchin, and L. H. Greene, Phys. Rev. B **57**, 14416 (1998).
244. R. A. Kaindl, M. A. Carnahan, J. Orenstein, D. S. Chemla, H. M. Christen, H.-Y. Zhai, M. Paranthaman, and D. H. Lowndes, Phys. Rev. Lett. **88**, 027003 (2002).
245. J. Demsar, R. D. Averitt, A. J. Taylor, V. V. Kabanov, W. N. Kang, H. J. Kim, E. M. Choi, and S. I. Lee, Phys. Rev. Lett. **91**, 267002 (2003).
246. D. N. Basov and T. Timusk, Rev. Mod. Phys. **77**, 721 (2005).
247. L. Ozyuzer, A. E. Koshelev, C. Kurter, N. Gopalsami, Q. Li, M. Tachiki, K. Kadowaki, T. Yamamoto, H. Minami, H. Yamaguchi, T. Tachiki, K. E. Gray, W.-K. Kwok, and U. Welp, Science **318**, 1291 (2007).
248. S. V. Dordevic, S. Komiya, Y. Ando, and D. N. Basov, Phys. Rev. Lett. **91**, 167401 (2003).
249. J. Corson, R. Mallozzi, J. Orenstein, J. N. Eckstein, and I. Bozovic, Nature **398**, 221 (1999).
250. P. L. Gammel, D. J. Bishop, G. J. Dolan, J. R. Kwo, C. A. Murray, L. F. Schneemeyer, and J. V. Waszczak, Phys. Rev. Lett. **59**, 2592 (1987).
251. B. Parks, S. Spielman, J. Orenstein, D. T. Nemeth, F. Ludwig, J. Clarke, P. Merchant, and D. J. Lew, Phys. Rev. Lett. **74**, 3265 (1995).

Index